Advances in Sustainability S and Technology

Series Editors

Robert J. Howlett, Bournemouth University & KES International, Shoreham-by-sea, UK

John Littlewood, School of Art & Design, Cardiff Metropolitan University, Cardiff, UK

Lakhmi C. Jain, University of Technology Sydney, Broadway, NSW, Australia

The book series aims at bringing together valuable and novel scientific contributions that address the critical issues of renewable energy, sustainable building, sustainable manufacturing, and other sustainability science and technology topics that have an impact in this diverse and fast-changing research community in academia and industry.

The areas to be covered are

- Climate change and mitigation, atmospheric carbon reduction, global warming
- Sustainability science, sustainability technologies
- Sustainable building technologies
- Intelligent buildings
- Sustainable energy generation
- Combined heat and power and district heating systems
- Control and optimization of renewable energy systems
- Smart grids and micro grids, local energy markets
- Smart cities, smart buildings, smart districts, smart countryside
- Energy and environmental assessment in buildings and cities
- Sustainable design, innovation and services
- Sustainable manufacturing processes and technology
- Sustainable manufacturing systems and enterprises
- Decision support for sustainability
- Micro/nanomachining, microelectromechanical machines (MEMS)
- Sustainable transport, smart vehicles and smart roads
- Information technology and artificial intelligence applied to sustainability
- Big data and data analytics applied to sustainability
- Sustainable food production, sustainable horticulture and agriculture
- Sustainability of air, water and other natural resources
- Sustainability policy, shaping the future, the triple bottom line, the circular economy

High quality content is an essential feature for all book proposals accepted for the series. It is expected that editors of all accepted volumes will ensure that contributions are subjected to an appropriate level of reviewing process and adhere to KES quality principles.

The series will include monographs, edited volumes, and selected proceedings.

More information about this series at http://www.springer.com/series/16477

P. Muthukumar · Dilip Kumar Sarkar · Debasis De ·
Chanchal Kumar De
Editors

Innovations in Sustainable Energy and Technology

Proceedings of ISET 2020

Editors
P. Muthukumar
Department of Mechanical Engineering
Indian Institute of Technology Guwahati
Assam, India

Dilip Kumar Sarkar
Department of Applied Sciences
University of Quebec at Chicoutimi
Chicoutimi, QC, Canada

Debasis De
Energy Institute, Bengaluru
Rajiv Gandhi Institute of Petroleum
Technology, Bengaluru, India

Chanchal Kumar De
Department of Electronics
and Communication Engineering
Haldia Institute of Technology
Haldia, West Bengal, India

ISSN 2662-6829 ISSN 2662-6837 (electronic)
Advances in Sustainability Science and Technology
ISBN 978-981-16-1121-6 ISBN 978-981-16-1119-3 (eBook)
https://doi.org/10.1007/978-981-16-1119-3

This Springer imprint is published by the registered company Springer Nature Singapore Pte Ltd.
The registered company address is: 152 Beach Road, #21-01/04 Gateway East, Singapore 189721, Singapore

Committee

Chief Patron

Prof. A. S. K. Sinha, Director, Rajiv Gandhi Institute of Petroleum Technology, Jais, Amethi, India

Convener

Dr. Debasis De, Energy Institute, Bengaluru (Centre of R.G.I.P.T.), India

Organizing Secretary

Dr. Roopa Manjunatha, Energy Institute, Bengaluru (Centre of R.G.I.P.T.), India

Program Committee Members

Technical Committee

Dr. S. Sundaram, Energy Institute, Bengaluru (Centre of R.G.I.P.T.), India
Dr. Debasis De, Energy Institute, Bengaluru (Centre of R.G.I.P.T.), India

PC Members

Dr. Hakeem Niyas U. S., Energy Institute, Bengaluru (Centre of R.G.I.P.T.), India
Dr. Roopa Manjunatha, Energy Institute, Bengaluru (Centre of R.G.I.P.T.), India

Finance Committee

Dr. Arjun Deo, Energy Institute, Bengaluru (Centre of R.G.I.P.T.), India
Dr. Bhaskor Jyoti Bora, Energy Institute, Bengaluru (Centre of R.G.I.P.T.), India

International Advisory Committee

Prof. G. Tamizhmani, Arizona State University
Prof. Dilip Kumar Sarkar, University of Quebec a Chicoutimi, Canada
Prof. Sukanta Roy, Curtin University, Malaysia

National Advisory Committee

Dr. Shamasundar, Rajiv Gandhi Institute of Petroleum Technology, Jais, India
Prof. A. K. Choubey, Rajiv Gandhi Institute of Petroleum Technology, Jais, India
Dr. Nagahanumaiah, CMTI, Bengaluru, India
Prof. K. Rajanna, IISc, Bengaluru, India
Prof. S. Bandyopadhyay, IIT Bombay, India
Prof. R. P. Saini, IIT Roorkee, India
Prof. V. V. Goud, IIT Guwahati, India
Prof. Dinesh Rangappa, Visvesvaraya Technological University, Karnataka, India
Prof. Amit Chakraborthy, NIT Durgapur, India
Prof. Agnimitra Biswas, NIT Silchar, India
Dr. Nripen Chanda, CSIR-CMERI, Durgapur, India
Dr. Srinivasarao Naik B., CSIR-CBRI, Roorkee, India
Prof. Uma Maheshwar, NIT Surathkal, India
Mr. A. Balasubramanian, General Manager (CGM), GAIL, Bengaluru, India

Technical and Review Committee Members

Dr. Y. Rajasekhar, School of Mechanical Engineering, VIT Vellore, India
Dr. M. A. Asha Rani, Electrical Engineering, NIT Silchar, India
Dr. R. Ramya, School of Electrical Engineering, SRM University, Chennai, India
Dr. P. Sankar, Electrical Engineering, NIT Andhra Pradesh, India
Dr. Binu Ben Bose, VIT Chennai, India
Dr. B. Satya Sekhar, IIT Jammu, India
Dr. S. Vedharaj, NIT Tiruchirappalli, India
Dr. Manish Kumar Rathod, Sardar Vallabhai NIT Surat, India
Dr. Koushik Das, NIT Meghalaya, Shillong, India
Dr. Bukke Kiran Naik, Simon Fraser University, Canada
Dr. Mubarak Ali M., TKM College of Engineering, Kollam, India
Dr. Hemant Ahuja, ABES Engineering College Ghaziabad, India
Dr. Mohit Bansal, GL Bajaj Institute of Technology and Management, Greater Noida
Dr. V. S. K. V. Harish, PDPU, Gandhi Nagar, India
Dr. Sanjay Agarwal, IGNOU, Delhi, India
Dr. Ankita Gaur, IIT Guwahati, India
Dr. Sumit Tiwari, Shiv Nadar University, Noida, India
Dr. Chanchal Kumar De, Haldia Institute of Technology, Haldia, India
Dr. Santigopal Pain, Haldia Institute of Technology, Haldia, India
Dr. Kalisadhan Mukharjee, PDPU, Gandhi Nagar, India
Dr. A. A. Prasanna, Malnad College Engineering, Hassan, Karnataka, India
Dr. K. M. Hossain, Dr. B. C. Roy Engineering College, Durgapur, India
Dr. Subhasis Roy, University of Kolkata, Kolkata, India
Prof. M. M. Nayak, Indian Institute of Science, Bengaluru, India
Prof. D. Roy Mahapatra, Indian Institute of Science, Bengaluru, India
Dr. Veeranna B. Nasi, Ramaiah Institute of Technology, Bengaluru, India
Dr. M. Jyothirmayi, Ramaiah Institute of Technology, Bengaluru, India
Prof. Sai Siva Gorthi, Indian Institute of Science, Bengaluru, India
Dr. D. Mallick, Girijananda Chowdhury Institute of Technology, Guwahati, India
Dr. Achinta Sarkar, KIIT, Orissa, India
Dr. Ashok Siddharamanna, Dayananda Sagar University, Bengaluru, India
Dr. K. Fathima Patham, VIT Vellore, India
Dr. Vedalakhsmi, CSIR-CECERI, Karaikudi, India
Dr. Saranyan Vijayaraghavan, CSIR-CECERI, Karaikudi, India
Dr. Karthick Krishnan, CSIR-CECERI, Karaikudi, India
Dr. Mayavan Sundar, CSIR-CECERI, Karaikudi, India
Dr. Subbiah Aawarappan, CSIR-CECERI, Karaikudi, India

Additional Reviewers

Reviewer, Organization

Dr. Ajit Dattatray Phule, Gyeongsang National University, Jinju, Korea
Dr. Dola Sinha, Dr. B. C. Roy Engineering College, Durgapur, India
Dr. Kamalika Tiwari, Dr. B. C. Roy Engineering College, Durgapur, India
Dr. Sudipta Bhadra, Heritage Institute of Technology, Kolkata, India
Dr. Animesh Mondal, Gayeshpur Government Polytechnic College, West Bengal, India

Preface

The 1st National Conference on Innovations in Sustainable Energy and Technology (ISET) India 2020 was organized by Energy Institute, Bengaluru (Centre of Rajiv Gandhi Institute of Petroleum Technology, Jais, Amethi), India, during December 3 and 4, 2020, through virtual mode. Energy Institute, Bengaluru, is dedicated to the objectives of creating highly trained professional manpower in various disciplines of engineering related to renewable and sustainable energy and technologies. It has gained reputation through institutional dedication to teaching and research.

In response to call for papers of ISET India 2020, a total of 47 papers were submitted for presentation and inclusion in the proceedings of conference. These papers were evaluated and ranked based on their novelty, significance and technical quality by at least two reviewers per paper. After a careful and blind refereeing process, 32 papers were selected for inclusion in the proceedings. These papers cover current research in renewable energy, sustainable energy, smart energy systems and E-mobility. The conference hosted one plenary talk by Prof. P. Muthukumar (IIT Guwahati, India) and five keynote talks by Dr. Saji Salkalachen (Former General Manager, Bharat Heavy Electricals Limited, Bengaluru, India), Prof. Amit Kumar Chakraborty (NIT Durgapur, India), Prof. Zakir Hussin Rather (IIT Bombay, India), Prof. Sudip Ghosh (IIEST, Shibpur, India) and Prof. S. Iniyan (Anna University, Chennai, India). Also, Prof. K. Rajanna, Emeritus Professor, IISc, Bengaluru, India, was the chief guest on this occasion.

A conference of this kind would not be possible without the full support from different committee members. The organizational aspects were looked after by the organizing committee members who spent their time and energy in making the conference a reality. We also thank all the technical program committee members and additional reviewers for thoroughly reviewing the papers submitted to the conference and sending their constructive suggestions to improve the quality of papers. Our hearty thanks to Springer for agreeing to publish the conference proceedings.

We are indebted to Mangalore Refinery and Petrochemicals Limited (MRPL) and Rajiv Gandhi Institute of Petroleum Technology for sponsoring and supporting the event. Last but not least, our sincere thanks go to all speakers, participants and all authors who have submitted papers to the conference ISET India 2020. We sincerely hope that the readers will find the proceedings stimulating and inspiring.

Assam, India	P. Muthukumar
Chicoutimi, Canada	Dilip Kumar Sarkar
Bengaluru, India	Debasis De
Haldia, India	Chanchal Kumar De

Message from the Volume Editors

It is a great pleasure for us to organize the 1st National Conference on *Innovations in Sustainable Energy and Technology (ISET) India 2020* held during December 3 and 4, 2020. Our main goal is to provide an opportunity to the participants to learn about contemporary research in renewable energy, sustainable energy systems, smart energy systems and E-mobility and exchange ideas among themselves and with experts present in the conference as plenary and keynote speakers. It is our sincere hope that the conference helps the participants in their research and training and open new avenues for work for those who are either starting their research or are looking for extending their area of research to a different direction of current research in *renewable energy, smart energy systems and electric vehicle technology*.

After an initial call for papers, 47 papers were submitted for presentation at the conference. All submitted papers were sent to external referees, and after refereeing, 32 papers were recommended for publication for the conference proceedings that will be published by Springer in its Book Series: Advances in Sustainability Science and Technology (ASST).

We are grateful to the speakers, participants, referees, organizers, sponsors and Energy Institute, Bengaluru, for their support and help, without which it would have been impossible to organize the conference. We express our gratitude to the organizing committee members who work behind the scene tirelessly in taking care of the details in making this conference a success.

Contents

About the Editors

P. Muthukumar is a Professor in the Department of Mechanical Engineering in IIT Guwahati, India. He is the Fellow of Institute of Engineers (India). He served as the President, Indian Society of Heating, Refrigerating and Air Conditioning Engineers (ISHRAE), Guwahati sub-chapter. He is also the recipient of Fulbright-Nehru Academic and Professional Excellence Award 2017 from USIEF and also received Mechanical Engineering Design National Award from NDRF. He is a reviewer for more than 50 international journals. He has successfully completed 11 research projects funded by various government agencies and 5 consultancy projects funded by industries. Currently, he is handling 5 research and 1 consultancy projects. His specialization includes refrigeration, hydrogen storage, metal hydride based thermal machines, porous medium combustion and thermal energy storage.

Dilip Kumar Sarkar is a Professor at University of Quebec at Chicoutimi (UQAC), Quebec, Canada, in the Applied Sciences Department. He is an active member of American Chemical Society (ACS), Canadian Association of Physicist (CAP) as well as that of the aluminum research cluster of Quebec (REGAL). He holds several important research grants from the Government of Quebec and the Government of Canada. He holds a research group at the UQAC comprising International and National graduate students and post-doctoral fellows working on various aspects of Materials Science, mainly concentrated on thin films fabrication and various applications that includes, energy storage, corrosion, as well as antibacterial nanomaterials.

Debasis De is an Assistant Professor in Energy Institute, Bengaluru (Centre of R.G.I.P.T, Jais), India. He is an active member of the International Solar Energy Society and The Institute of Engineers (India). He was selected as one of the fourteen Indian student delegates to attend the JSPS-DST Asia Science Seminar held at Yokohama, Japan. He has received Quebec Merit Scholarship for foreign student in 2012. He has published more than 30 research papers in various journals and conferences. He is a reviewer for more than 5 international journals. His current research interests include different types of solar cells, OER/HER, bio-inspired surfaces, nanostructures for electronics and advanced materials, energy materials.

Chanchal Kumar De is a Professor and Head in the Department of Electronics and Communication Engineering, Haldia Institute of Technology, Haldia, India. His research interests include cognitive radio networks, cooperative communication, energy harvesting, wireless ad hoc and sensor networks, etc. He is a reviewer of IEEE, Springer and Elsevier journals. He has published more than 25 research articles in different journals and conferences. He is an editor of proceeding of the 2nd International Conference on Communication, Devices and Computing ICCDC 2019, Lecture Notes in Electrical Engineering, Springer.

WattCastLSTM—Power Demand Forecasting Using Long Short-Term Memory Neural Network

V. Vijay Sankar, P. Chitra, and B. Poonkuzhali

Abstract Power demand forecasting has a substantial role in smart meter technology to alleviate the analyzing and decision-making process in power consumption, demand, and scheduling. As the observance of electricity load is nonlinear, recurrent neural networks (RNN) fit well for understanding and erudition of the nonlinear behavior which can be efficiently used for power demand forecasting. In this paper, "WattcastLSTM," a stacked long short-term memory (LSTM) architecture is devised to perform demand forecasting of the electricity consumption in an institution on hourly basis. Model is developed such that it captures inconsistent trends and patterns of power demand. This helps organizations to manage the timings of laboratories, machinery usage and also helps in capacity planning and preparing budget. Thus, proper forecasting of power demand in an inconsistent environment helps in reducing financial burden followed by efficient usage of resources. The performance of the proposed architecture of WattCastLSTM is compared with the other standard models used in forecasting of time series data. The optimized solution is achieved with the least root mean squared error (RMSE) value. The model resulted in better forecasting of the power demand of an organization.

Keywords Deep learning · LSTM · Forecasting · Smart meter technology · Power consumption data · Scheduling · Decision making

1 Introduction

One of the major integral and indispensable parts of the advanced metering infrastructure solution (AMI) is the smart meters [1]. Consumers are provided with a finer

V. Vijay Sankar (✉) · P. Chitra · B. Poonkuzhali
Department of Computer Science and Engineering, Thiagarajar College of Engineering, Madurai, Tamil Nadu, India

P. Chitra
e-mail: pccse@tce.edu

P. Muthukumar et al. (eds.), *Innovations in Sustainable Energy and Technology*,
Advances in Sustainability Science and Technology,
https://doi.org/10.1007/978-981-16-1119-3_1

access to the informatical data and which capacitate them to make more sophisticated decisions in power usage. An effectual demand forecasting method can be used to maintain a balance between power supply and demand chain, thus curtailing the production cost, estimating the pragmatic energy prices, management of scheduling the resources and future capacity planning.

Deep learning also called as deep neural learning is an arrangement of neural networks, which is used to perform the machine learning task with maximum efficiency. Some of the related works include working algorithms like long short-term memory (LSTM) [2, 3], machine learning-based switching model [4], autoregressive integrated moving average (ARIMA) [5], support vector regression (SVR) [6] and artificial neural network (ANN) [7]. This paper proposes a new deep learning model WattcastLSTM, which has a long short-term memory neural network architecture [2, 3, 8–10] to forecast the electricity consumption in an organization. Since many organization schedule/follow events on an hourly basis, forecasting demand on an hourly basis provides greater insight for resource planning and decision making.

The rest of the proposed paper is compiled as follows. Section 2 comes up with the necessary literature study for the various forecasting techniques with different configurations. Section 3 provides the background and proposed WattcastLSTM framework. Section 4 introduces case study which deals with data preparation and preprocessing, exploratory data analysis, implementation and performance comparison of the proposed model with the other benchmark candidates. Section 5 concludes the paper.

2 Related Works

Many experimentation and exploration works have been carried out in the field of demand and load forecasting. Wei-Jen Lee adopted switching model for energy load forecasting using machine learning [4]. Mack Robinson J proposed an ARIMA Model [5] and compared it to ANN's performance. Ping-Huan-Kuo adopted a forecasting technique based on short-term load forecasting [7] using smart grid technology and accurate deep neural network algorithm for STLF [7].

Luca Ghelardoni proposed a long-term forecasting problem [6], which predicts the energy consumption 25 months ahead. Kermanshahi developed a RNN model [11] for forecasting with three-layered backpropagation. Motahar Reza performed a research in spark cluster to forecast energy load using gated recurrent unit (GRU) and LSTM [3]. Zheng adopted empirical mode decomposition (EMD-LSTM) neural networks for short-term load forecasting and used Xgboost algorithm for the evaluation and selection of features based on importance [9]. Jian proposed electric load forecasting in smart grids using LSTM [10].

Precise and error-free forecasting of short-term electric load is intriguing due to the non-stationary and non-seasonal nature of the power load. The proposed WattcastLSTM model uses the modified LSTM to forecast the power demand on an hourly basis.

3 The Demand Forecasting Model Based on LSTM

3.1 The LSTM Neural Network Model

LSTM [2] is a subset of RNN [12, 13] introduced to increase the capability of learning the long-term dependencies, which is used for categorizing and foresight predictions in the time series data. A LSTM cell can either store or remove the information which are authorized by a time controlled and inner looped regulators (also called as gates). A LSTM cell/unit has an input gate, an output gate and a forget gate.

Input gate layer: It has a sigmoid function which decides whether the new information should be stored and tanh function creates new vector which is added to the state.

Output gate layer: This decides whether the current state of the cell affects the other. A Non linear sigmoid activation function is executed, followed by a tanh function, which is multiplied with the output of the sigmoid gate.

Forget gate layer: It has a sigmoid function which helps to decide what information has to be stored.

LSTM model should be developed such that it compromises both modeling capability and the performance efficiency. Higher the capability would lead to the problem of over fitting of data.

3.2 WattCastLSTM—Architecture

WattCastLSTM is a stacked LSTM with three hidden layers having 200, 150 and 100 neurons. Dropout regularization technique is used to reduce over fitting while stacking LSTM layers. The dense layer provides the output of the stacked LSTM layers. Linear activation function [11] is applied on the output layer, since the values are unbounded. The stacking of LSTM layers for building neural network in WattcastLSTM is shown in Fig. 1.

3.3 Performance Metrics

Mean squared error (MSE) is the mean of the squared error (estimated value—original value). RMSE is the square root of MSE. The mean absolute percentage error (MAPE) is the prediction measure to check how accurate a forecasting method is in trend estimation and prediction. Cross-validation root mean squared error—CV (RMSE) is used to measure of variability of the errors.

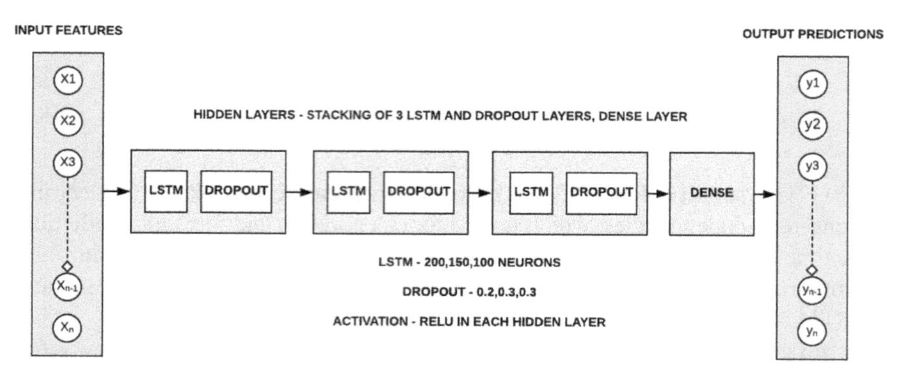

Fig. 1 WattcastLSTM architecture—outline

4 Case Study

The data used in this research study is collected from the smart meters installed in the different laboratories of Thiagarajar College of Engineering, Madurai. The dataset consists total power consumed and timestamp in fixed time interval. The data from one of the laboratories in the institution for a span of six months (April 2018 to September 2018) in hourly interval is collected and used for the analysis. This period is selected since it covers both high consumption (working days) and low consumption (Semester holidays). The design methodology for WattcastLSTM is shown in Fig. 2.

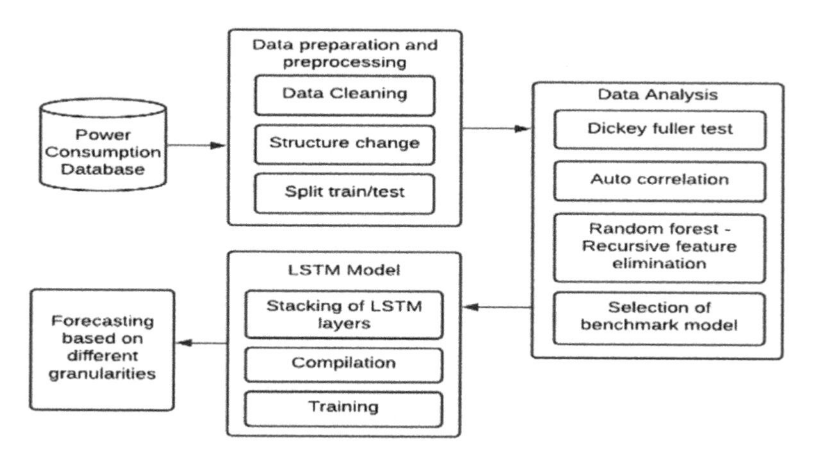

Fig. 2 WattcastLSTM design methodology

4.1 Data Preparation and Preprocessing

Data preparation and preprocessing is used to obtain better performance and accuracy [3]. It consists of methods to resolve inconsistent, missing and noisy input data. The preprocessing tasks for forecasting include,

- Data cleaning—Conversion of time and date to a single date-time format.
- Structure change—Normalization of data in the range [−1, 1] by scaling.

Generally, data fed to the machine learning models will be splitted into train and test data in the ratio of 80:20 or 70:30 [4]. The WattcastLSTM model is trained on the 80% of the data and the remaining 20% of the data is used for testing the forecasting accuracy. Training data includes data from April 2018 to August 2018 and test data includes data of September 2018 in hourly interval. Thus, there are totally 3672 (153 days * 24) data points for training and 720 (30 days * 24) data points for testing.

4.2 Data Analysis

The power load of a laboratory from April to September 2018 is shown in Fig. 3.

It is observed that the hourly load demand varies for every day in the month and does not follow a same sequence throughout the period. Day-wise plot of a month April is shown in Fig. 4a.

In a monthly plot of April, there is no or very minimal consumption of power due to holidays as shown in Fig. 4a. Weekly electricity load is shown in Fig. 4b.

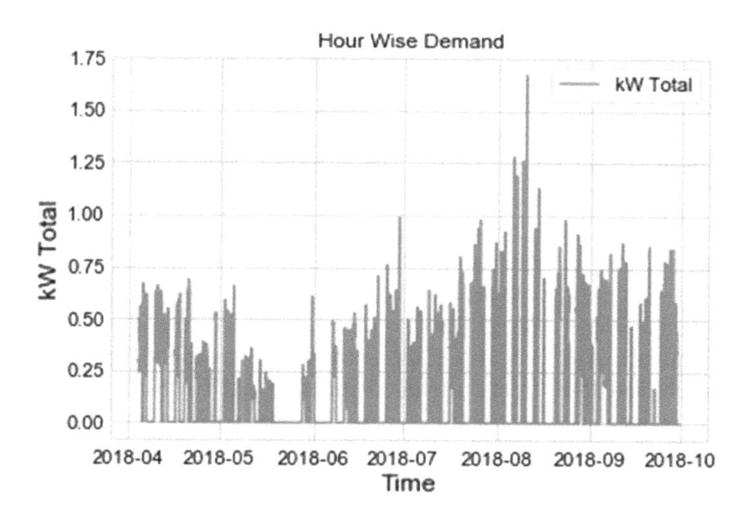

Fig. 3 Hourly electricity load versus time (April to September 2018)

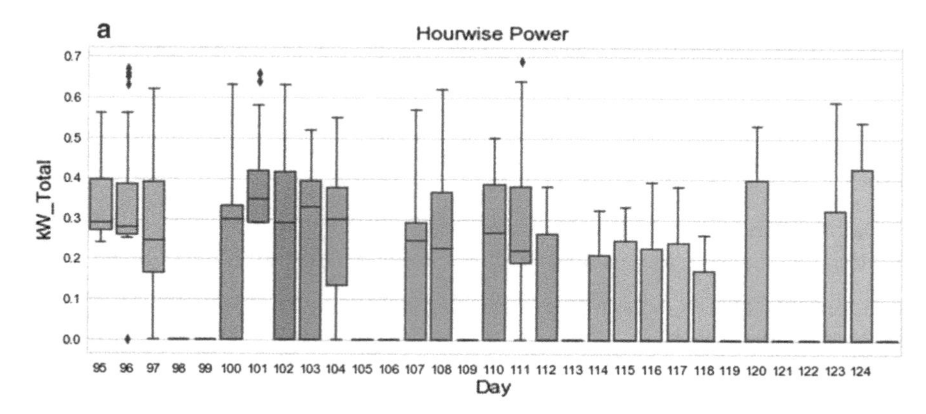

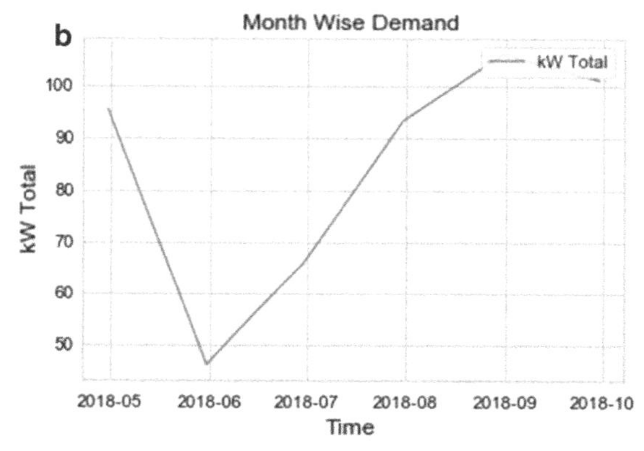

Fig. 4 **a** April month electricity load versus time. **b** Weekly electricity load versus time (April to September 2018)

Weekly plot shows low consumption in the month of June as compared to others since power demand decreases due to semester holidays. Hence, week demand of one month is not the same as the week demand of another month.

Dickey–Fuller Test. Dickey–Fuller test was carried out to check stationarity if any exists in the given dataset [14]. If the resulting *p*-value from the test for a time series data is less than 0.05, then that data is said to be stationary. Dickey–Fuller test on the respective dataset is achieved with *p*-value less than 0.05; thus, it is inferred that the time series data is stationary. Plot of the test is shown in Fig. 5.

The randomness in the lag plot of the data also implies that the selected data is stationary in nature. And thus, it becomes easier for the model to fit the data.

Autocorrelation Plot. Autocorrelation function (ACF) [5] is implemented to discover the optimal number of time lags. The time correlation plot is shown in Fig. 6.

Value of p is minimum (0.009) at the time lag 30. Hence, 30 is chosen as optimal time lag for the forecasting model.

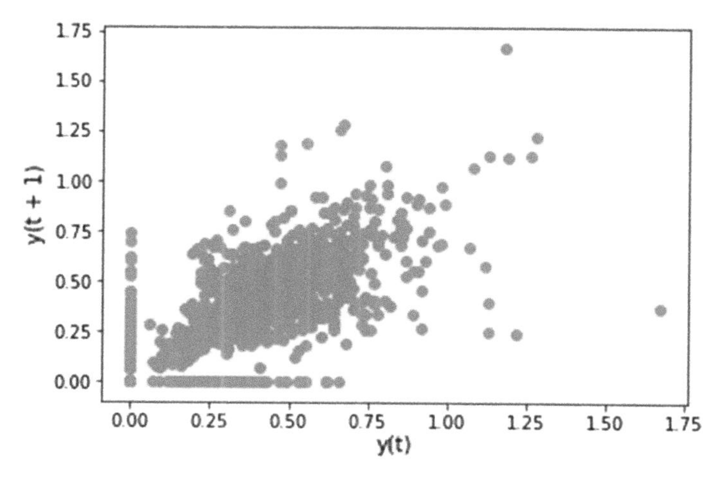

Fig. 5 Dickey–Fuller test

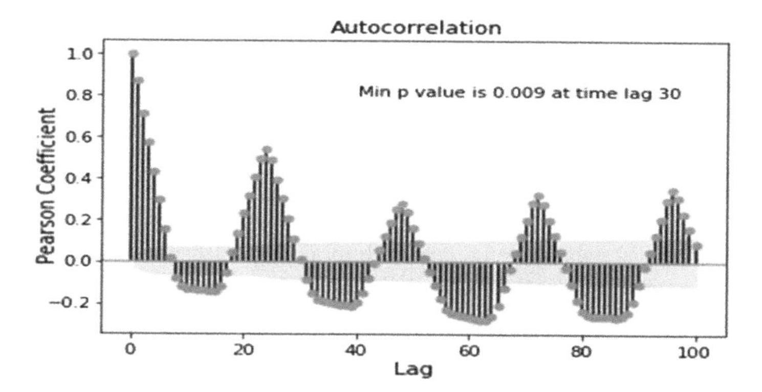

Fig. 6 Autocorrelation plot

Random Forest-Recursive Feature Elimination (RF-RFE). RF-RFE is used for determining the important subset of features. RF-RFE is used on those 30 features to obtain the optimal subset. Figure 7 represents feature selection rank for features obtained using RF-RFE [2].

From this, t-24, t-23, t-7, t-1 are selected as important features for the model since it has smaller rank (smaller ranks are the better ones). Hence, four features are selected as the input for the model (optimal features [t-24, t-23, t-7, t-1]).

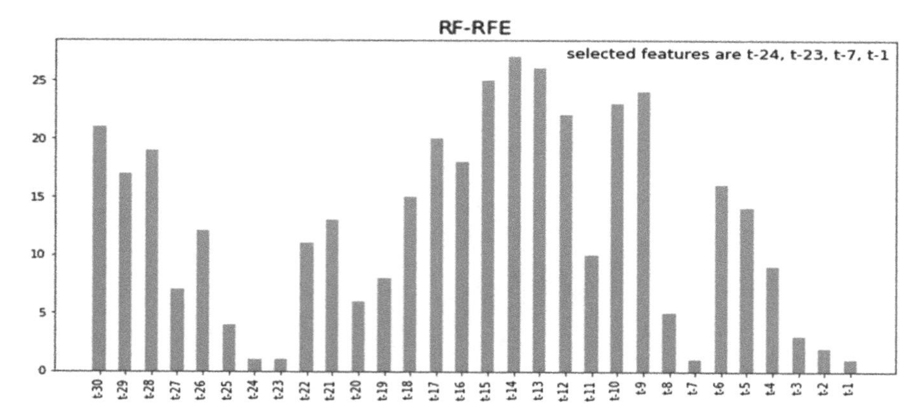

Fig. 7 Feature selection using RF-RFE

Table 1 Performance metrics of WattCastLSTM

Algorithm/metrics	MSE	RMSE	MAPE	CV(RMSE)
WattCastLSTM	0.014	0.118	0.056	0.784
WattCastLSTM (Optimal 4 features)	0.005	0.069	0.033	0.529

4.3 Implementation and Results

The WattCastLSTM was implemented on Intel i7-9750H @ 2.60 GHz with 16 GB RAM. The average time for training WattCastLSTM was around 100 s for 200 epochs of batch size 150. To assess the performance and importance of feature selection, base model with all the features and the fine-tuned model with the selected optimal features are trained separately. Performance of the models are calculated via MSE, RMSE, MAPE and CV(RMSE) [8, 10], and the results are depicted in Table 1.

Thus, WattCastLSTM with optimal feature performed better than WattCastLSTM with all features. Hence, RMSE is reduced at rate of 41.5%. Model with optimal features is tested for September 2018. Prediction result for hourly basis is shown in Fig. 8a.

The prediction result shows a stable prediction of the power demand and captures inconsistent trends and patterns.

4.4 Comparison with Baseline Models

Benchmarking is an approach to test the proposed model by comparing the results with the existing candidate models. The candidate models and its corresponding parameters are summarized in Table 2.

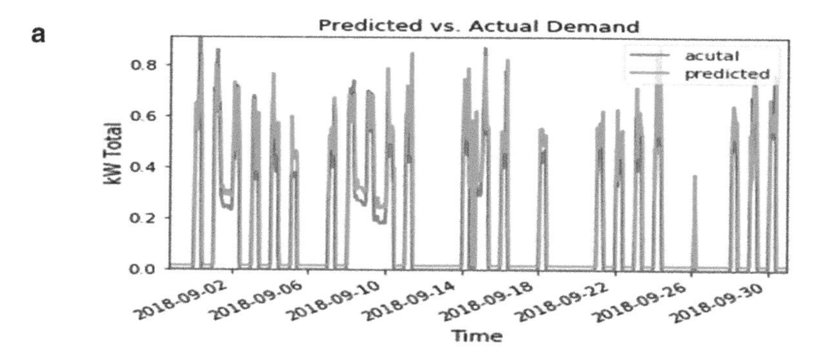

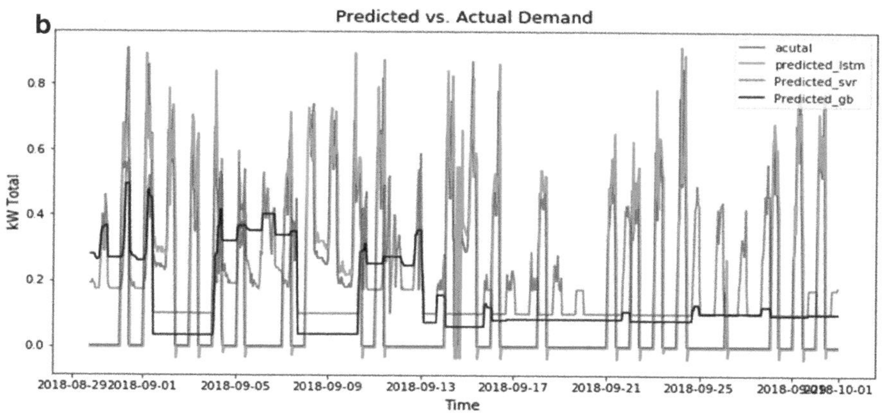

Fig. 8 Prediction result of WattCastLSTM. **b** Performance of the fine-tuned models

Table 2 Parameters for models

Algorithm	Parameters
ARIMA	No. of lag observation = 5, degree of differencing = 1, order of moving average = 0
Triple exponential smoothing	Default alpha, beta and gamma values. Trend = add, seasonal = add, periods = 720
Random forest	No of trees = 1000, max_depth = default
Support vector regression (SVR)	Kernel = rbf, regularization parameter c = 1e3, gamma = 0.1
Gradient boosting	n_estimators = 200, max_depth = 2, learning rate = 0.01, min_samples_split = 5, loss = ls

Table 3 Performance metrics of models

Algorithm/metrics	MSE	RMSE	MAPE	CV(RMSE)
ARIMA	0.02	0.141	0.05	0.768
Exponential smoothing	0.4	0.6	0.5	4.6
Random forest	0.122	0.349	0.322	2.459
Support vector regression (SVR)	0.009	0.095	0.095	0.739
Gradient boosting (GB)	0.017	0.129	0.090	1.278

Table 4 Performance metrics of fine-tuned models

Algorithm/metrics	MSE	RMSE	MAPE	CV(RMSE)
Fine-tuned SVR	0.008	0.089	0.045	0.723
Fine-tuned gradient boosting (GB)	0.015	0.124	0.089	1.267
WattCastLSTM	0.005	0.069	0.033	0.529

All candidate models are trained and tested on the same selected dataset, the results are compared with the performance metrics—MSE, RMSE, MAPE, CV(RMSE) [8, 10]—and the inferences are given in Table 3.

From the performance metric analysis in the above table, two models—GB [2, 9] and SVR [6]—are selected as benchmark candidates for the current model.

For GB and SVR, three parameters are tuned using grid search technique [2, 9], with parameter grid for GB as n_estimators {300,400,500}, maximum depth {3,4,5} and learning rate {0.001, 0.1} and parameter grid for SVR as kernel {rbf, linear, sigmoid}, C {0.001, 0.01, 0.1} and gamma {0.001, 0.01, 0.1}. Comparison of results of the fine-tuned SVR (0.001, rbf, 0.001) and fine-tuned GB (500, 5, 0.001) with WattCastLSTM is shown in Table 4. Performance of the fine-tuned baseline models and WattcastLSTM model is shown in Fig. 8b.

RMSE is reduced at rate of 22.4% comparing with SVR and 44.3% in case of GB. Thus, the proposed WattCastLSTM performs better than fine-tuned baseline models—GB and SVR—in fitting the time series data and predicting it with high accuracy.

5 Conclusion

WattCastLSTM for power demand forecasting is presented in this paper with suitable figures and tables. LSTM model was enhanced using autocorrelation function and RF-RFE method which performs better than the other benchmark models. The results of WattCastLSTM with optimal features provide an accurate prediction mechanism for power demand forecasting in organizational level. Thus, the forecasting of smart meter data results in identification of exact power demand using WattCastLSTM.

References

1. Selvam, C., Srinivas, K., Ayyappan, G.S., Sarma, M.V.: Advanced metering infrastructure for smart grid applications. In: 2012 International Conference on Recent Trends in Information Technology, pp. 145–150. IEEE (2012)
2. Bouktif, S., Fiaz, A., Ouni, A., Serhani, M.A.: Optimal deep learning LSTM model for electric load forecasting using feature selection and genetic algorithm: comparison with machine learning approaches. Energies **11**(7), 1636 (2018)
3. Kumar, S., Hussain, L., Banarjee, S., Reza, M.: Energy load forecasting using deep learning approach-LSTM and GRU in spark cluster. In: 2018 Fifth International Conference on Emerging Applications of Information Technology (EAIT), pp. 1–4. IEEE (2018)
4. Fan, S., Chen, L., Lee, W.J.: Machine learning based switching model for electricity load forecasting. Energy Convers. Manage. **49**(6), 1331–1344 (2008)
5. Zhang, G.P.: Time series forecasting using a hybrid ARIMA and neural network model. Neurocomputing **50**, 159–175 (2003)
6. Ghelardoni, L., Ghio, A., Anguita, D.: Energy load forecasting using empirical mode decomposition and support vector regression. IEEE Trans. Smart Grid **4**(1), 549–556 (2013)
7. Kuo, P.H., Huang, C.J.: A high precision artificial neural networks model for short-term energy load forecasting. Energies **11**(1), 213 (2018)
8. Zhao, Z., Chen, W., Wu, X., Chen, P.C., Liu, J.: LSTM network: a deep learning approach for short-term traffic forecast. IET Intel. Transp. Syst. **11**(2), 68–75 (2017)
9. Zheng, H., Yuan, J., Chen, L.: Short-term load forecasting using EMD-LSTM neural networks with a Xgboost algorithm for feature importance evaluation. Energies **10**(8), 1168 (2017)
10. Zheng, J., Xu, C., Zhang, Z., Li, X.: Electric load forecasting in smart grids using long-short-term-memory based recurrent neural network. In: 2017 51st Annual Conference on Information Sciences and Systems (CISS), pp. 1–6. IEEE (2017)
11. Kermanshahi, B., Iwamiya, H.: Up to year 2020 load forecasting using neural nets. Int. J. Electr. Power Energy Syst. **24**(9), 789–797 (2002)
12. Rahman, A., Srikumar, V., Smith, A.D.: Predicting electricity consumption for commercial and residential buildings using deep recurrent neural networks. Appl. Energy **212**, 372–385 (2018)
13. Kong, W., Dong, Z.Y., Jia, Y., Hill, D.J., Xu, Y., Zhang, Y.: Short-term residential load forecasting based on LSTM recurrent neural network. IEEE Trans. Smart Grid **10**(1), 841–851 (2017)
14. Fathi, O.: Time series forecasting using a hybrid ARIMA and LSTM model. In: Velvet Consulting (2019)

Thermodynamic Analysis of Solar Photovoltaic/Thermal System (PVT) for Air-Conditioning Applications

A. Sai Kaushik and Satya Sekhar Bhogilla

Abstract In this paper, the study of a photovoltaic/thermal system (PVT) module with an integrated thermal energy storage system in the form of phase change material (PCM) to supply energy for the vapor absorption cycle to provide air-conditioning is carried out. The PV module characteristics have been studied, and the increase in the overall efficiency of the system due to the thermal extraction cycle coupled with the thermal energy storage (TES) system has been calculated. The heat absorption characteristics of various PCMs have been compared for the selection of the best optimum material to obtain the maximum performance of the system. The operating parameters of the vapor absorption cycle have been analyzed and the amount of cooling obtained from the vapor absorption cycle at different time intervals based on the input power obtained from the dissipated heat from the panel has been calculated to determine the COP of the vapor absorption cycle.

Keywords Photovoltaic · Energy storage · Solar energy · Air-conditioning

Nomenclature

Thermal efficiency	ηth (%)
Mass flow rate	$\dot{m}$ (kg/s)
Specific heat capacity of the coolant fluid	$c_{\text{cp(cf)}}$ (kJ/kg K)
Electrical efficiency	η_{e} (%)
Heat absorbed by the evaporator	Q_{e} (W)
Heat supplied to the generator	Q_{g} (W)
Mass of PCM	m_{PCM} (kg)
Latent heat of fusion of lauric acid	L_{PCM} (kJ/kg)

A. Sai Kaushik
IIITDM Kurnool, Kurnool, India

S. S. Bhogilla (✉)
Indian Institute of Technology Jammu, Jammu, India
e-mail: satya.bhogilla@iitjammu.ac.in

P. Muthukumar et al. (eds.), *Innovations in Sustainable Energy and Technology*, Advances in Sustainability Science and Technology,
https://doi.org/10.1007/978-981-16-1119-3_2

Heat transferred to the PCM	$\dot{Q}_{\mathrm{PCM}}$ (W)
Circulation ratio	γ
Concentration ratio	$\in$

1 Introduction

Hybrid systems for the co-generation of both electricity and thermal energy from the same system are becoming more common and are playing a pivotal role in extracting the maximum output from the given system under consideration and making them more diverse and flexible in terms of the different applications they can be utilized. Several studies have been conducted on analyzing the various operational parameters of the different types of the photovoltaic/thermal system (PVT) systems to observe their influence on the thermal and electrical performance of the total system [1]. Optical devices such as plane reflectors and parabolic dish collector are utilized as solar concentrators for capturing a large area of incident sunlight and focusing it onto the solar panel for operating the panel at its peak performance throughout its working cycle. As the efficiency of the solar cell decreases with increase in its temperature, the implementation of the thermal energy extraction cycle to the panel absorbs the excess heat and maintains the solar panel at its optimum temperature for best performance and to avoid any damage to any of the components. This leads to an increased lifespan of the solar panel in addition to the extra thermal energy that is being harnessed. Various studies have been done in modeling the PVT module with thermal energy storage (TES) system to increase the overall efficiency of the system in addition to extracting the excess thermal energy dissipated by the solar panel [2]. Tubes having very good thermal conductivity such as copper are attached under the solar panel for the coolant fluid to circulate underneath along with a thermal energy storage system for storing energy in the form of latent heat to act as a reservoir for energy storage. The thermal energy system ensures the continuous supply of the required energy for the thermal extraction cycle even during periods of insufficient supply of solar radiation. Organic materials such as hydrocarbons and inorganic materials such as various salt hydrates are primarily used as the phase change materials for the thermal energy storage system. A detailed analysis has been performed on water PV/T collectors [3] for calculating its performance characteristics. The extent of utilizing this system for various applications is determined by the amount of heat that is being absorbed by the thermal extraction cycle or the maximum outlet temperature that is obtained from the circulating coolant fluid [4–6].

2 System Modeling

The concept of the total system is shown in Fig. 1 and explained in reference [7]. The main objective of employing a thermal energy storage system is to store energy produced by the PV module during the day so as to provide energy output for the required application during night hours or when the solar module is operating below the required power production rate due to cloudy conditions or because of bad weather. The energy supplied to the vapor absorption cycle is utilized in providing air-conditioning or for refrigeration applications in the sample space under consideration.

2.1 Thermal Energy Storage System

Latent heat storage systems using phase change material as the medium for storing thermal energy are utilized in our system since it can help in retaining large amounts of stored energy for longer durations of time and can supply energy for the desired application during night hours or bad weather conditions. PCMs have high energy density and can store a large amount of energy for given unit mass of the substance. Furthermore, the relative constant temperature of heat exchange between the working fluid and the PCMs makes it a simpler yet effective process for energy storage. The operating principle of PCM is shown in Fig. 2.

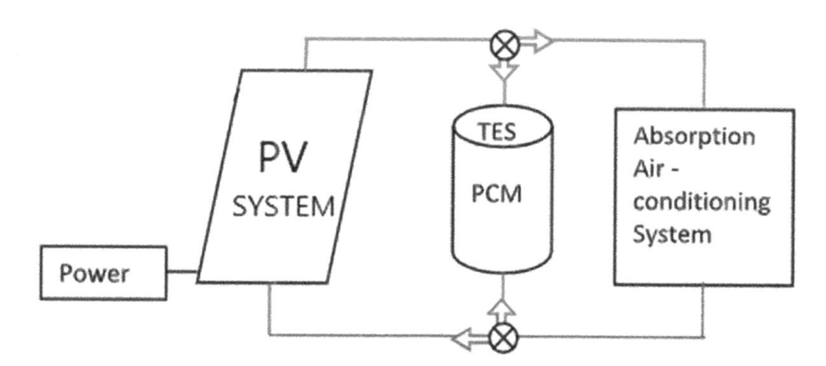

Fig. 1. Concept of total system [7]

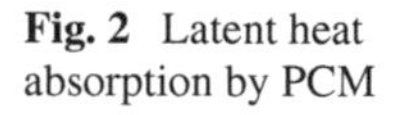

Fig. 2 Latent heat absorption by PCM

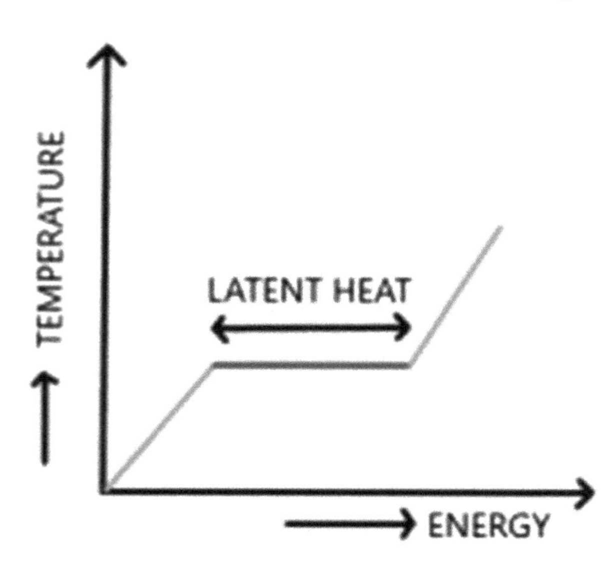

2.2 *Vapor Absorption Cycle*

A vapor absorption cycle is similar to that of the vapor compression cycle except that it utilizes a heat source to drive the cooling process. The mechanical compression employed in a vapor compression cycle is replaced by a thermal compression process wherein the external heat supplied to the generator of the vapor absorption cycle provides the necessary compression for the absorbent–refrigeration mixture for condensation to take place. The vapor absorption cycle is shown in Fig. 3. The vapor

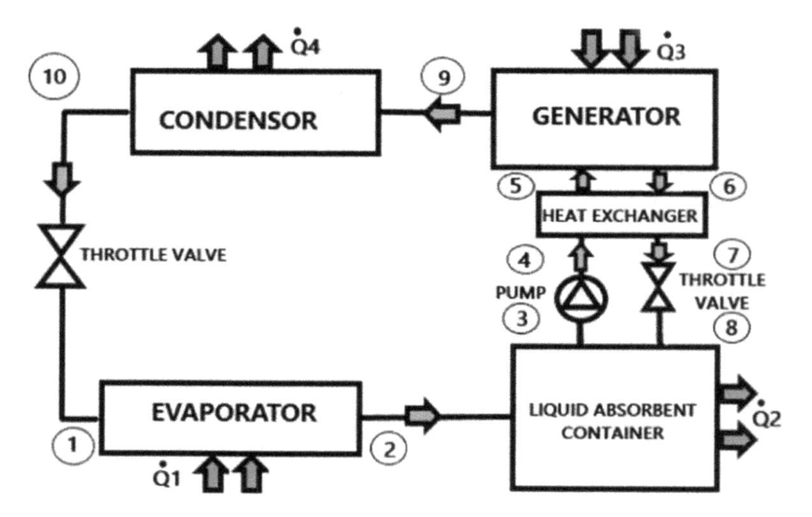

Fig. 3 Vapor absorption cycle

absorption cycle in our analysis is designed to provide optimum cooling from an initial temperature of 30 °C (room temperature) to a final temperature value of 4 °C. lithium–bromide/water system of vapor absorption cycle is selected for this analysis because of the high heat of vaporization of the water that acts as the refrigerant for efficient cooling and also because of the non-volatile nature of the absorbent (Li–Br). The equations used in this analysis are available in the references [7, 8].

3 Model Analysis and Observations

The irradiation data and temperature data are taken from the reference [8].

The electrical power and electrical efficiency with respect to time are shown in Fig. 4, and the electrical power of the system increases as time progresses reaching a maximum value of 152 W at 2.00 PM and gradually decreases toward the evening hours. This is because of higher levels of solar irradiation that is incident on the panel that produces a higher electrical output during the hotter afternoon hours as compared to the early and evening hours of the case study. It is the opposite in the case of the electrical efficiency wherein it decreases as the time progresses toward a hotter climate and then shows a gradual increase toward the evening hours with a maximum electric efficiency obtained being 13.7% at 4.00 PM, and the minimum electrical efficiency obtained being 9.89% at 2.00 PM. This is because the efficiency of the solar module is inversely proportional to its working temperature and as the climate becomes hotter, the efficiency of the panel decreases with optimum efficiency being obtained at relatively cooler morning and evening hours.

The maximum temperature difference obtained in the coolant fluid at different flow rates is shown in Fig. 5; it is observed that the temperature difference decreases

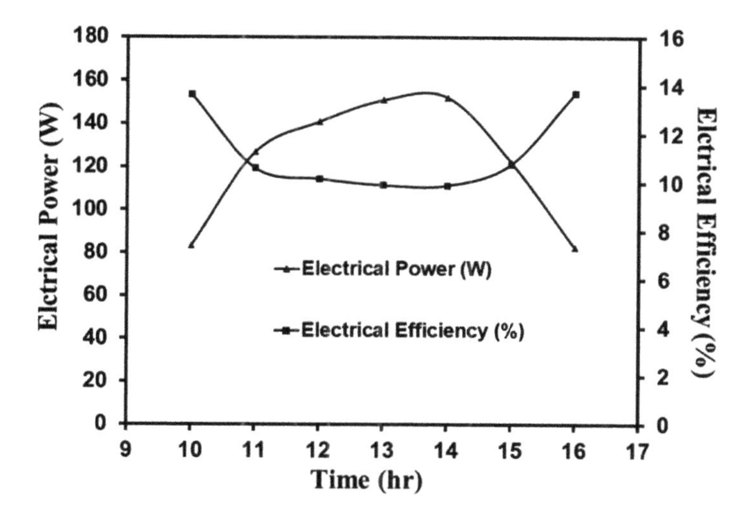

Fig. 4 Hourly variation of electrical performance

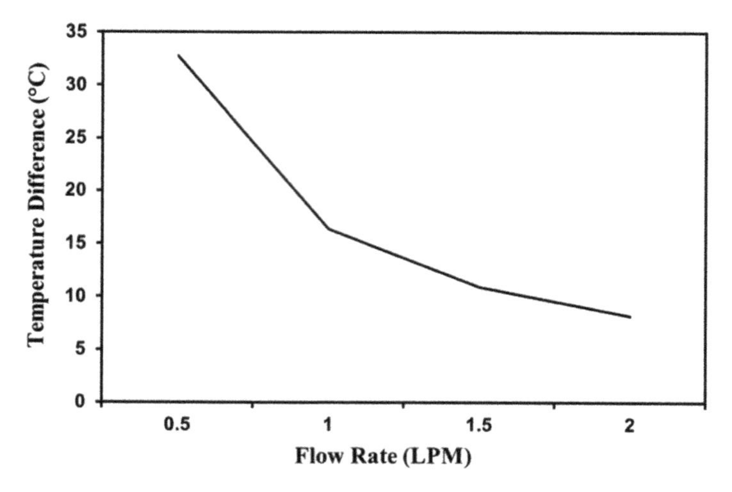

Fig. 5 Effect of water flow rate on fluid temperature difference

as the volume flow rate of the coolant fluid is increased gradually. As more amount of coolant fluid is passed through the circuit, there is a slower rate of temperature gain as the specific heat capacity of the coolant fluid is now increased due to the increased flow rate through the system. A maximum temperature difference of 32.75 °C is obtained with a volume flow rate of 0.5 LPM at 1.00 PM.

The thermal output with respect to time is shown in Fig. 6. The thermal output of the system increases as time progresses, reaching a maximum value of 1143 W at 1.00 PM and gradually decreases toward the evening hours. This is because as the climate becomes progressively hotter, there is more heat dissipation from the system to obtain higher amounts of thermal energy as compared to the morning and evening hours.

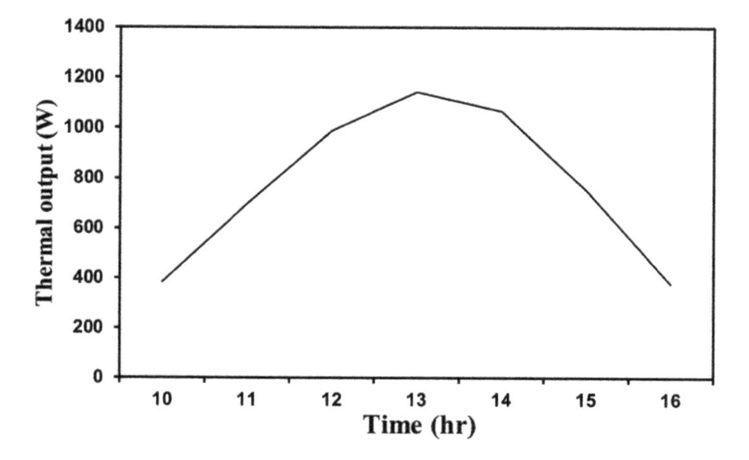

Fig. 6 Hourly variation of thermal output

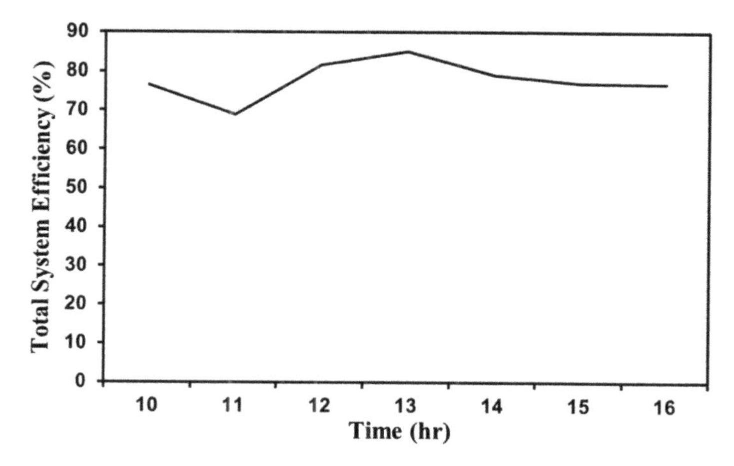

Fig. 7 Hourly variation of total system efficiency

The maximum overall efficiency of the system as plotted in Fig. 7 is higher than that of the individual thermal and electrical systems with a maximum efficiency of 85.13% obtained at 1.00 PM. As the thermal efficiency and the electrical efficiency show a contrasting trend as the time progresses, the overall system efficiency shows a fluctuating pattern with values in the range of 68.89–85.13%. The quality of the energy has not been taken into consideration in determining the maximum overall efficiency but gives an indication of the extra performance that can be harnessed from the hybrid system.

4 Thermal Energy Storage System Analysis

Average solar radiation intensity over a time period of 6 h (21,600 s): 730.84 W/m^2. Average thermal efficiency of the system: 73.7%. Average waste heat absorbed by the coolant fluid = (730.84 × 1.623) × 0.73 = 874.195 W. Assuming 50% distribution to the PCM: $\dot{Q}_{\mathrm{PCM}}$ (avg) = 437.1 W. Density of lauric acid = 880 kg/m^3. Amount of heat transferred to the PCM = $\dot{Q}_{\mathrm{PCM}}$ (avg) × 21,600 s = 9.441 MJ

$$\dot{Q}_{\mathrm{PCM}}(\mathrm{avg}) = m_{\mathrm{PCM}} x\, L_{\mathrm{PCM}} \tag{1}$$

where m_{PCM} is the mass of PCM required for the analysis and L_{PCM} is the latent heat of fusion of lauric acid = 228.29 kJ/kg. Hence, m_{PCM} = 41.245 kg.

4.1 Vapor Absorption Cycle Analysis

The operating parameters and conditions of the vapor absorption cycle along with the thermodynamic properties at different states of the cycle considered in the below tables are taken from the reference [9]: Assuming 50% distribution to the vapor absorption cycle: $\dot{Q}_{\text{VAC}}(\text{avg}) = 437.1$ W. This heat is provided as the input to the generator of the cycle: $\dot{Q}_{\text{VAC}}(\text{avg}) = \dot{Q}_{\text{g}}(\text{avg}) = \dot{Q}_3(\text{avg})$. The enthalpies of the working medium at various points of the cycle are given in Table 1.

Considering the energy balance equation of the generator:

$$\dot{Q}_{\text{g}} = \dot{Q}_3 = \dot{m}_{\text{ref}} h_9 + \dot{m}_6 h_6 - \dot{m}_5 h_5 \tag{2}$$

where
$\dot{m}_{\text{ref}}$ = mass flow rate of refrigerant (kg/s),
$\dot{m}_6$ = mass flow rate of rich absorbent solution (kg/s),
$\dot{m}_5$ = mass flow rate of refrigerant–absorbent mixture (kg/s).

The circulation ratio (γ) of the cycle is calculated from the relation:

$$\gamma = \frac{\epsilon_5}{\epsilon_6 - \epsilon_5} = 6 \tag{3}$$

where
ϵ_5 = concentration ratio of the refrigerant–absorption mixture, ϵ_6 = concentration ratio of the rich absorbent solution.

The relation between the mass flow rates of the three distinct channels mentioned above is given by:

$$\dot{m}_5 = (\gamma + 1)\dot{m}_{\text{ref}} = 7\dot{m}_{\text{ref}} \tag{4}$$

Table 1 Thermodynamic properties of the vapor absorption cycle

State points	Temperature (°C)	Pressure (mm of Hg)	Enthalpy (kJ/kg)	Concentration (ϵ)
1	30.00	6.1	125.7	–
2	4.00	6.1	2508.7	–
3	20.00	6.1	−180.0	0.48
4	20.00	32	−180.0	0.48
5	53.85	32	−115.7	0.48
6	64.00	32	−120.0	0.56
7	20.00	32	−195.0	0.56
8	20.00	6.1	−195.0	0.56
9	64.00	32	2616.5	–
10	30.00	32	125.7	–

Table 2 Energy input and refrigeration effect of the vapor absorption cycle

Time	$\dot{Q}_g$ (W)	$\dot{m}_{ref}$ (kg/s)	$\dot{Q}_e$ (W)
10 AM	191.95	0.071	169.367
11 AM	348.78	0.1288	307.75
12 PM	495.22	0.0183	436.96
1 PM	571.878	0.2113	504.6
2 PM	532.89	0.197	470.2
3 PM	376.36	0.14	332.1
4 PM	265.456	0.1	234.22

$$\dot{m}_6 = (\gamma)\dot{m}_{ref} = 6\dot{m}_{ref} \tag{5}$$

By substituting the above two relations in the energy balance equation of the generator along with the enthalpy values at different states:

$$\dot{Q}_g = \dot{Q}_3 = \dot{m}_{ref}h_9 + 6(\dot{m}_{ref}h_6) - 7(\dot{m}_{ref}h_5)) \tag{6}$$

$$\dot{Q}_g = \dot{Q}_3 = 2706.4(\dot{m}_{ref}) \tag{7}$$

Hence, the mass flow rate of the refrigerant required for different amounts of generator heat that is supplied can be calculated based on the above equation:

$$\dot{Q}_e = \dot{Q}_1 = \dot{m}_{ref}(h_2 - h_1) \tag{8}$$

The average mass flow rate of the cycle obtained from the generator heat input: $\dot{m}_{ref} = 0.1711$ kg/s. The heat absorbed by the evaporator: $\dot{Q}_e(\text{avg}) = 409.2$ W. COP of the vapor absorption cycle = 0.88. Table 2 shows the energy input and refrigeration effect of the vapor absorption cycle. The amount of heat absorbed is increasing as time increases toward hotter afternoon climate with a maximum absorber heat = 571.8 W at 1.00 PM and gradually decreases toward evening hours of our case study. This is due to that the maximum amount of thermal dissipation from the system as the climate gets hotter and hence, a higher value of generator heat can be supplied to the vapor absorption cycle at this time for the best cooling performance.

During night hours or bad weather conditions, the PCM thermal energy storage releases its stored energy to provide the necessary input for our application:

$$\dot{Q}_{PCM}(\text{avg}) = 437.1 = \dot{Q}_g, \dot{m}_{ref} = .1611 \text{ kg/s},$$

$$\dot{Q}_e = \dot{Q}_1 = 385.68 \text{ W}, \text{COP} = 0.882$$

Thus, the PCM provides the necessary input energy to absorb 385.6 W of energy during bad weather conditions or when the system is switched off, i.e., when its duty cycle has been completed.

5 Conclusions

In this analysis, solar photovoltaic/thermal system (PVT) for air-conditioning applications has been studied. It is observed that with increasing mass flow rate of the liquid water circulating fluid, the lesser change in the temperature at the outlet of the coolant flow and the maximum change in the temperature of 32.75 °C is observed when the flow rate of the water is set at 0.0834 kg/s and at a time of 1.00 PM. The maximum electrical efficiency of the PV module obtained is 13.7% obtained at 4 PM with the electrical efficiency showing a slight decline in value as the incident intensity of light and temperature of the system keep increasing. The electrical efficiency obtained was higher during morning and evening hours of the case study. The maximum thermal efficiency obtained from the heat extraction cycle is 75.21% at 1 PM due to the higher heat dissipation obtained from the PV module during afternoon hours. The overall efficiency of the system has been calculated, and the maximum overall efficiency of the system obtained is 85.13% at 1.00 PM, with values showing a fluctuating pattern as the time progressed during the case study. Also, an amount of 385.68 W of energy can be absorbed by the evaporator when the PCM material provides the heat stored in it as the input to the generator.

References

1. Riffat, S.B., Cuce, E.: A review on hybrid photovoltaic/thermal collectors and systems. Int. J. Low-Carbon Technol. **6**, 212–241 (2011)
2. Javed, M., Leila, L.: Optimal management of a solar power plant equipped with a thermal energy storage system by using dynamic programming method. Proc. Mech. Eng. Power Energy **230**, 219–233 (2016)
3. Nick, J., Thomas, S.: Concentrated solar power: recent developments and future challenges. Proc. Mech. Eng. Power Energy **229**, 693–713 (2015)
4. Arkin, H., Navon, R., Burg, I.: HVAC with thermal energy storage: optimal design and optimal scheduling. Proc. Mech. Eng. Power Energy **18**, 31–38 (1997)
5. Cocco, D., Cau, G.: Energy and economic analysis of concentrating solar power plants based on parabolic trough and linear Fresnel collectors. Proc. Mech. Eng. Power Energy **229**, 677–688 (2015)
6. Shouquat, H.M.D., Abd, R.N.B., Selvaraj, J.A.L., Kumar, P.A.: Experimental Investigation on energy performance of hybrid PV/T-PCM system. In: ICEES 2019 Fifth International Conference on Electrical Energy Systems (2019)
7. Saikaushik, A., Pulla Rao, M., Bhogilla, S.S.: Solar photovoltaic/thermal system (PVT) for air-conditioning applications. In: Theoretical, Computational, and Experimental Solutions to Thermo-Fluid Systems -Select Proceedings of ICITFES (2020)
8. Hossain, M.S., Pandey, A.K., Selvaraj, J., Rahim, N.A., Islam, M.M., Tyagi, V.V.: Two side serpentine flow based photovoltaic-thermal-phase change materials (PVT-PCM system): energy, exergy and economic analysis. Renew. Energy **136**, 1320–1336 (2019)
9. Kaushik, S., Singh, S.: Thermodynamic analysis of vapor absorption refrigeration system and calculation of COP. Int. J. Res. Appl. Sci. Eng. Technol. **2**, 1–8 (2014)

PbS Nanoparticle Sensitized Fe-Doped Mesoporous TiO_2 Photoanodes for Photoelectrochemical Water Splitting

Somoprova Halder, Soumyajit Maitra, and Subhasis Roy

Abstract Utilization of solar energy to generate hydrogen by photoelectrochemical water splitting using nanostructured semiconductor electrodes has gained a lot of attention due to its promising potential to solve the energy crisis and related environmental issues. The design of ideal heterojunctions based on the proper selection of materials and tuning nanostructure morphology for efficient charge separation has gained ground as an important research area for the above application. This work demonstrates the fabrication of PbS nanoparticle sensitized Fe doped mesoporous TiO_2 photoanodes for solar radiation-activated water splitting applications. Fe-Doped TiO_2 photoanodes were fabricated by the doctor blading technique followed by PbS sensitization. The PbS nanoparticles were deposited by a successive ionic layer adsorption and reaction (SILAR) method whereby the repetition cycles were varied to tune the layer thickness. The fabricated cells were analyzed using a potentiostat with the platinum counter electrode and Ag/AgCl reference electrode with 0.5M Na_2S and 0.5M Na_2SO_3 as an electrolyte. Upon illumination with simulated sunlight (AM 1.5G, 100 mW/cm^2), an optimized ABPE of 0.6% was observed at 0.35 V for the fabricated photoelectrode having 1.5 wt% Fe-doped mesoporous TiO_2 with 20 cycles of PbS deposition.

Keywords Photoanode · P-n junction · Mesoporous · Nanoparticles

1 Introduction

The Earth receives almost 1,000,000 TW of radiation every day from the sun. With the depletion of conventional energy resources, technologies are being developed for the generation of electricity and fuel using solar energy. Photocatalytic water splitting revolves around evolution of hydrogen from water, on photoexcitation of semiconductor material to generate electrons and a potential difference, evolving

S. Halder · S. Maitra · S. Roy (✉)
Department of Chemical Engineering, University of Calcutta, 92 A. P. C. Road, Kolkata 700009, India
e-mail: srchemengg@caluniv.ac.in

P. Muthukumar et al. (eds.), *Innovations in Sustainable Energy and Technology*, Advances in Sustainability Science and Technology,
https://doi.org/10.1007/978-981-16-1119-3_3

hydrogen at cathode and oxygen at the anode. Although platinum is the conventional catalyst for hydrogen evolution reaction through electricity, over recent years, semiconductor oxides and sulphides have been experimented with to effectively utilize the solar spectra for photoexcitation generation, lower the potential costs and increase inavailability. The transition from the usage of bulk metal oxides and chalcogenides to nanostructuring has led to enhanced photocatalytic activity due to exposure of several catalytic sites, magnificently increased surface area, and thus higher efficiency. Various solution-based routes have been employed for the synthesis of nanostructures using spin coating [1], hydrothermally synthesized powders [2] or thin films [3, 4], sputtering [5], atomic layer deposition [6], electrodeposition [7], spray pyrolysis, etc. Hematite, α-Fe_2O_3, has shown good potential as a photoanode in water splitting because it possesses several desirable properties, including a bandgap between 1.9 and 2.2 eV that maximizes absorption from the solar spectrum as the bandgap is excited by visible radiation, stability in a humid environment under typical operating conditions, abundance in the Earth's crust, and affordability. Despite these advantages, hematite exhibits a low value of electrical mobility and a short hole diffusion length (2–4 nm), and requires an applied overpotential to overcome energy band misalignment [8]. TiO_2, on the other hand, has a bandgap of 3.0–3.2 eV for polycrystalline anatase phases, which covers a very small region of visible spectra but a significant region of the UV spectra. Despite low absorbance in the visible region, TiO_2 has certain remarkable advantages such as an excellent compact structure, good electron conductivity, and requirement of a low overpotential in oxygen evolution reaction (OER) [9, 10]. In order to overcome these disadvantages, scientists have used doping methods to enhance absorbance in TiO_2 thin films [11] and conductivity of Fe_2O_3 thin films separately [12]. The introduction of states or sub-bandgap energy levels in the photocatalyst is necessary for tuning its absorption bandwidth and increase electron mobility and charge separation. However, other than doping methods, control over nanostructure morphology [13] also affects photocatalytic activity due to exposure of certain facets, electron transport in specific dimensions, and the occurrence of defect sites. There have been many publications on charge carrier transfer and collection, diffusion of electrons, and electron–hole recombination in titanium dioxide and hematite thin films [14]. Changes in activity with different synthesis methods also have an effect on photoelectrochemical (PEC) activity [15]. Passivation of haematite surfaces states that promote electron recombination by TiO_2 and Al_2O_3 has improved efficiency due to better charge separation and transfer processes from the photoanode to the electrolyte.

Presently, the fabrication of heterojunctions in photoelectrodes has been a fascinating area of research. Not only does it incorporate two different semiconductors to harness the solar spectrum efficiently, but it majorly involves the creation alignment of energy bands in the two materials to ensure charge separation. This might include n-n (type-2) [16] p-n (type-1) [17], p-n-p [18], n-n-p heterojunctions [19]. Although semiconductor material modification has been researched to a large extent, and creation of heterojunctions of n-type semiconductor oxides with CdS has been a much researched one [20], there are fewer reports on use of lead sulphide in PEC photoanodes [21] despite it having a good absorption band in the far-visible or near-infrared regions. In this work, we have fabricated photoanodes for OER at a low cost by designing a p-n junction of n-type Fe-doped mesoporous TiO_2 [22] and p-type PbS nanoparticles on FTO glass substrate. The use of mesoporous TiO_2, a widely used component in dye-sensitized solar cells, is also a novel aspect of the work.

2 Experimental

FTO glasses were sonicated in soap water, ethanol, and distilled water for 30 min each. A $FeCl_3$ solution of 0.2M in N-Methyl Pyrrolidone was prepared with a drop by drop addition of 35% concentration HCl till the solution became transparent. Equal weights, 0.2 g of TiO_2 nanopaste (Dyesol) were weighed individually for 5 sets of FTO, and different volumes of the $FeCl_3$ solution were added to the separately weighed mesoporous TiO_2 pastes and mixed thoroughly to create 5, 10, 25, 30, 40 wt% Fe-doped TiO_2. These were labelled as (Fe–Ti–O(1), Fe–Ti–O(2), Fe–Ti–O(3), Fe–Ti–O(4), and Fe–Ti–O(5)). The prepared nanopaste was applied to the conductive surface of the FTO substrates using doctor blading [23] technique. With different doping concentrations, the paste colour changed from light orange to yellowish-orange and became more difficult to blade the slurry due to its changing consistency. The cells were annealed at 400 °C for 1 h followed by 550 °C for another hour. A mesoporous film of Fe-doped TiO_2 was obtained. A set of cells with the same weight percentages of Fe doping was also created by spin coating a TiO_2 sol followed by annealing. Both the set of films were loaded with PbS in a SILAR approach to study the effect of the use of mesoporous paste. For the deposition of PbS nanoparticles, the chemical bath two solutions: Solution1—0.001M $Pb(CH_3COO)_2$, 0.1 ml acetic acid in 30 ml distilled water, and Solution 2—0.001 Na_2S in 30 ml water was prepared. Chemical bath deposition was carried out on each of the cells for varied dipping cycles, dipping time in each solution lasting for 30 s. The cells were tested for PEC water splitting, and the concentration of Fe in the cell showing best performance was selected for further analysis. Fe–Ti–O(3)was selected due to its best photoresponse out of all the remaining cells, and different dipping cycles of PbS for 5, 10,15, 20, 25 cycles were applied. The cell with 25 cycles was almost entirely black and rarely allowed the passage of light through it. The one with fewer concentrations of PbS, however, showed decreased light absorption. The cell with 20 dipping cycles showed the best performance (Fig. 1).

3 Results and Discussions

The use of mesoporous TiO_2 paste hugely increased the surface area for PbS loading. Reference films made by spin coating, on the other hand, yielded very compact films with much less thickness. PbS deposition on compact thin films reached almost

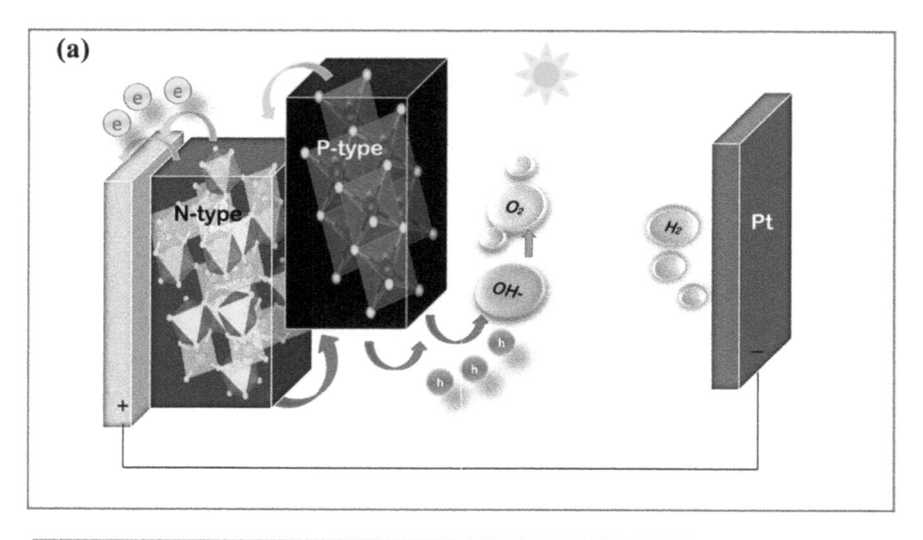

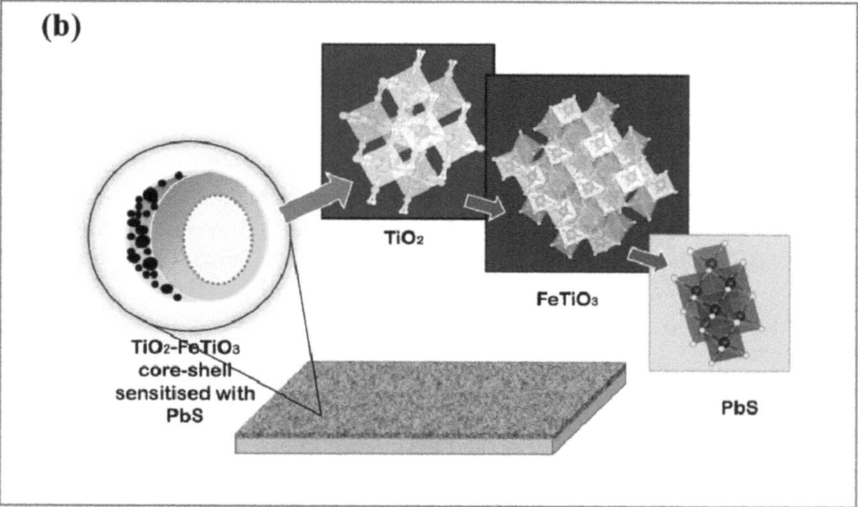

Fig. 1 Schematic representation of PbS-sensitised photoanode **a** electron–hole transport mechanism **b** insight of the p-n junction on FTO glass

saturation after 8 dipping cycles, and PbS could not load further. Fe–Ti–O(5) with the highest Fe concentration in the first set showed poor bindability to the FTO substrate, which can be ascribed to lower the relative concentration of binder molecules ethyl cellulose α-terpeniol in the TiO_2 paste. The opaqueness of the films increased with increasing Fe content. Since Fe–Ti–O(3) showed the best photocurrent, as shown in Fig. 5, it was selected for further sensitization by PbS. After the chemical bath deposition, uniform thin films of Fe doped TiO_2 with black PbS deposition was observed. The films allowed good optical passage of light, and the deposition of

PbS was smooth uniform. Among the second set of cells, the one with 25 cycles was almost completely black and rarely allowed the passage of light. The one with less concentrations of PbS, however, showed decreased light absorption in the visible range. Figure 2a shows the change in the thin film's absorbance with different dopant concentrations of Fe in TiO_2. In Fig. 2b the absorbance of the thin films with different numbers of PbS loading cycles on Fe–Ti–O(3) has been depicted.

From the absorbance results obtained by UV–Vis Spectroscopy, we made the Tauc plots as shown in Fig. 3 to find out the band gaps. The film's thickness was assumed to be 0.06 mm, which is the thickness of the Scotch tape used for doctor blading. In the second case, the thickness was considered to be the sum of the thickness of Scotch tape and the diameter of the PbS nanoparticles.

The nature of the Tauc plots reveals the direct bandgap nature of the n-type Fe-doped TiO_2 thin films. The bandgap is maximum for the lowest doping percentage and gradually shifts towards shorter band gaps in 2.5–2.8 eV for higher doping amounts. This can be attributed to the formation of $FeTiO_3$ phases validated by

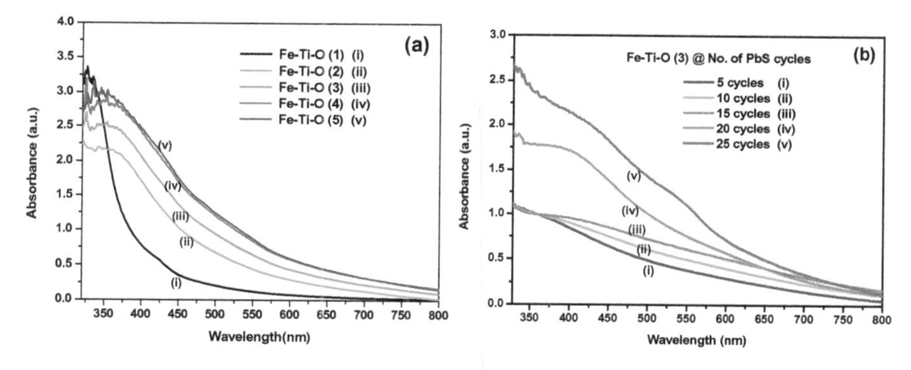

Fig. 2 UV–Vis absorption spectra; **a** absorbance versus wavelength for different dopant concentrations, Fe–Ti–O(x) where x = 1–5; **b** absorbance graph for different loading cycles of PbS on Fe–Ti–O(3)

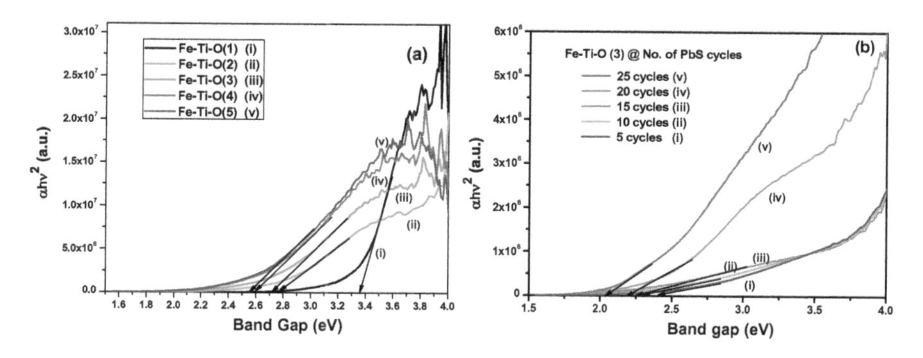

Fig. 3 Tauc plots; **a** Tauc plot for Fe–Ti–O(x), where x = 1–5; **b** Tauc plot for different loading cycles of PbS on Fe–Ti–O(3)

an X-ray diffraction pattern. The PbS loaded cells show wide absorbance range throughout the entire length of the solar spectra, the absorbance values increasing with an increase in number of dipping cycles. On the PbS loaded cells, the band gap was found to be ranging from 2.0 to 2.5 eV. This variable band gap of the PbS nanoparticles can be attributed to their dimensions which fall in the quantum confinement regime. The particles deposited are too small for lower number of dipping cycles, leading to larger band gaps. When the deposition cycles are increased, the size of PbS nanoparticles grows, and the bandgap reduces. PbS being p-type is a good conductor of holes [24, 25]; therefore, photoexcited electrons from PbS make their way out to the conduction band minimum edge of Fe-doped TiO_2, which lies just below that of PbS. Photogenerated holes in Fe-doped TiO_2 move to the valence band maximum edge of PbS from VB_{max} of TiO_2. This leads to enhanced separation of photogenerated electron–hole pairs, preventing recombination. The holes are responsible for OER at the photoanode while the electrons are transported from FTO through the external circuit to the platinum cathode, where hydrogen evolution occurs.

The crystalline phase of materials used in photoanodes was analyzed by XRD analysis shown in Fig. 4. The fabricated cells were analyzed at 0.20/s using Cu K-α rays (8.04 keV, 1.5406 Angstrom). Characteristic peaks of anatase TiO_2 were observed (JCPDS Card no. 21-1272) [26]. Iron doping resulted in Fe atoms' diffusion into the anatase TiO_2 lattice forming $FeTiO_3$perovskite phase (JCPDS Card No. 071-1140) where distinct peaks of Iron Titanate has been observed alongside anatase TiO_2. PbS has been identified as its distinct peaks (JCPDS Card No. 78-1901.) [27] to be

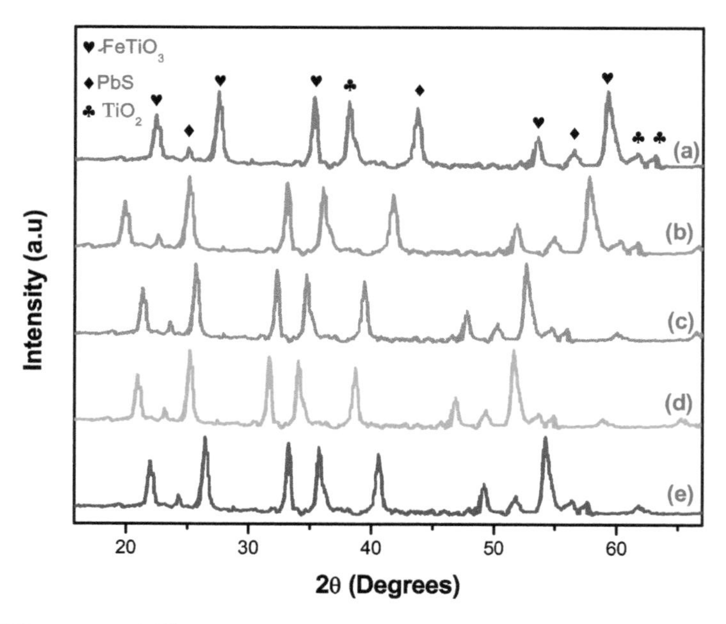

Fig. 4 XRD patterns of different number of cycles PbS loading on Fe–Ti–O(3) **a** 5 cycles, **b** 10 cycles, **c** 15 cycles, **d** 20 cycles, **e** 25 cycles

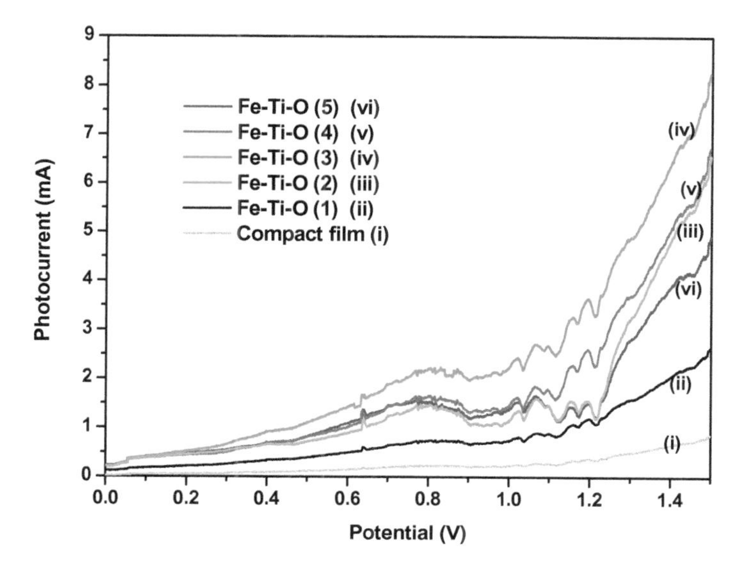

Fig. 5 Linear Sweep Voltammetry (LSV) curves under illumination for Fe–Ti–O(*x*)

in its face-centered cubic structure. All the distinct peaks are present in the XRD data showing $TiO_2/FeTiO_3/PbS$ heterostructure. In the XRD pattern, we can observe significant peak shifts in both the $FeTiO_3$ and PbS phases due to high lattice strains, which can be accounted to insertion of dopant atoms in the anatase TiO_2 lattice and small size of PbS nanoparticles leading to excessive development of strain.

Linear Sweep Voltammetry (LSV) analysis was carried out using a potentiostat (Autolab) with Ag/AgCl reference electrode under simulated solar illumination (100 mW/cm^2), yielding an output as shown in Fig. 5. The electrolyte used was 0.5M solution of Na_2S and Na_2SO_3 [28]. In this set, Fe doped TiO_2 cells were analyzed under illumination. It is observed that Fe–Ti–O(3) yielded the maximum value of current. Current density gradually increased with an increasing amount of iron, while excess deposition at 0.5 ml reduced the current value. Excess iron results in the formation of excess $FeTiO_3$ on the surface which results in band misalignment and hinders electron flow by acting as a sink to pin electrons from the solution. Controlling the amount of Fe diffused into TiO_2 lattice is very important to form a homojunction between the anatase TiO_2 and $FeTiO_3$. This strategy is employed to promote efficient charge transfer of photoexcited electrons from low bandgap $FeTiO_3$ into the wideband TiO_2, which acts as an electron collector to reduce recombination. Excess formation of $FeTiO_3$ is thus not desirable as hole electron recombination increases due to its inefficient charge separation and intrinsic poor electron conduction, and longer diffusion length [29] as described in Fig. 6. The peak obtained at 0.8 V is due to oxidation of Fe^{2+} to Fe^{3+} [30] since in the perovskite phase, Fe is present in its 2 + state. At 0 V, there is a current density of 0.17μA/cm^2, indicating that there is a

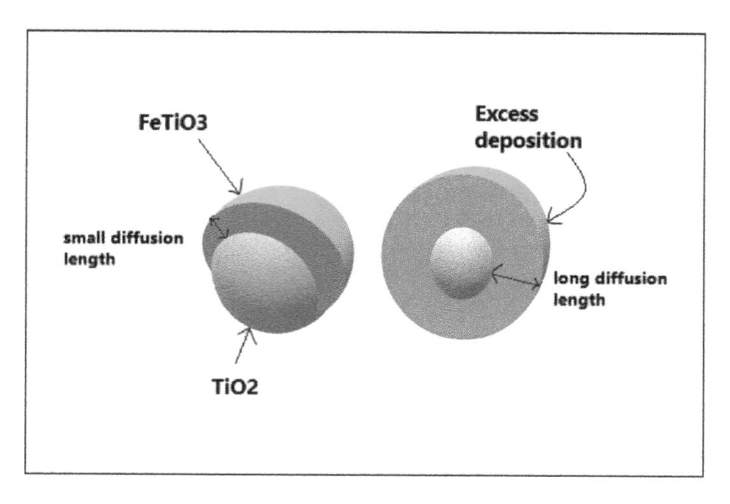

Fig. 6 Diagram representing the doping effect on TiO_2 nanoparticles

photo-induced electron transfer due to the built-in electric field of the materials and the absence of deep-lying trap states requiring extra bias.

PbS was deposited by a Successive Ionic Layer Adsorption and Reaction (SILAR) method to create a p–n junction and increase light absorbance and photoexcited electron generation. The prepared cells' SEM image in Fig. 7 shows that the nanoparticles are evenly distributed throughout the surface. The underlayer containing Fe doped TiO_2 has a few ridges, which are observable, causing the LSV data fluctuations in Fig. 5. It can be seen that the PbS nanoparticles have filled up the spaces, thereby reducing the defects by acting as a photosensitive filler to maintain grain continuity. The particle size of the PbS nanoparticles has been observed to be in the range of 20–30 nm on average. From the SEM microstructural imaging, it becomes clear that the bandgap values that we have obtained from the Tauc plots occur due to quantum confinement due to the extremely small dimensions of the PbS nanoparticles.

The chopped LSV data shows an increase and decrease of current as light is switched on and off. It is observed that the roughness of the LSV curve of Fe doped TiO_2 disappears in the case of PbS sensitization due to the fact that any crystal defects produced during annealing and diffusion of Fe into the TiO_2 lattice that resulted in current disruption disappeared due to space-filling by PbS nanoparticles. The SILAR deposited nanoparticles under SEM analysis showed a size of 20–30 nm. Thus any fissures produced in the film were healed by PbS filling, resulting in a smoother curve. It was observed that PbS sensitization increased current density by 1000 times indicating the effectiveness of its sensitization. An increase in PbS cycling increased the current density value until 20 cycles after which the current dropped. The decrease in photocurrent can be attributed to the fact that excess PbS shields light, thus obstructing transmittance into lower layers. Higher particle size and layer thickness result in lower interfacial contact between PbS and Fe doped TiO_2 underlayer required to produce an efficient p-n junction. ABPE% value of the

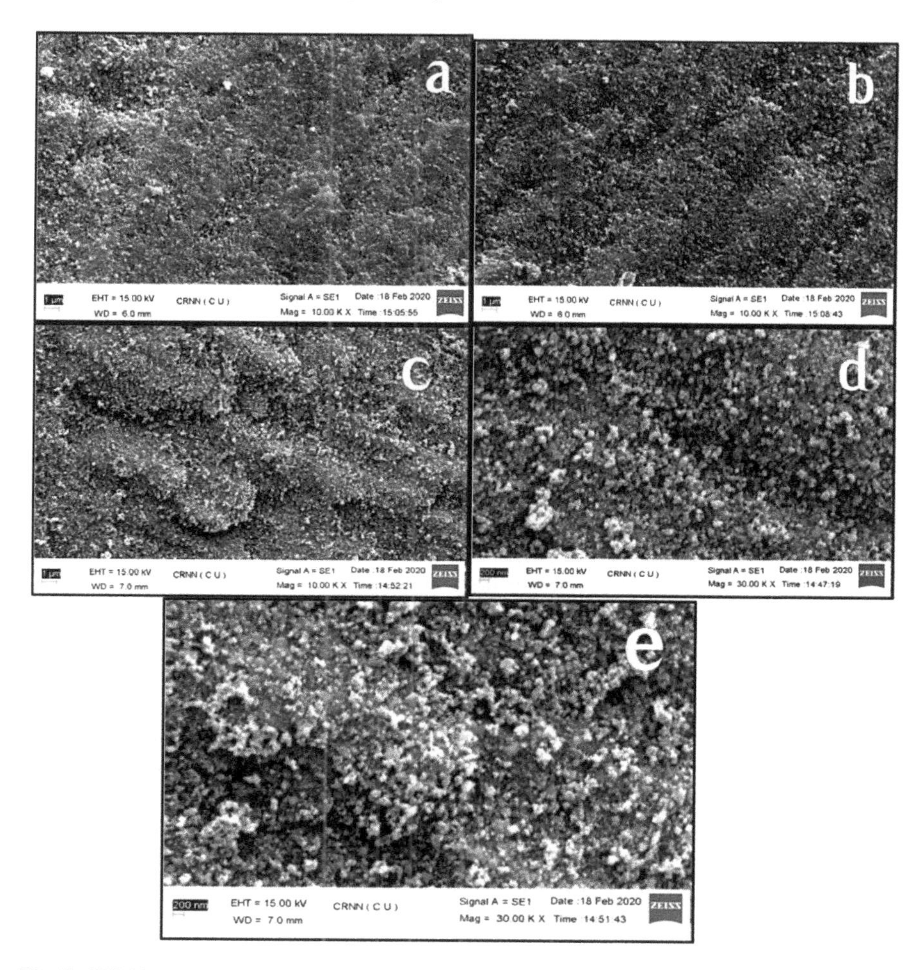

Fig. 7 SEM images of the PbS sensitized mesoporous Fe doped-TiO_2 photoanodes with dipping cycles of **a** 5, **b** 10, **c** 15, **d** 20, **e** 25 dipping cycles

fabricated photoanodes was calculated, and the results showed a maximum 0.58% ABPE at 0.35 V for 20 cycles of PbS deposition, as shown in Fig. 7.

The overpotential indicates that the SILAR method produces asymmetric disordered structures without preferential crystal facets and morphology for efficient hole electron separation and charge transport. Thus additional bias is needed. Advanced fabrication techniques like chemical vapour deposition (CVD) or atomic layer deposition (ALD) can be explored to preferentially fabricate ordered nanostructure arrays to solve this problem (Fig. 8).

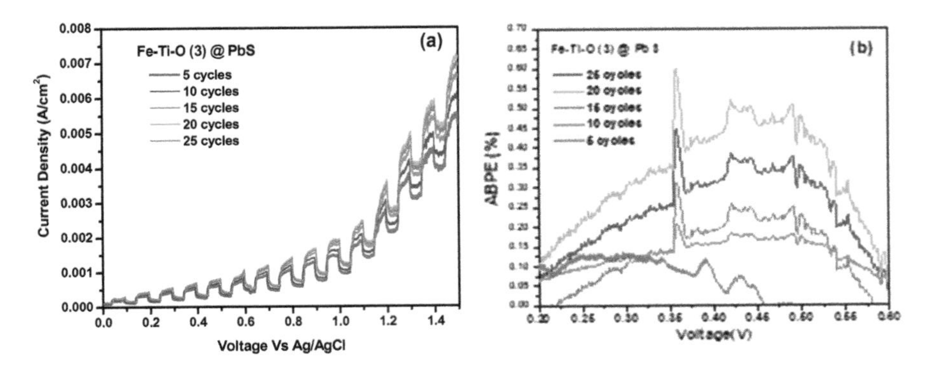

Fig. 8 **a** Chopped LSV performance on the alternate dark and light illumination, **b** ABPE% with the change in PbS cycling on the cell Fe–Ti–O (3)

4 Conclusion

The effect of employing a mesoporous layer as photoanode followed by creation of a p-n junction in a facile method was studied. Doping with Fe in different proportions was carried out to alter the bandgap of TiO_2 and introduce intermediate energy levels for photoexcitation. The deposition of a layer of PbS helped in achieving a p–n junction for better activity in water splitting. Maximum ABPE of 0.6% at 0.35 V was obtained for PbS deposited after 20 cycles on Fe doped TiO_2 photoanodes prepared by 0.3 ml addition of $FeCl_3$ solution to 0.2 g of mesporous nanopaste. The above work opens doors to the development of mesoporous photoanodes for PEC water splitting applications which have evitably shown better performance than compact thin films. The performance of these mesoporous electrodes can be further enhanced by creation of one or more heterojunctions.

Acknowledgements This work was supported by Science and Engineering Research Board (SERB) grants funded by Department of Science and Technology (DST) Central, Government of India through Teachers Associateship for Research Excellence (TAR/2018/000195) (Subhasis Roy).The authors would like to acknowledge Dr. Achintya Singha, Associate Professor, Department of Physics Main Campus Bose Institute Kolkata, for his valuable comments and ideas, which helped to improve the manuscript.

References

1. Souza, F.L., Lopes, K.P., Nascente, P.A., Leite, E.R.: Nanostructured hematite thin films produced by spin-coating deposition solution application in water splitting. Sol. Energy Mater. Sol. Cells **93**(3), 362–368 (2009)
2. Khan, I., Ali, S., Mansha, M., Qurashi, A.: Sonochemical assisted hydrothermal synthesis of pseudo-flower shaped Bismuth vanadate ($BiVO_4$) and their solar-driven water splitting application. Ultrason. Sonochem. **36**, 386–392 (2017)

3. Roy, S., Han, G.S., Shin, H., Lee, J.W., Mun, J., Shin, H., Jung, H.S.: Low temperature synthesis of rutile TiO_2 Nanocrystals and their photovoltaic and photocatalytic properties. J. Nanosci. Nanotechnol. **15**, 6 (2015)
4. Maitra, S., Sarkar, A., Maitra, T., Halder, S., Roy, S., Kargupta, K.: Cadmium sulphide sensitized crystal facet tailored nanostructured nickel ferrite @ Hematite core-shell ternary heterojunction photoanode for photoelectrochemical water splitting. MRS Adv. **5**(50), 2585–2593 (2020)
5. Chen, L., Alarcón-Lladó, E., Hettick, M., Sharp, I.D., Lin, Y., Javey, A., Ager, J.W.: Reactive sputtering of bismuth vanadate photoanodes for solar water splitting. J. Phys. Chem. C **117**(42), 21635–21642 (2013)
6. Wang, T., Luo, Z., Li, C., Gong, J.: Controllable fabrication of nanostructured materials for photoelectrochemical water splitting via atomic layer deposition. Chem. Soc. Rev. **43**(22), 7469–7484 (2014)
7. Phuan, Y.W., Chong, M.N., Zhu, T., Yong, S.T., Chan, E.S.: Effects of annealing temperature on the physicochemical, optical and photoelectrochemical properties of nanostructured hematite thin films prepared via electrodeposition method. Mater. Res. Bull. **69**, 71–77 (2015)
8. Roy, S., Botte, G.: Perovskite solar cell for photocatalytic water splitting with a TiO_2/Co-doped hematite electron transport bilayer. RSC Adv. **8**(10), 5388–5394 (2018)
9. Dey, A., Karan, P., Sengupta, A., Moyez, A., Sarkar, P., Majumder, S., Pradhan, D., Roy, S.: Enhanced charge carrier generation by dielectric nanomaterials for quantum dots solar cells based on CdS-TiO_2 photoanode. Solar Energy **158**, 83–88 (2017)
10. Kumari, S., Singh, A.P., Deva, D., Shrivastav, R., Dass, S., Satsangi, V.R.: Spray pyrolytically deposited nanoporous Ti4+ doped hematite thin films for efficient photoelectrochemical splitting of water. Int. J. Hydrogen Energy **35**(9), 3985–3990 (2010)
11. Kim, J.Y., Youn, D.H., Kim, J.H., Kim, H.G., Lee, J.S.: Nanostructure-preserved hematite thin film for efficient solar water splitting. ACS Appl. Mater. Interfaces. **7**(25), 14123–14129 (2015)
12. Wang, C.W., Yang, S., Fang, W.Q., Liu, P., Zhao, H., Yang, H.G.: Engineered hematite mesoporous single crystals drive drastic enhancement in solar water splitting. Nano Lett. **16**(1), 427–433 (2016)
13. Walter, M.G., Warren, E.L., McKone, J.R., Boettcher, S.W., Mi, Q., Santori, E.A., Lewis, N.S.: Solar water splitting cells. Chem. Rev. **110**(11), 6446–6473 (2010)
14. Hussain, S., Hussain, S., Waleed, A., Tavakoli, M.M., Wang, Z., Yang, S., Fan, Z., Nadeem, M.A.: Fabrication of $CuFe_2O_4/\alpha\text{-}Fe_2O_3$ composite thin films on FTO coated glass and 3-D nanospike structures for efficient photoelectrochemical water splitting. ACS Appl. Mater. Interfaces **8**(51), 35315–35322 (2016)
15. Leng, W.H., Barnes, P.R., Juozapavicius, M., O'Regan, B.C., Durrant, J.R.: Electron diffusion length in mesoporous nanocrystalline TiO_2 photoelectrodes during water oxidation. J. Phys. Chem. Lett. **1**(6), 967–972 (2010)
16. Kim, J.Y., Magesh, G., Youn, D.H., Jang, J.W., Kubota, J., Domen, K., Lee, J.S.: Single-crystalline, wormlike hematite photoanodes for efficient solar water splitting. Sci. Rep. **3**, 2681 (2013)
17. Feng, W., Lin, L., Li, H., Chi, B., Pu, J., Li, J.: Hydrogenated TiO_2/ZnO heterojunction nanorod arrays with enhanced performance for photoelectrochemical water splitting. Int. J. Hydrogen Energy **42**(7), 3938–3946 (2017)
18. Liu, S., Luo, Z., Li, L., Li, H., Chen, M., Wang, T., Gong, J.: Multifunctional TiO_2 overlayer for p-Si/n-CdS heterojunction photocathode with improved efficiency and stability. Nano Energy **53**, 125–129 (2018)
19. Liu, Q., Wu, F., Cao, F., Chen, L., Xie, X., Wang, W., Tian, W., Li, L.: A multijunction of ZnIn2S4 nanosheet/TiO_2 film/Si nanowire for significant performance enhancement of water splitting. Nano Res. **8**(11), 3524–3534 (2015)
20. Ikram, A., Dass, S., Shrivastav, R., Satsangi, V.R.: Integrating PbS quantum dots with hematite for efficient photoelectrochemical hydrogen production. Phys. Status Solidi A **216**, 1800839 (2019)
21. Hartmann, P., Doh-Kwon, L., Smarsly, B. M., Janek, J.: Mesoporous TiO_2 comparison of classical sol−gel and nanoparticle based photoelectrodes for the water splitting reaction. ACS Nano **4**(6), 3147–3154 (2010)

22. Tang, L., Deng, Y., Zeng, G., Hu, W., Wang, J., Zhou, Y., Wang, J., Tang, J., Fang, W.: CdS/Cu_2S co-sensitized TiO_2 branched nanorod arrays of enhanced photoelectrochemical properties by forming nanoscale heterostructure. J. Alloy. Compd. **662**, 516–527 (2016)
23. Li, L., Dai, H., Feng, L., Luo, D., Wang, S., Sun, X.: Enhance photoelectrochemical hydrogen-generation activity and stability of TiO_2 nanorod arrays sensitized by PbS and CdS quantum dots under UV-visible light. Nanoscale Res. Lett. **10**(1), 418 (2015)
24. Godea, F., Baglayan, O., Guneric, E.: P-type nanostructure PbS thin films prepared by the silar method. Chalcogenide Lett. **12**(10), 519–528 (2015)
25. Li, W., Liang, R., Hu, A., Huang, Z., Zhou, Y.: Generation of oxygen vacancies in visible light activated one-dimensional iodine TiO_2 photocatalysts. RSC Adv. **4**(70) (2014)
26. Parmar, M.D., Chaudhari, J.M., Patel, J.C., Chaudhary, M.D.: Structural analysis of PbS synthesized by different technique. Int. J. Innov. Res. Multidisc. Field **02**, 400–404 (2016)
27. Li, C., Hu, P., Meng, H., Jiang, Z.: Role of sulphites in the water splitting reaction. J. Solution Chem. **45**, 67–80 (2016)
28. Leng, W.H., Barnes, P.R.F., Juozapavicius, M., O'Regan, B.C., Durrant, J.R.: Electron diffusion length in mesoporous nanocrystalline TiO_2 photoelectrodes during water oxidation. J. Phys. Chem. Lett **1**(6), 967–997 (2010)
29. Shen, S., Lindley, S.A., Chen, X., Zhang, J.Z.: Hematite heterostructures for photoelectrochemical water splitting: rational materials design and charge carrier dynamics. Energy Environ. Sci. **9**, 2744–2775 (2016)

In Situ Ni-Doped Co_3O_4 Nanostructure: An Efficient Electrocatalyst for Hydrogen Evolution

Arijit Basu, N. Srinivasa, S. Ashoka, and Debasis De

Abstract The paper represents an environmentally strong facile one-pot wet chemical synthesis for the production of nickel (Ni) doped cobalt oxide (Co_3O_4) at different weight% (1, 3, 5, 7 and 10%) of Ni that demonstrates efficient electrocatalysis towards hydrogen evolution reaction (HER), in 1M KOH solution. The Ni doped Co_3O_4 were synthesized using cobalt nitrate hexahydrate $\{Co(NO_3)_2.6H_2O\}$, nickel nitrate hexahydrate $\{Ni(NO_3)_2.6H_2O\}$and citric acid $\{C_6H_8O_7\}$, among which 7% Ni-doped Co_3O_4 proved to be an optimal HER catalysis. Nickel foam electrode drop cast with Ni-doped Co_3O_4 displayed remarkable HER catalysis with an overpotential (η) of 212 mV (vs RHE) recorded at 50 mA/cm^2. The composite also exhibits Tafel slope of 48 mV/dec @ HER which is kinetically more efficient. 7% Ni-doped Co_3O_4 has the potential to replace precious-metal-based catalysts as the anodic/cathodic material within electrolyzers, reducing the associated costs of hydrogen production from water splitting.

Keywords Wet chemical synthesis · Ni-doped Co_3O_4 · Hydrogen evolution reaction (HER) · Ni foam · Water splitting

1 Introduction

With the increase in the demand for replacing fossil fuel reserves and to meet the impending need of energy crisis that has led to growing concern over global warming, greenhouse effect, and drastic climate change [1], there have been breakthrough attention paid towards storage systems with low cost and environment benignity, high efficiency and alternative form of energy. Hydrogen is considered to be the most asepsis, energy-efficient form of resource whose production relies on steam reforming of

A. Basu · D. De (✉)
Energy Institute, Bengaluru (Centre of Rajiv Gandhi Institute of Petroleum Technology, Jais, Amethi), Bengaluru, India
e-mail: debasisd@rgipt.ac.in

N. Srinivasa · S. Ashoka
Department of Chemistry, School of Engineering, Dayanand Sagar University, Bengaluru, India

P. Muthukumar et al. (eds.), *Innovations in Sustainable Energy and Technology*, Advances in Sustainability Science and Technology,
https://doi.org/10.1007/978-981-16-1119-3_4

hydrocarbons or gasification process of biomass substances leading to the generation of carbon dioxide. Eventually, several techniques have been developed such as that of electrochemical or photoelectrochemical water splitting reaction which has proven to be environmentally sustainable for the generation of hydrogen gas that is of high purity and high production [2]. Study based on Co_3O_4 catalyst materials have been made for a long time in numerous fields related degradation [3], Zn-air batteries [4], supercapacitor [5] and water splitting [6] because of their excellent properties. To make significant changes in morphology and structural notations up-gradation in methods have been implemented. These include direct high-temperature oxidation [7], hot plate combustion method [8], precipitation method [9], and hydrothermal reaction [10]. Recently, trends have gone in the synthesis of two (2D)-dimensional porous structures that give high specific surface area and more active sites for reaction. Following this trend doping based catalyst of specified groups includes carbon (C)-doping [6, 11], PdO-doping [12], La^{3+}-doping [13], sulfur (S)-doping [14], iron (Fe)-doping [15, 16], manganese (Mn)-doping [17, 18], chromium (Cr)-doping [19] and nickel (Ni)-doped Co_3O_4 is also reported for electrocatalytic property [20, 21]. So far, few reports depict that Ni-doped Co_3O_4 catalysts with low percentage of nickel contain shows effect of water splitting in both alkaline and acidic medium.

Over recent years it is observed that electrolysis of water is now economically viable for niche application and be deployed for the large-scale within a decade [22]. Therefore, the ability for hydrogen to be produced on a large-scale basis which has significant environmental, societal and economic value will come into play shortly as a global energy system [23]. Splitting action of water as a whole experimental setup is concerned is divided into two half reaction namely: Hydrogen Evolution Reaction (HER) and Oxygen Evolution Reaction (OER), they both are important as per as overall efficiency of water splitting. To enhance the reaction kinetics and lower overpotential several efficient catalyst or doped catalyst have been implemented to proceed towards water electrolysis. From industrial scale, there exist a strong rigidity towards the unavailability of cheap and highly efficient electrocatalyst. Few catalysts based on non-noble earth-abundant metal such as CoS [24], CoP [25, 26], FeP [27], WP_2 [28], Mo_2C [29], MoB [30], MoS_x [31], $Co(OH)_2$ [32] and H_2-Co [33] have shown low stability under working conditions. Recent literature survey reveals that currently catalyst like RuO_2 and IrO_2 [23] being used but are considered to be the rarest element on earth. To mitigate such circumstances, they have been replaced with non-precious catalysts which are abundant on earth like cobalt and its substituted cobaltite's $M_xCo_{3-\underline{x}}O_4$ (M = Ni, Fe, Cu or Zn) showing excellent activities for splitting of water [34, 35]. Due to electrochemical oxidation metallic cobalt gets transformed into $Co(OH)_2$, CoO, Co_3O_4 and CoOOH [35–37]. Through several studies, it is observed that Co_3O_4 has proven to be one of the superior catalysts to display efficient corrosion stability. Morphological activities show that to enhance the electrochemical behaviour it is important to increase the effective surface area (Table 1).

Herein we present a comparative study based on the facile one-pot wet chemical synthesis assisted by combustion for the catalyst used i.e., Ni-doped Co_3O_4 which is formed as a result of doping characteristic of Nickel Nitrate Hexahydrate: Ni

Table 1 Comparison of catalyst, overpotential and Tafel analysis of the present work for the reported Nickel and Cobalt oxides based electrocatalysts[a] for HER

Catalysts	Substrate	Electrolyte	Overpotential (mV)	Tafel plot (mV/dec)	References
NiCoP NSA	3D Ni foam	1M KOH	308	68	[38]
Porous Cobalt Phosphide/phosphate TF	Glass slide	1M KOH	430	–	[39]
FeCoO	Ni foam	1M KOH	205	117	[40]
$NiCo_2S_4$ @ NiFe LDH	Ni foam	1M KOH	200	101.1	[41]
Ni-doped Co_3O_4 porous nanoplates	–	1M KOH	120	62	[42]
Ni-doped CoS_2- NN	SS	1 M KOH	350	76	[43]
$Ni(OH)_2$	Ni foam	1M KOH	237	140	[44]
Sulphur doped Nickel Cobalt (S–LDH)	SS	6M KOH	380	69	[45]
Ni-doped Co_3O_4	Ni foam	1M KOH	212	48	This work

[a]*NSA*-Nanosheet arrays; *TF*-Thin film; *LDH*-Double-layer hydroxide; *NN*-Nanoneedle; *SS*-Stainless steel

$(NO_3)_2.6H_2O$ with Cobalt Nitrate Hexahydrate: Co $(NO_3)_2.6H_2O$ along with Citric Acid at the concentration level of 1, 3, 5, 7 and 10%. Using different characterisations of powder XRD, SEM, study Ni-doped Co_3O_4 has been examined to get a relationship among catalyst activity, morphology and electrochemical behaviour for the generation of hydrogen. Finally, this is accompanied by the preparation of electrode materials using a drop cast technique. Once the electrode is prepared it is directed for hydrogen evolution. The material worked as a highly stable and active electrocatalyst for HER in 1M KOH of pH 13.6. Amongst all, Ni-doped Co_3O_4 (7% doped) achieves 50 mA/cm^2 current densities at a low overpotential of 212 mV showing better performance than bare Ni foam electrode or undoped Co_3O_4 electrode. Results show that 7% Ni-doped Co_3O_4 achieves highly efficient, stable and robust electrocatalyst with superior activity as compared to that of previously documented reports.

2 Experimental Methodology

2.1 Reagents and Materials

Chemicals and solvents used for synthesis are of analytical grades cobalt nitrate hexahydrate ($Co(NO_3)_2.6H_2O$) of purity 97%; Laboratory Reagent—Spectrum Reagent & Chemicals; C 0117, nickel nitrate hexahydrate ($Ni(NO_3)_2.6H_2O$) of purity 99%; Analytical Reagent—Sd fine-CHEM Limited; UN 2725, citric acid ($C_6H_8O_7$) of purity 99.5%; Laboratory Reagent—Sd fine-CHEM Limited and potassium hydroxide (KOH) of purity 85%; Laboratory Reagent—MERCK; CAS 1310-58-3 were purchased from SD Fine Chemicals Ltd. and received without any purification.

2.2 *Material Synthesis*

Preparation of Doped Compound

Typical solution process in Fig. 1: Nickel nitrate hexahydrate (0.00290 g–1%, 0.00872 g–3%, 0.01453 g–5%, 0.020355 g–7%, 0.02907 g–10%) is dissolved in 10 ml of deionized water and stirred until the nitrate dissolved completely. To the above solution, Cobalt nitrate hexahydrate (0.28812 g, 0.282 g, 0.2764 g, 0.27066 g, 0.26193 g) and Citric acid (0.11673 g, 0.11661 g, 0.11670 g, 0.11673 g, 0.11674 g) was added and continually stirred with magnetic stirrer to get a uniform solution. This was followed by keeping the precursor solution in a preheated muffle furnace maintained at 510 °C. The reaction was rested to form graphitic carbon by taking out the reaction vessel after 4 min from preheated muffle furnace. Finally, the as-obtained powder of Ni-doped Co_3O_4 at a concentration of 1, 3, 5, 7 and 10% doping level was crushed in a mortar and pestle which was used for further study (Fig. 2).

Characterization

The synthesized product of Ni-doped Co_3O_4 crystal structure was examined using powder X-ray Diffraction (XRD), Zeiss Rigaku series, with Cu Kα radiation (λ =

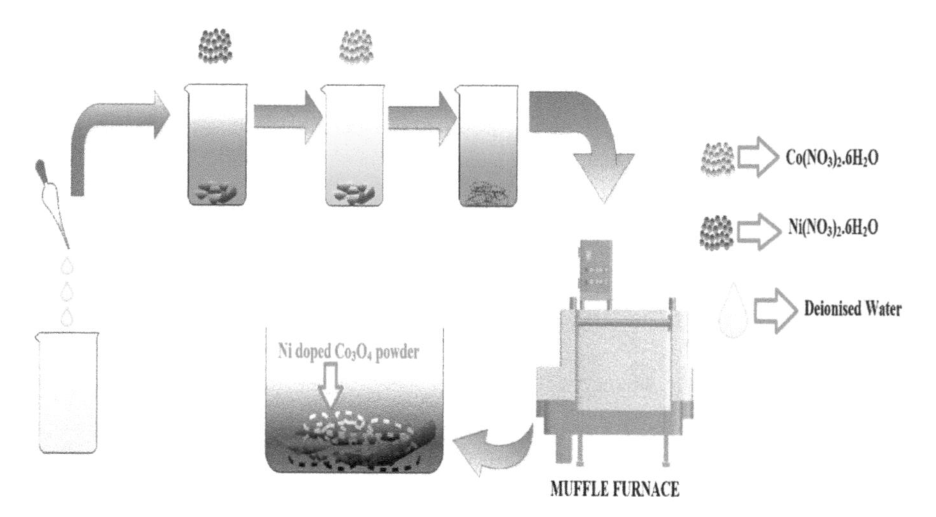

Fig. 1 Schematic diagram of Ni-doped Co_3O_4 synthesis

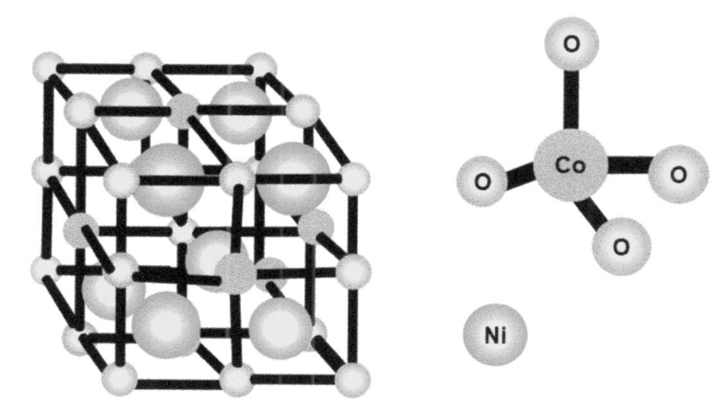

Fig. 2 Proposed structural model of Ni-doped Co_3O_4

1.5418 Å). Surface morphological behaviour of the samples was characterized using a scanning electron microscope (SEM, VEGA3 TESCAN).

Electrode Fabrication

The fabrication of the Ni-Foam electrode utilized within this study as illustrated in Fig. 3. The electrodes for all the samples were dipped in concentrated HCl under ultrasonication for 15 min. Then 5 mg of Co_3O_4, and Ni (1, 3, 5, 7 and 10%) doped Co_3O_4 were taken individually. To it 0.2 ml deionized water added and 0.2 ml ethanol was also added. To hold the sample in a semi-liquid state Nafion binder of 10 μl was added. All the samples were then sonicated for 30 min to get a homogenous dispersion. 3 mg of the resultant homogenous dispersion was drop cast onto the Ni-Foam electrode and subjected to drying for 2 h in a hot air oven. Thus, the electrocatalytic performance towards the HER of each modified Ni-Foam electrode was ready to be assessed.

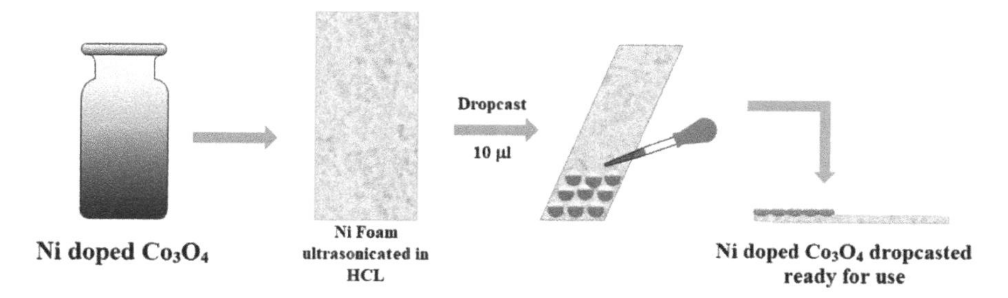

Fig. 3 Steps for the fabrication of Ni foam electrode with Ni-doped Co_3O_4 post sonication

Electrochemical Measurements

Cyclic Voltammetry (CV), Linear Sweep Voltammetry (LSV) and Electrochemical Impedance Spectroscopy (EIS) measurements have been carried out using a three-electrode system where the Cobalt Oxide or Nickel doped Cobalt Oxide electrodes drop cast on Ni-Foam, Pt and Ag/AgCl electrodes were used as working electrode, the counter electrode and reference electrode respectively. The catalytic activity of the Co_3O_4, Ni–Co_3O_4 (Ni-1%, 3%, 5%, 7% and 10%) towards the hydrogen evolution was studied in 1M KOH of pH = 13.6.

3 Results and Discussion

3.1 *Microstructural Characteristics of Ni–Co_3O_4*

The crystallographic science of the as-synthesized compound was analyzed with X-ray diffraction patterns in Fig. 4. As far the powder XRD of the prepared materials exhibit peaks at 31.11°, 36.648°, 44.75°, 59.101° and 65.120° corresponding to (220), (311), (400), (511) and (440) crystalline plane. All the diffraction peaks can be perfectly indexed and represented to that of spinel cubic structure of Co_3O_4 with Fd3m space group (JCPDS 74-2120) [40]. For 1%, Ni-doped Co_3O_4 shows a shift in peak as compared to other samples which is attributed strongly to the formation of phrase due to high-temperature reaction leading to change in structure. Further, it can also happen that due to low weight percentage of nickel doping at 510 °C some parts may have remained unreacted completely with cobalt oxide. The presence of NiO formation is denoted by *.

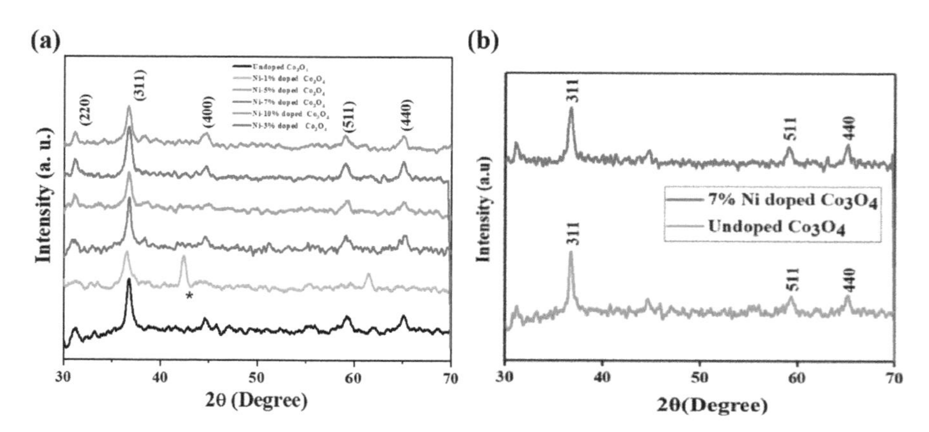

Fig. 4 **a** Powder XRD patterns of the undoped Co_3O_4 and Ni-doped Co_3O_4 (Ni-1%, 3%, 5%, 7% and 10%.). **b** Distinct peaks at the observed planes for 7% Ni-doped Co_3O_4 and undoped Co_3O_4

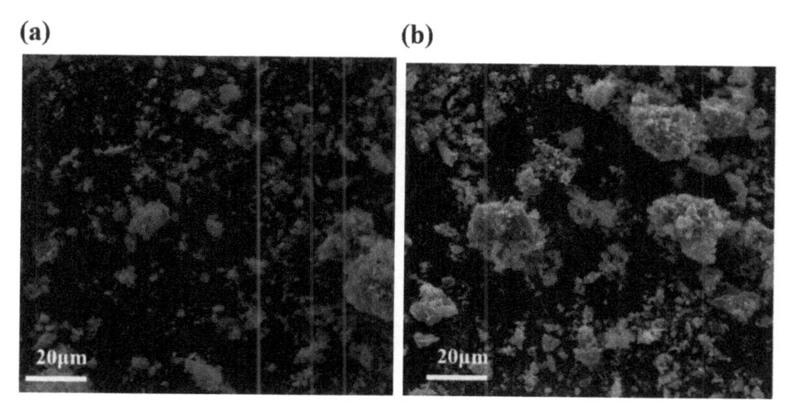

Fig. 5 SEM image of **a** Pure Co_3O_4 and **b** 7% Ni-doped Co_3O_4 prepared by wet chemical synthesis followed by combustion temperature of 510 °C

Apart from this size of the crystallite can be well evaluated using the Scherrer equation based on the full width of half maximum. This is generally represented by Eq. (1):

$$t = \frac{0.9\lambda}{\beta \mathrm{Cos}\theta} \tag{1}$$

Here t is the crystallite size, λ is the wavelength, β is the full width at half maximum of the peak and θ is the peak diffraction angle. The average crystallite size of undoped and doped sample is ~20 nm.

Figure 5 represents SEM images that narrate surface morphology of the particles in (a) pure Co_3O_4 and (b) 7% Ni-doped Co_3O_4 formation synthesized by wet chemical method at 510 °C in a muffle furnace. From the SEM image of pure Co_3O_4 formation has small agglomerated elliptical structures in the range of 8 μm through 15 μm. SEM image of 7% Ni-doped Co_3O_4 shows a comparatively large agglomerated compacted structure with the formation of Ni active sites over the surface of Co_3O_4 particles. They ranged from 19 to 32 μm. The agglomerated structure is attributed to high-temperature combustion for doping. Apart from these images give a clear notion of active sites being formed over doped surfaces for further high-performance HER activity.

3.2 Electrochemical Applications of Nickel-Doped Cobalt Oxide

The Ni doped Co_3O_4 synthesized using citric acid for estimation of HER was investigated in 1M KOH electrolyte. An amount of 3 mg Undoped Co_3O_4, Ni-doped Co_3O_4

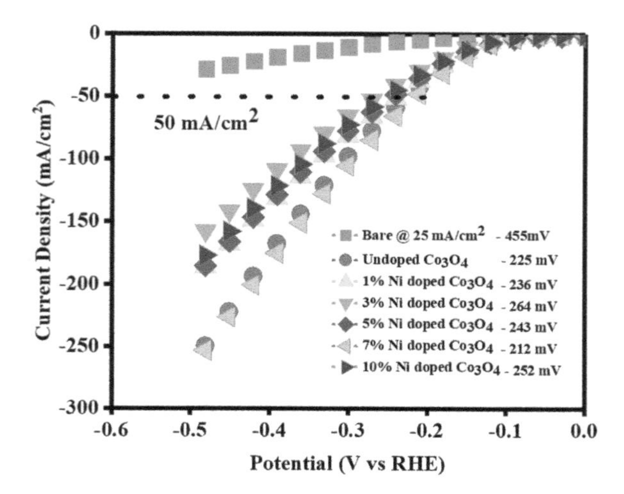

Fig. 6 LSV curve for HER of pure Co_3O_4, and Ni-doped Co_3O_4(Ni-1, 3, 5, 7, and 10%) Co_3O_4 in 1M KOH electrolyte

(Ni-1, 3, 5, 7, 10%) and bare nickel foam (NF) was loaded using drop cast with Nafion binder as a working electrode. From the linear sweep voltammetry (LSV) curve, it depicts the onset potential at −0.15 V due to reduction of Co^{4+} to Co^{3+} suggesting that the Co^{4+} containing octahedral sites responsible for this onset potential. This is accompanied by a sudden increase in current density after −0.15 V for the hydrogen evolution to proceed. The LSV curve depicts the overpotential required for HER at 50 mA/cm^2 for the all the doped Co_3O_4, undoped Co_3O_4 and bare electrodes presented in the curve.

From Fig. 6 it is inferred that 7% of Ni-doped Co_3O_4 exhibits the lowest overpotential of 212 mV required to deliver a current density of 50 mA/cm^2. For the purpose of comparison bare nickel foam and Ni-doped Co_3O_4 (Ni-1, 3, 5, 10%) was conducted which showed an overpotential of 455, 236, 264, 243, 252 mV to deliver current density of 25 mA/cm^2 (bare nickel foam) and 50 mA/cm^2 for doped materials. The main drawback lies with the higher overpotential, there is the evolution of larger H_2 bubbles on the active surface of catalyst resulting in low mass transport and less active surface area by attaching to the hydrophobic surface of the catalyst [17, 18] Such surfaces can be altered using rotating electrodes but here our material as-synthesized contains oxide which makes it hydrophilic resulting in easy release of H_2 gas molecules in the form of small tiny bubbles. Hence, the 7% Ni-doped Co_3O_4 shows very low overpotential of 212 mV to drive a current density of around 50 mA/cm^2 which makes this an efficient electrocatalyst for practical applications. Furthermore, from the Tafel slope (Fig. 7) of the bare nickel foam, undoped Co_3O_4 and all Ni-doped Co_3O_4 shows 160 mV/dec, 122 mV/dec, 57 mV/dec, 68 mV/dec, 36 mV/dec, 48 mV/dec and 50 mV/dec respectively (Fig. 7). Here, at 48 mV/cm^2 considering 7% Ni-doped Co_3O_4 adsorption of H^+ led to the formation of H2 bubbles. The reaction involved in H^+ adsorption mechanism given by Eqs. (2) and (3):

$$H_2O + e^- \rightarrow H_{(absorbed)} + OH^- \quad (2)$$

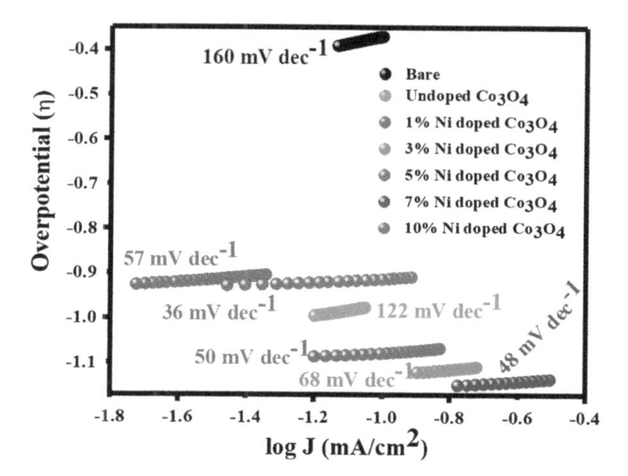

Fig. 7 Tafel plot of pure Co_3O_4 and Ni-doped Co_3O_4 (Ni-1, 3, 5, 7 and 10%)

$$H^+ + H_{(absorbed)} + e^- \rightarrow H_2 \quad (3)$$

The kinetic behaviour of adsorption is attributed to the graphitic carbon of citric acid which results in increasing the electronic conductivity and active sites to enhance the charge and mass transfer process. Thereby, making 7% Ni-doped Co_3O_4 modest and efficient electrocatalyst with low overpotential and Tafel slope for water splitting.

For further investigation, the reaction kinetics of the as-synthesized electrocatalysts, electrochemical impedance spectroscopy (EIS) measurements were carried out in 1M KOH solution. Based on the Nyquist plots which are exhibited within the frequency range of 0.01 Hz–100 KHz an outstanding performance is observed by 7% Ni-doped Co_3O_4 agglomerated catalyst showing significantly low interfacial charge transfer resistance. The results indicate that 7% Ni-doped Co_3O_4 promoting effective amount of HER kinetics compared to those of pure Co_3O_4 or another Ni-doped Co_3O_4. All the doped catalyst including that of pure Co_3O_4 consist of the semicircle. The measurement of these semicircle i.e. diameter indicates the charge transfer resistance (R_{ct}) of the hydrogen evolution reaction; when the value of this R_{ct} is a small potential for faster charge transfer is possible. Figure 8 represents the Nyquist plots of all Ni-doped Co_3O_4 composites, pure Co_3O_4, bare electrode and RuO_2 for comparison and measures impedance spectra using Zsimpwin software by fitting with an electrical equivalent circuit. An equivalent circuit that is composed of Rct, constant phase element Q, and Warburg impedance (W) corresponds to different electrochemical processes that occur at the electrode/electrolyte interface. The Nyquist plot consists of an intercept and straight sloping line at high and low-frequency regions indicates the Rct and W respectively. From the results R_{ct} value of 7% Ni-doped Co_3O_4 to be 20.67Ω which is much lower than the value of bare electrode (732.8Ω), 3% Ni-doped Co_3O_4 (634.3Ω) and pure Co_3O_4 (269.6Ω). The enhanced charge transfer rate is attributed to the change in morphological and

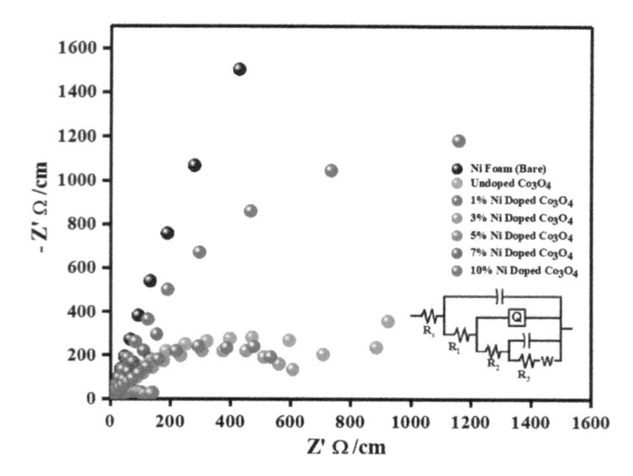

Fig. 8 Electrochemical impedance spectroscopy of Ni-foam, undoped Co_3O_4 and Ni-doped Co_3O_4 (Ni- 1, 3, 5, 7, and 10%)

structural orientation after the doping effect of nickel over cobalt oxide at different percentages.

4 Conclusions

In summary, we have grown a rapid one-pot wet chemical synthesis assisted with combustion at significantly 1, 3, 5, 7 and 10% weight % of Ni doped Co_3O_4 for the fabrication of doped catalysts. Here, bare electrode and pure Co_3O_4 has been used for comparison to show the best result for hydrogen evolution with doped electrocatalysts. 7% Ni-doped Co_3O_4 shows exactly in favour of significant HER generation for a current density of 50 mA/cm^2 at a low overpotential of 212 mV. Moreover, due to the small charge transfer resistance of 20.67Ω, this catalyst exhibits the most efficient one among all other synthesized catalysts. Based on this result as a part of interface development material, it is expected to be a promising candidate for hydrogen evolution reaction. Thus, it also proposes that adoption to transition metal oxides with small concentration doping inspires in generating bifunctional electrocatalysts.

References

1. Abdullah, M.I., Hameed, A., Zhang, N., Ma, M.: Nickel doped cobalt - hollow nanoparticles as an efficient electrocatalyst for hydrogen evolution from neutral water. J. Hydrogen Energy **44**, 14869–14876 (2019)
2. Glenk, G., Reichelstein, S.: Economics of converting renewable power to hydrogen. Nat. Energy **4**, 216–222 (2019)

3. Gao, R., Yan, D.: Fast formation of single-unit-cell-thick and defect-rich layered double hydroxide nanosheets with highly enhanced oxygen evolution reaction for water splitting. Nano Res. **11**, 1883–1894 (2018)
4. Liu, Q., Wang, L., Liu, X., Yu, P., Tian, C., Fu, H.: N-doped carbon-coated Co_3O_4 nanosheet array/carbon cloth for stable rechargeable Zn-air batteries. Sci. China Mater. **62**, 624–632 (2018)
5. Meher, S.K., Rao, G.R.: Ultra-layered Co_3O_4 for high-performance supercapacitor applications. J. Phys. Chem. C **115**, 15646–15654 (2011)
6. Liu, R., Wang, Y., Liu, D., Zou, Y., Wang, S.: Water-plasma-enabled exfoliation of ultrathin layered double hydroxide nanosheets with multivacancies for water oxidation. Adv. Mater. **29**, 1701546 (2017)
7. Ren, S., Guo, Y., Ju, L., Xiao, H., Hu, A., Li, M.: Facile synthesis of petal-like nanocrystalline Co_3O_4 film using direct high-temperature oxidation. J. Mater. Sci. **54**, 7922–7930 (2019)
8. Kombaiah, K., Vijaya, J.J., Kennedy, L.J., Kaviyarasu, K.: Green synthesis of Co_3O_4 nanorods for highly efficient catalytic, photocatalytic, and antibacterial activities. J. Nanosci. Nanotechnol. **l**(19), 2590–2598 (2019)
9. Wang, Y., Wei, X., Hu, X., Zhou, W., Zhao, Y.: Effect of formic acid treatment on the structure and catalytic activity of Co_3O_4 for N_2O decomposition. Catal. Lett. **149**, 1026–1036 (2019)
10. Chen, C., Wang, B., Liu, H., Chen, T., Zhang, H., Qiao, J.: Synthesis of 3D dahlia-like Co_3O_4 and its application in superhydrophobic and oil-water separation. Appl. Surf. Sci. **471**, 289–299 (2019)
11. Xu, A., Dong, C., Wu, A., Li, R., Wang, L., Macdonald, D.D., Li, X.: Plasma-modified C-doped Co_3O_4 nanosheets for the oxygen evolution reaction designed by Butler–Volmer and first-principle calculations. J Mater. Chem. A **7** 4581- 4595 (2019)
12. Zhong, H., Liu, T., Zhang, S., Li, D., Tang, P., Alonso-Vante, N., Feng, Y.: Template-free synthesis of three-dimensional NiFe-LDH hollow microsphere with enhanced OER performance in alkaline media. J. Energ. Chem. **33**, 130–137 (2019)
13. Mahmoudi-Moghaddam, H., Tajik, S., Beitollahi, H.: Highly sensitive electrochemical sensor based on La3+-doped Co_3O_4 nanocubes for determination of sudan I content in food samples. Food Chem. **286**, 191–196 (2019)
14. Liang, W., Fan, K., Luan, Y., Tan, Z., Al-Mamun, M., Wang, Y., Liu, P., Zhao, H.: Sulfur-doped cobalt oxide nanowires as efficient electrocatalysts for iodine reduction reaction. J. Alloys Compd. **772**, 80–91 (2019)
15. Deng, J., Xu, M., Feng, S., Qiu, C., Li, X., Li, J.: Iron-doped ordered mesoporous Co_3O_4 activation of peroxymonosulfate for ciprofloxacin degradation: Performance, mechanism and degradation pathway. Sci. Total. Environ. **658**, 343–356 (2019)
16. Gong, M., Li, Y., Wang, H., Liang, Y., Wu, J.Z., Zhou, J., Wang, J., Regier, T., Wei, F., Dai, H.: An advanced Ni–Fe layered double hydroxide electrocatalyst for water oxidation. J. Am. Chem. Soc. **135**, 8452–8455 (2013)
17. Li, L., Zhang, Y., Li, J., Huo, W., Li, B., Bai, J., Cheng, Y., Tang, H., Li, X.: Facile synthesis of yolk–shell structured $ZnFe_2O_4$ microspheres for enhanced electrocatalytic oxygen evolution reaction. Inorg. Chem. Front. **6**, 511–520 (2019)
18. G. Li, M. Chen, Y. Ouyang, D. Yao, L. Lu, L. Wang, X. Xia, W. Lei, S. Chen, D. Mandler, Q. Hao.: Manganese doped Co3O4 mesoporous nanoneedle array for long cycle-stable supercapacitors; Appl. Surf. Sci (469), 941–950, 2019.
19. Ravi Dhas, C., Venkatesh, R., Sivakumar, R., Dhandayuthapani, T., Subramanian, B., Sanjeev Raja, C.A.: Electrochromic performance of chromium-doped Co_3O_4 nanocrystalline thin films prepared by nebulizer spray technique. J. Alloys Compd. **784**, 49–59 (2019)
20. Singhal, A., Bisht, A., Irusta, S.: Enhanced oxygen evolution activity of $Co_{3-x}Ni_xO_4$ compared to Co_3O_4 by low Ni doping. J. Electroanal. Chem. **823**, 482–491 (2018)
21. Banerjee, S., Debata, S., Madhuri, R., Sharma., P.K.: Controlled hydrothermal synthesis of graphene supported $NiCo_2O_4$ coral-like nanostructures: an efficient electrocatalyst for overall water splitting. Appl. Surf. Sci. **449**, 660–668 (2018)

22. Staffell, I., Scamman, D., Velazquez Abad, A., Balcombe, P., Dodds, P.E., Ekins, P., Shah, N., Ward, K.R.: The role of hydrogen and fuel cells in the global energy system. En. Environ. Sci. **12**, 463–491 (2019)
23. Petrykin, V., Macounova, K., Shlyakhtin, O.A., Krtil, P.: Tailoring the selectivity for electrocatalytic oxygen evolution on ruthenium oxides by zinc substitution. J. Angew. Chem. Int. Ed. **28**, 4813–4815 (2010)
24. Sun, Y., Liu, C., Grauer, D.C., Yano, J., Long, J.R., Yang, P., Chang, C.J.: Bioinspired iron sulfide nanoparticles for cheap and long-lived electrocatalytic molecular hydrogen evolution in neutral water. J. Am. Chem. Soc. **135**, 17699–17702 (2013)
25. Tian, J., Liu, Q., Asiri, A.M., Sun, X.: Self-supported nanoporous cobalt phosphide nanowire arrays: an efficient 3D hydrogen-evolving cathode over the wide range of pH 0–14. J. Am. Chem. Soc. **136**, 7587–7590 (2014)
26. Pu, Z., Liu, Q., Jiang, P., Asiri, A.M., Obaid, A.Y., Sun, X.: CoP nanosheet arrays supported on a Ti plate: an efficient cathode for electrochemical hydrogen evolution. Chem. Mater. **26**, 4326–4329 (2014)
27. Tian, J., Liu, Q., Liang, Y., Xing, Z., Asiri, A.M., Sun, X.: Strong metal-phosphide interactions in core-shell geometry for enhanced electrocatalysis. ACS Appl. Mater Interfaces **6**, 20579–20584 (2014)
28. Pu, Z., Liu, Q., Asiri, A.M., Sun, X.: Tungsten phosphide nanorod arrays directly grown on carbon cloth: a highly efficient and stable hydrogen evolution cathode at all pH values. ACS Appl. Mater Interfaces **6**, 21874–21879 (2014)
29. Xile, H.: Molybdenum boride and carbide catalyze hydrogen evolution in both acidic and basic solutions. Angew Chem. Int. Ed **51**, 12703–12706 (2012)
30. Merki, D., Vrubel, H., Rovelli, L., Fierro, S., Hu, X.: Fe Co, and Ni ions promote the catalytic activity of amorphous molybdenum sulfide films for hydrogen evolution. Chem Sci **3**, 2515–2525 (2012)
31. Cobo, S., Heidkamp, J., Jacques, P.-A., Fize, J., Fourmond, V., Guetaz, L., Jousselme, B., Ivanova, V., Dau, H., Palacin, S.: A Janus cobalt-based catalytic material for electro-splitting of water. Nat. Mater. **11**, 802 (2012)
32. Li, Y., Hasin, P., Wu, Y.: Magnetic-field-assisted preparation of one-dimensional (1-D) wire-like NiO/Co_3O_4 composite for improved specific capacitance and cycle ability. J. Adv. Mater. **17**, 1926–1929 (2010)
33. Kumar, B., Saha, S., Basu, M., Ganguli, A.K.: J. Mater. Chem. A, **11**, 4728–4734 (2013)
34. Chi, B., Li, J., Yang, X., Lin, H., Wang, N.: Electrophoretic deposition of $ZnCo_2O_4$ spinel and its electrocatalytic properties for oxygen evolution reaction. J. Electrochim. Acta **10**, 2059–2064 (2005)
35. Yu, Y., Zhang, J., Zhong, M., Guo, S.: J. Springer publication 2018, **9**, 653–661 (2018)
36. Chanda, D., Hnát, J., Paidar, M., Schauer, J., Bouzek, K.: Synthesis and characterization of $NiFe_2O_4$ electrocatalyst for the hydrogen evolution reaction in alkaline water electrolysis using different polymer binders. J. Power Sourc. **285**, 217, (2015)
37. Yang, Y., Fei, H., Ruan, H., Tour, J.M.: Porous cobalt-based thin film as a bifunctional catalyst for hydrogen generation and oxygen generation. Adv. Mater. **27**, 3175–3180 (2015)
38. Bandal, H.A., Jadhav, A.R., Tamboli, A.H., Puguan, J.M.C.: Electrochim. Acta **249**, 253–262 (2017)
39. Liu, J., Wang, J., Zhang, B.: Hierarchical $NiCo_2S_4$@NiFe LDH heterostructures supported on nickel foam for enhanced overall-water-splitting activity. ACS Appl. Mater. Interfaces **9**, 15364 – 15372 (2017)
40. Zhang, Y., Li, J., Fang, J., Dai, Y.: Low Ni-doped Co_3O_4 porous nanoplates for enhanced hydrogen and oxygen evolution reaction. J. Alloys Compounds **823**, 153750 (2019)
41. He, G., Zhang, W., Deng, Y.: Engineering pyrite-type bimetallic Ni-doped CoS_2 nanoneedle arrays over a wide compositional range for enhanced oxygen and hydrogen electrocatalysis with flexible property. Catalysts **7**, 366 (2017)
42. Rao, Y., Wang, Y., Ning., Li, P.: Hydrotalcite-like $Ni(OH)_2$ nanosheets in situ grown on nickel foam for overall water splitting. ACS Appl. Mater. Interfaces **8**, 33601–33607 (2016)

43. Karthik, N., Atchudan, R., Xiong, D.: Electro-synthesis of sulfur doped nickel cobalt layered double hydroxide for electrocatalytic hydrogen evolution reaction and supercapacitor applications. J. Electroanalyt. Chem. **833**, 105–112 (2018)
44. Jia, Q., Liu, E., Jiao, L., Li, J., Mukerjee, S.: Current understandings of the sluggish kinetics of the hydrogen evolution and oxidation reactions in base. Curr. Opin. Electrochem. **12**, 209 (2018)
45. Li, Y., Zhang, H., Jiang, M.: Ternary NiCoP nanosheet arrays: an excellent bifunctional catalyst for alkaline overall water splitting; Nano Res. **9**, 2251–2259 (2016)

Aerodynamic Performance Enhancement and Optimization of a H-rotor Vertical Axis Wind Turbine

Abhilash Nayak, Sukanta Roy, Sivasankari Sundaram, Arjun Deo, and Hakeem Niyas

Abstract The aerodynamic performance of a H-rotor vertical axis wind turbine (VAWT) is dependent on a number of factors, with the shape of aerofoil being the crucial one. So, in this research study, an attempt is made to incorporate three such aerofoils like DU 06-W-200, NACA 0015, NACA 64-415. These aerofoils exhibit varied properties viz. camber, chord length and thicknesses, which have variable effects on the performance of the wind turbine. The aerofoils show a decent performance with an optimal value of coefficient of performance when incorporated in the wind turbine. Further, simulations are executed with these aerofoils being implemented in a single H-rotor VAWT. The flow governing equations around the VAWT using these aerofoils are solved using ANSYS-Fluent software. The performance of VAWTs is studied in terms of power and moment coefficients to determine its efficacy. As an outcome of the research, the best wind turbine structures [C1C2C2] and [C2C2S1] are used for optimum performance and self-starting characteristics, owing to the perks associated with them.

Keywords Vertical axis wind turbine · Aerodynamic performance · Aerofoil · Self-starting characteristic · Stability

1 Introduction

Wind is generated by the uneven heating of the atmosphere by the sun, variations in the earth's surface, and rotation of the earth. Moreover, mountains, water bodies and vegetation also influence the wind flow patterns. Wind power is the technique of utilizing the kinetic energy of wind to generate electricity for which wind turbines are implemented to harness the energy. A wind turbine is defined as a rotating machine

A. Nayak (✉) · S. Sundaram · A. Deo · H. Niyas
Energy Institute, Centre of Rajiv Gandhi Institute of Petroleum Technology, Bengaluru, Karnataka, India
e-mail: abhilash.nayak2@gmail.com

S. Roy
Curtin University, Miri, Sarawak, Malaysia

P. Muthukumar et al. (eds.), *Innovations in Sustainable Energy and Technology*, Advances in Sustainability Science and Technology,
https://doi.org/10.1007/978-981-16-1119-3_5

that converts the kinetic energy of wind into mechanical energy. This mechanical energy is then converted to electricity, thus naming it as a wind generator, wind energy converter, or aerogenerator. Wind turbines can be divided into two types based on the axis in which the turbine rotates such as horizontal axis wind turbine (HAWT) and vertical axis wind turbine (VAWT). Among all the renewable energy sources, wind energy is a promising research area for which a lot of effort and resources have been invested in order to upgrade the technology of power generation. It has a huge potential of energy with worldwide production of 650.8 GW as per September 2020. The global wind power market is mostly centered in countries like China, the USA, Germany, India, Spain, etc. Considering the global scenario, as shown in Fig. 1, wind energy has its huge potential and production at China with installed capacity of 237 GW followed by the USA, Germany, India, Spain and many other countries.

HAWTs are involved with a lot of advantages to it such as variable blade pitch that gives the turbine blades the optimum angle of attack, taller height having an access to stronger winds, high efficiency, etc. These merits have led to its preference over other wind turbines. Besides, it also incurs difficulty in transportation of large blades and towers, installation issues, requirement of additional yaw control mechanism. Further technological advancement has led to the development of VAWT.

VAWTs have the main rotor shaft arranged vertically with its generator and gearbox at its base. The advantages of these structures include yaw less mechanism, small structure, lower wind start-up speeds, low noise, etc. It also has some disadvantages such as low efficiency, maintenance, etc. Inspite of low coefficient of performance, VAWTs are recognized to be an alternative to HAWTs particularly in urban areas for small-scale power generation. As discussed above, VAWTs have been implemented depending on the drag and lift force-based structures, which has led to

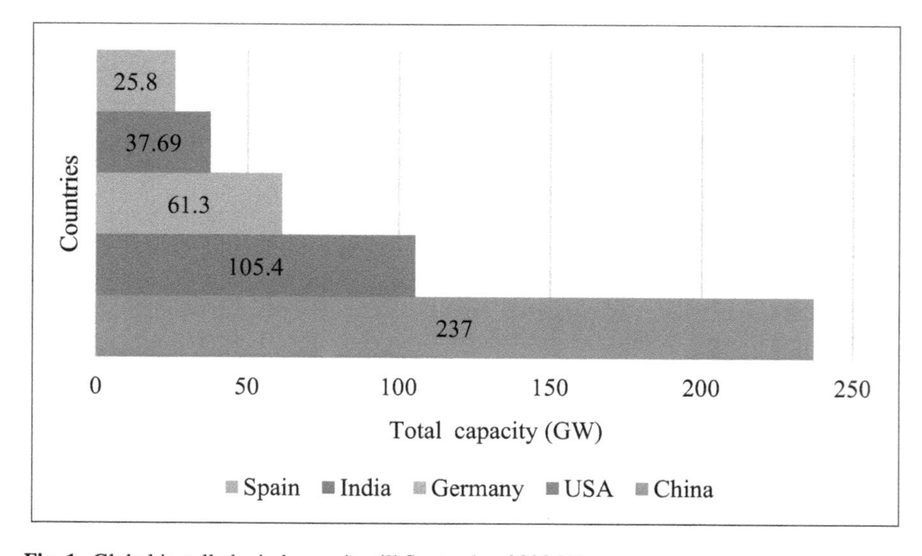

Fig. 1 Global installed wind capacity till September 2020 [1]

different design configurations such as Darrieus, Savonius, H-rotor, Helical structure, etc. Considering the Darrieus and H-rotor turbines, which are lift-based structures, the shortcomings in self-starting issues is compensated by giving an initial rotation to the turbine to help it get acquainted with the wind flow. According to Mohamed et al. [2], these turbines are generally preferred among all VAWTs because of its comparatively better efficiency and the wide range of Tip Speed Ratio (TSR) (~2–8). Zemamou et al. [3] commented on the performance of Savonius rotors, the drag-based turbines, which has its minimal preference due to its low efficiency (~30%) and low TSR range (~0–1.8). The merit of VAWTs being omni-directional in nature, allows the turbine to receive wind from any direction so that a yawing mechanism is not required. Yet, VAWTs possess certain minor limitations in the operational efficiency and self-starting issues in comparison with HAWTs. Irrespective of the limitations of VAWTs with respect to the operational parameters like TSR, self-starting, poor efficiency, etc., these incorporate a broader scope of research because of its lower maintenance, lesser pre-installation procedures, acceptance of low wind speeds, etc. Also, VAWTs can be an attractive concept in peak clipping in demand response strategies serving a limited residential area. Hence, researchers have made an attempt to explore the possibilities of addressing the operational problems of VAWTs.

In the current study, a 3 bladed H-rotor VAWT has been considered, where the objective is to enhance and optimize the aerodynamic performance using multiple aerofoil structured blades in a single turbine. Being in ease with its installation, the turbines are practically implemented in road medians, rooftops, etc. The flow governing equations around VAWT are solved using ANSYS Fluent software. This 2D numerical study aimed to investigate the performance and efficiency using various configurations of aerofoils in the wind turbine.

2 Literature Survey

As we know, the optimization of VAWT has guided in the exploration of possibilities and played a major role in increasing the operational range of TSR, further deciding the optimum coefficient of performance. Literatures reported in this area are classified based on the type of aerofoil used and on the techniques of simulations performed.

In the design approach, the discussion regarding aerofoil was done on the basis of designs. According to Bhutta et al. [4], crossflex turbine was found to be more effective, when installed across high rise buildings experiencing wind speed of at least 14 m/s. A hybrid design of Savonius and Darrieus with C_p of 0.35 was also taken into account which showed variations in its values when plotted against TSR. Moreover, mathematical equations related to exergy losses were also incorporated during the process of irreversibility. According to Batista et al. [5], a new blade EN0005 was further developed with self-starting ability and good performance at high TSRs with C_p of 0.375 at TSR 7.5. The blade had its ability to self-start at 1.25 m/s and remained stable till 25 m/s, without any production of noise. Battisti et al. [6] led the further study by comparing different VAWT structures on the basis of blade no., design, inclination, TSR and azimuthal position. As the 2 bladed rotor had issues with its stability, so a 3 bladed H-rotor DU 06-W-200 was chosen which had a C_p of 0.4 at starting TSRs. The study also proposed helical bladed rotors because of its wide TSR range (1–6.5). Some modified aerofoils were also taken into consideration in Chen et al. [7] with parameters like TCR (thickness to chord ratio), LEN (leading edge thickness), MTITOC (maximum thickness in terms of chord). The research work for optimizing the aerofoil was based on C_p using 2 algorithms: OFAAT (one factor at a time) and Orthogonal algorithm. According to the report, NACA0015-64 was chosen as the best with C_p of 0.413975 through Orthogonal Algorithm while NACA0018-64, with C_p of 0.4585, was opted through OFAAT algorithm. Again a study was done by Qamar and Janajreh [8] comparing 3 aerofoils: NACA 4512, NACA 7512, NACA 0012, where they found that the cambered foils performed better than the symmetrical ones. They also came to a conclusion that moderate cambered aerofoil, i.e. NACA 4512 was much more effective than NACA 7512 with regards to the ability to self-start. A rigorous study by Mohamed et al. [2] was done by performing CFD simulations on 25 different aerofoils. It was concluded that LS(1)-0413 (C_p ~0.415) performed better than standard NACA0021 aerofoil. In a study for the aerodynamic enhancement of the horizontal axis wind turbines by Sedighi et al. [9], proposal of spherical dimples on the suction sides of the turbine blades was executed using RANS solver and k-ω Shear Stress Transport turbulent model. The parameters such as radius, location, quantity of dimples, blade pitch angle, wind speed happen to be of prime importance for the torque and power generation. The improvised model with 150 dimples effective at a pitch angle of 4°–11° and wind speed range of 12–16 m/s was found out to be lucrative in enhancing the torque by about 16.08%. In a reviewed study based on the performance analysis of VAWT by Kaushik et al. [10], experiments were conducted by wind tunnel under steady flow and uniform conditions and field test setup under wind conditions and dynamic flows. The field tests were executed to determine the 3D characteristics and performance under natural wind conditions whereas the wind tunnel aided in the aerodynamic design with the help of parameters viz. TSR, C_p, C_T, R_e, blade number, overlapping effects, blockage effects, etc.

In the simulation approach, the discussion regarding aerofoil was done based on simulations performed. In the study by Jin et al. [11], various techniques of research like computational aerodynamics, CFD, experimental methods were carried out. The 2D CFD model was preferred because of its fastest computational time. Similar methods and techniques were observed in the study by Bedon et al. [12] where 2D URANS CFD model was used to optimize the aerofoil shape. Among the works done, WUP 1615 structure was found to have a higher C_p than standard NACA 0018 at TSR of 4. The study by Balduzzi et al. [13] conducted the 3D simulations considering losses, stall, vorticity, wake interaction. The simulations were compared and power reduction between 2D and 3D was noticed. Moreover, emphasis on finer grid was given as simulations with coarser grid led to ambiguous results. Hasan et al. [14] proposed a new way of simulation approach that included BEM and CFD analyses. Though the value of C_p was higher in the case of BEM rather than in CFD, simulations using CFD was preferred because of its proximity to experimental values. The CFD simulations were also performed taking into account the factors like TSR, solidity, wind speed, number of blades, blade profile, stalling, self-start, etc. by Ghasemian et al. In this paper, the reduction of noise was projected wherein TSR and solidity were the influencing factors for noise. Hence, a solution was proposed by Ghasemian et al. [15] by improving the solidity and implementing the double spacing theory for noise reduction. A further new development was found in the study by Zhang et al. [16] with the concept of winglet designing which aids in vortex reduction, thereby enhancing the performance. The study implied that twist angle is the most crucial parameter for design. An improvement in the aerodynamic performance of the VAWTs was proposed by Wang et al. [17] using the CFD and Taguchi method at a low TSR of 2. A novel design of leading edge wavy serrations and helical blade structures with a twist angle of 60° formed the basis for the optimized wind turbine with an enhancement in the efficiency by 18.3%.

The consideration of a single type of aerofoil in a H-rotor VAWT, according to the previous research works has impelled the objective of this current effort to consider multiple design aerofoils in a single wind turbine to enhance the performance.

3 Proposed Model

According to the inference from reported research study, the unemployability of different wind turbine blade configurations in a single topology was identified. This led to the proposition of an unique model of the turbine incorporating different aerofoil structured blades in a H-rotor VAWT. The three best structures were chosen based on TSR, C_p, aerofoil thickness, stability, self-start, etc. consisting of 1 symmetrical and 2 cambered aerofoils. The combinations include three types of aerofoils: DU 06-W-200 (C1) ($C_p = 0.4$ at starting TSRs, 20% thickness and 0.8% camber) [6], NACA 64-415 (*C2*) ($C_p = 0.405$ at TSR range of 2–8), NACA 0015 (S1) ($C_p = 0.41$) [2] out of which two are cambered (*C*) and one is symmetrical (*S*). The modelling of the H-rotor wind turbine is done via. AutoCAD with simulations in ANSYS. The proposed turbine model for the further research to be carried out included 2 different configurations, i.e. [C1C2C2] and [C2C2S1]. These structures were found out to be efficient enough for practical implementation, yet with some more focus to be considered on its stability issues.

4 Numerical Investigation

A number of aerofoils were considered and simulations were performed based on the TSR, C_p, thickness, stability, self-start, etc. For the current work, 6 best aerofoils, i.e. DU 06-W-200 [6], NACA 63-215, NACA 0015, LS(1)-0413, NACA 64-415, S1046 [2] were chosen among which, best 3 structures were opted owing to characteristics like high C_p (~0.4) and a wider TSR range (~2–8) as shown in Fig. 2.

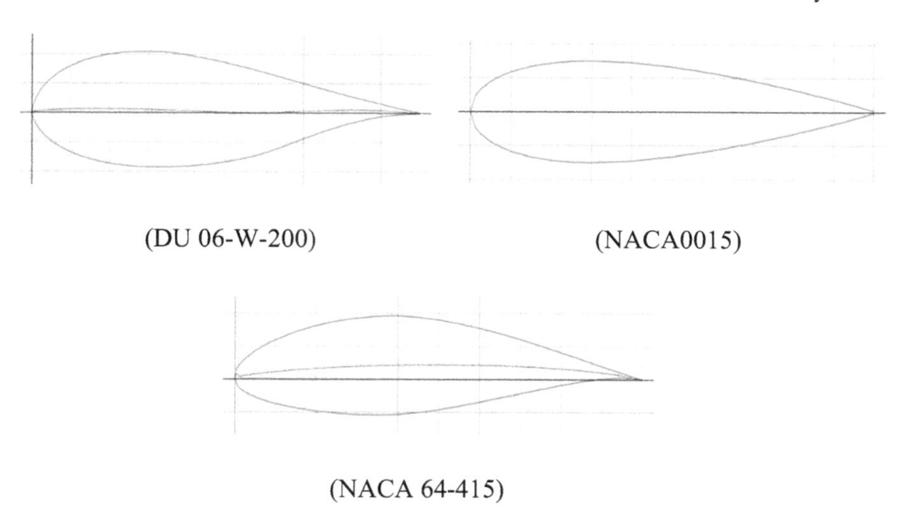

Fig. 2 Aerofoils selected for simulations [2, 6, 17]

These aerofoils have defined chord length and thicknesses which is generated with ease [18]. With the current scenario of implementing three different structures in the same turbine, possible combinations are examined through simulations. Ten combinations as shown in Table 1 incorporating 30 simulations are considered, thus generating variations of C_{m} (moment coefficient), C_{L} (lift coefficient), C_{d} (drag coefficient) with respect to the rotational time of turbine.

The modelling of a 2D H-rotor VAWT is done by generation of aerofoils from NACA website [18], which is then further imported in AutoCAD software for designing. The blades are positioned in an equilateral profile with 120° angular distance among them. A rotational hub of diameter 2000 mm and blade domains of diameter 400 mm each are constructed for identifying the variations in its vicinity. A boundary layer with 18,540 mm × 8240 mm zone is also considered with respect

Table 1 Combination of aerofoils [2, 6]

Aerofoil types	Combinations
DU 06-W-200 (C1)	C1C1C2 (1)
	C1C1S1 (2)
	C1C2C2 (3)
NACA 64-415 (C2)	C1C2S1 (4)
	C1S1S1 (5)
	C2C2S1 (6)
NACA 0015 (S1)	C2S1S1 (7)
	S1S1S1 (8)
	C1C1C1 (9)
	C2C2C2 (10)

to the wind speed and the wake effect. The designed model is hence transferred to the ANSYS to further carry out the required simulations. It inculcates numerous mathematical equations within it to determine the flow of a process. The parameters such as Refinement, Inflation, Sizing, etc. are taken into consideration around the domains and rotating hub for the generation of mesh. The simulations in ANSYS generate contours with information related to velocity, turbulence, pressure, etc. which make use of k-ω SST model along with Navier Stokes equation (both continuity and momentum equations), as indicated in Eqs. (2, 3, 4). The location of the blades along with the boundary conditions for inlet and outlet are pre-determined and mentioned with respect to the wind velocity and pressure. A sample modelled view of the meshing using the ANSYS software is shown in Fig. 3.

The GIT (grid independence test) is a kind of accuracy determining test done by using successively smaller cell sizes for the calculations, thus modifying the coarser mesh to a much finer mesh. The test commenced with the number of cell elements increasing gradually from 80,000 till 200,000, where the fineness of the mesh was achieved at 180,000 elements with an acceptable error with magnitude less than 2%. The representation of the GIT along with the error reduction is projected as shown in Fig. 4. Gradually with the increase in the number of cells, the error settles down thus confirming the convergence of the test. Some of the main features of the turbine are as tabulated in Table 2.

Certain basic equations regarding the power performance of the wind turbine in terms of power coefficient & moment coefficient are as mentioned in Eqs. (1), (2) & (3). The equation of power output from a wind turbine is formulated as in Eq. (4). The simulations in the modelling incorporated SST k-ω turbulence model along with Navier Stokes equations. The SST k-ω governing equations were formulated based on turbulent kinetic energy, as represented in Eq. (5) and specific dissipation rate,

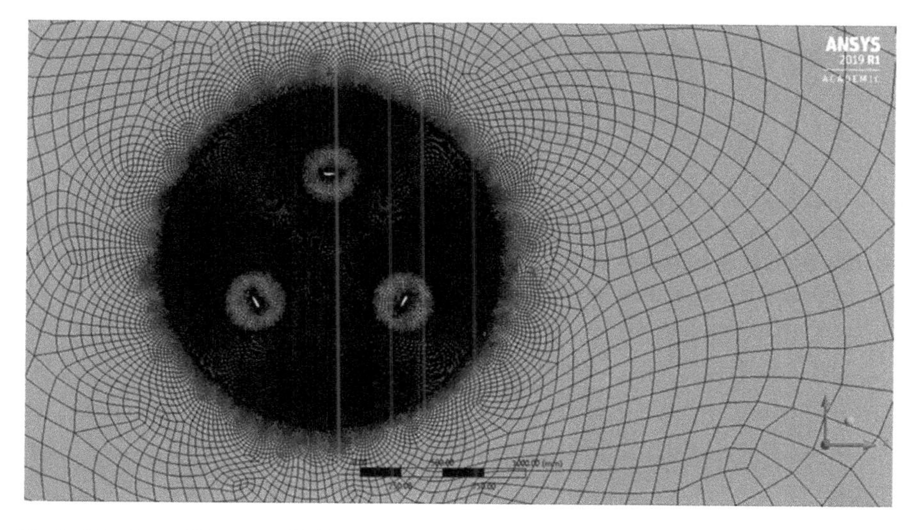

Fig. 3 Meshing of H-rotor VAWT

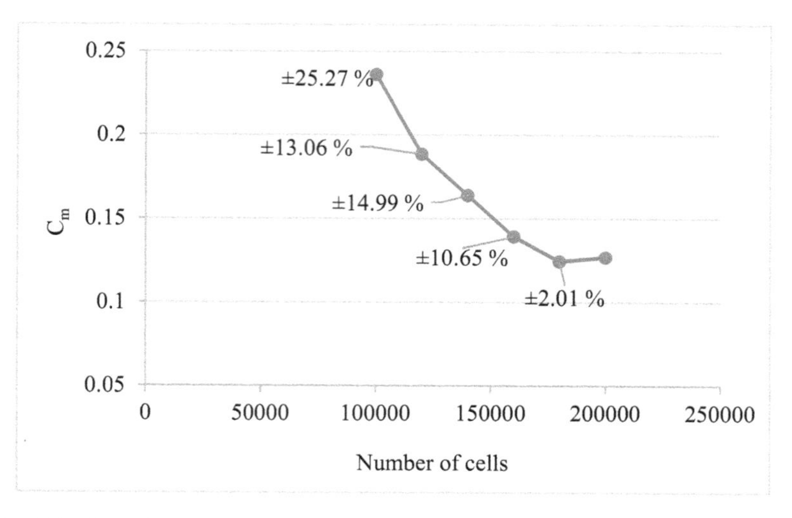

Fig. 4 Grid independence test (GIT)

Table 2 Features of the wind turbine

Specifications	Values
Rotor diameter (D) (mm)	1030
Rotor height (H) (mm)	1 (2D simulation)
Chord length (c) (mm)	85.8
No. of blades (N)	3
Wind speed (v) (m/s)	9
Rotational speed (ω) (rad/s)	73.398
Solidity (σ)	0.25
Tip speed ratio (λ) (TSR)	4.18
Viscosity (μ) (kg/m s)	1.7894×10^{-5}
Reynolds no. (R_e)	6.346×10^{5}

as represented in Eq. (6) [19]. The value of F_1 in Eq. (6) ranges from 0 in the free stream to 1 inside the boundary layer. **The Navier–Stokes equation of Newtonian viscous fluid is denoted as in** Eq. (7). When the flow is assumed as incompressible (i.e. constant density), then the equation gets simplified to Eq. (8).

$$C_p = \frac{P_{\text{rated}}}{\frac{1}{2}\rho A v^3} = \frac{T\omega}{\frac{1}{2}\rho A v^3} \tag{1}$$

$$C_m = \frac{T}{\frac{1}{2}\rho A \frac{D}{2} v^2} \tag{2}$$

$$C_p = \lambda \cdot C_m \tag{3}$$

$$P = \frac{1}{2}\rho A v^3 C_p(\lambda, \beta) \tag{4}$$

$$\frac{\partial k}{\partial t} + U_j \frac{\partial k}{\partial x_j} = P_k - \beta^* k\omega + \frac{\partial}{\partial x_j}\left[(\nu + \sigma_k \nu_T)\frac{\partial k}{\partial x_j}\right] \tag{5}$$

$$\frac{\partial \omega}{\mathrm{d}t} + U_j \frac{\partial \omega}{\partial x_j} = \alpha S^2 - \beta\omega^2 + \frac{\partial}{\partial x_j}\left[(\nu + \sigma_\omega \nu_T)\frac{\partial \omega}{\partial x_j}\right] + 2(1 - F_1)\sigma_\omega \frac{1}{\omega}\frac{\partial k}{\partial x_j}\frac{\partial \omega}{\partial x_j} \tag{6}$$

$$\rho\frac{\partial v}{\partial t} = -\nabla p + \rho g + \mu\nabla^2 v - \rho(v \cdot \nabla)v \tag{7}$$

$$\rho\frac{\partial v}{\partial t} = -\nabla p + \rho g + \mu\nabla^2 v \tag{8}$$

$\rho\frac{\partial v}{\partial t}$: momentum convection ($kg/m^2s^2$)
∇p: surface force, pressure gradient (kg/m^2s^2)
ρg: mass force, external forces (such as gravity, electro-magnetic) (kg/m^2s^2)
$\mu\nabla^2 v - \rho(v \cdot \nabla)v$: viscous force ($kg/m^2s^2$)
ρ: air density (kg/m^3)
v: wind speed (m/s)
C_p: Coefficient of performance
C_m: Coefficient of moment
P_{rated}: Rated power (kW)
A: Area of the VAWT (m^2)
D: Diameter of the VAWT (m)
λ: Tip speed ratio of the VAWT
β: Blade pitch angle of the turbine (°)
ω: Rotational speed (rad/s)
T: Torque (N m)
ν_T: kinematic eddy viscosity (m^2/s).

5 Results and Discussion

5.1 *Influence on Static Pressure*

The static pressure contours resulted from simulations are shown in Fig. 5. According to the pictorial representation of the contours, the maximum value of the static pressure is represented in red with a gradual reduction in the values towards the minimum, represented in blue. Similar representations depicting the variations of the property through different colours as mentioned above is applicable for the consecutive

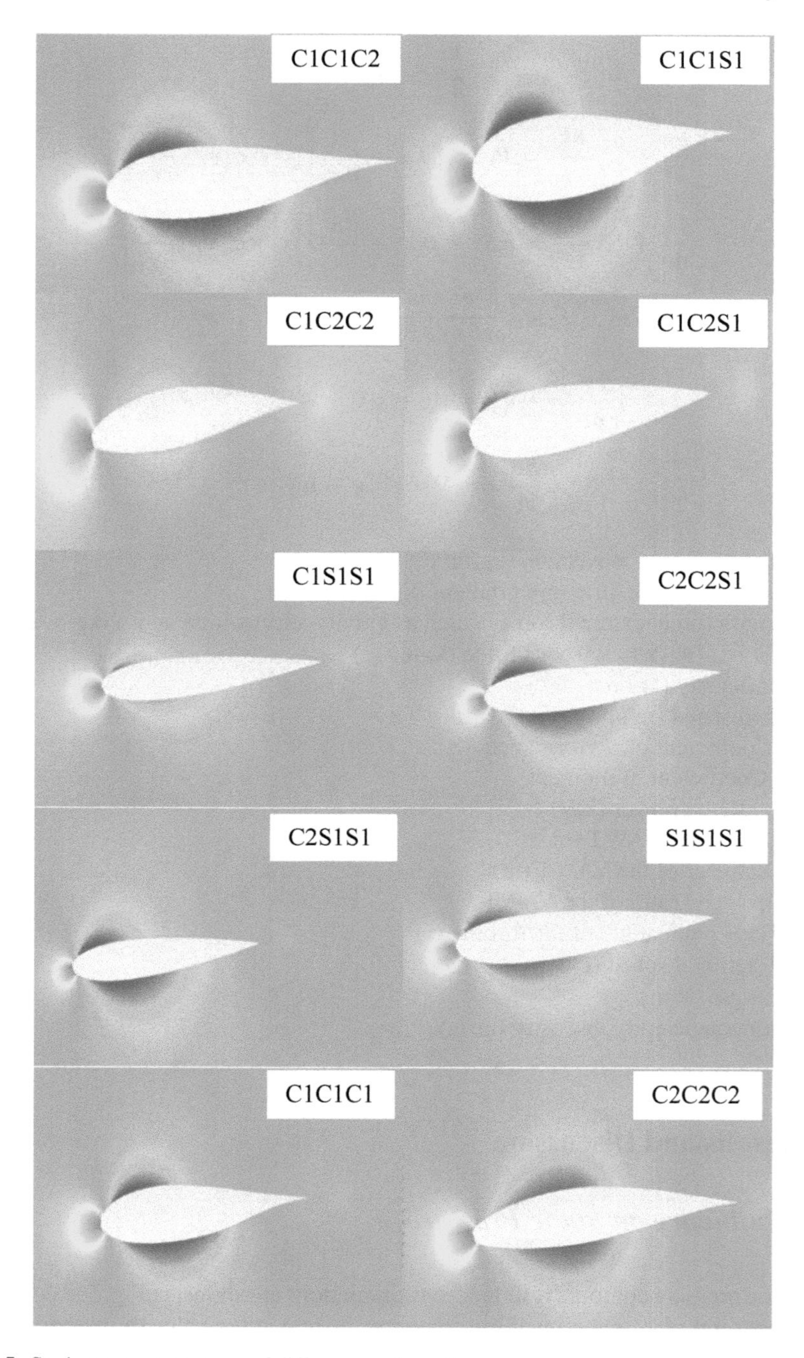

Fig. 5 Static pressure contours of different configurations

figures. The static pressure contours signify and pose a possibility to implement multiple aerofoil configurations with respect to the pressure gradients as compared to the standard symmetrical and cambered ones. The mixed configurations perform well as the symmetrical ones which can be deciphered from the illustrations of [C2C2S1] and [S1S1S1]. The magnitude of pressure at the lower surface of the aerofoil in the case of [C2C2S1] is a little bit high (around −843 Pa) compared to that of [S1S1S1] which is (around −922 Pa), posing for an efficient lift thereby assisting in an efficient turbine rotation.

5.2 Influence on Total Pressure

The total pressure contours resulted from simulations are shown in Fig. 6. The total pressure contours show us the pressure variations across the surfaces at certain positions. If we consider the [C2C2C2] structure, which consists of all cambered aerofoils, the pressure magnitude across the outer surface of the blade, around (−556 Pa) is uniformly distributed thus stabilizing the turbine during rotation. The situation is the same in the case of [C2C2S1] where the outer pressure with the value of (−604 Pa) is uniformly low thus enabling the structure to experience lift. Moreover, the pressure in the trailing edge of [S1S1S1] with the magnitude of (−615 Pa) is a bit lower compared to [C2C2S1], consequently generating a lesser force to make a forward move.

5.3 Effects on Velocity

The velocity contours resulted from simulations are as shown in Fig. 7. The velocity contours give us an insight about the flow around the blades. In the case of [C1C1S1] and [C1S1S1], the velocity at the trailing edge seems to be at its higher peak with around 48 m/s, thus defining the speed of the blade. The vigorous rotation of [C1C1S1] and [C1S1S1] can lead to instability thereby damaging the turbine or mitigating the aerodynamic performance. The velocity contours of [C1C2C2] and [C2C2S1] with a speed of around 40 m/s seem not so violent at the trailing edge of its blades. The selection can be further justified by its comparison with [S1S1S1] where the velocity magnitude is the highest at its tail with 48 m/s. It is the stability issue of the turbine which leads to such variations, even though the blade is facing wind directly.

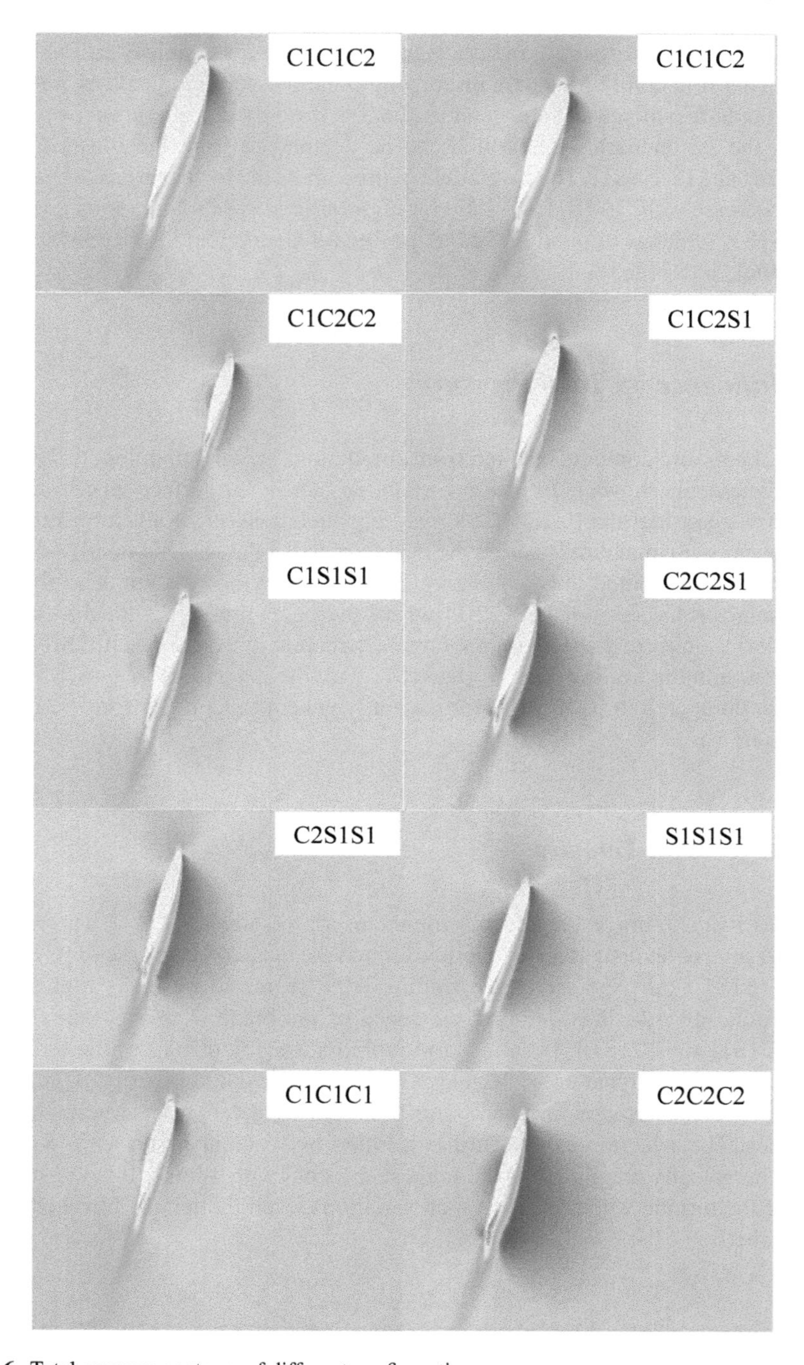

Fig. 6 Total pressure contours of different configurations

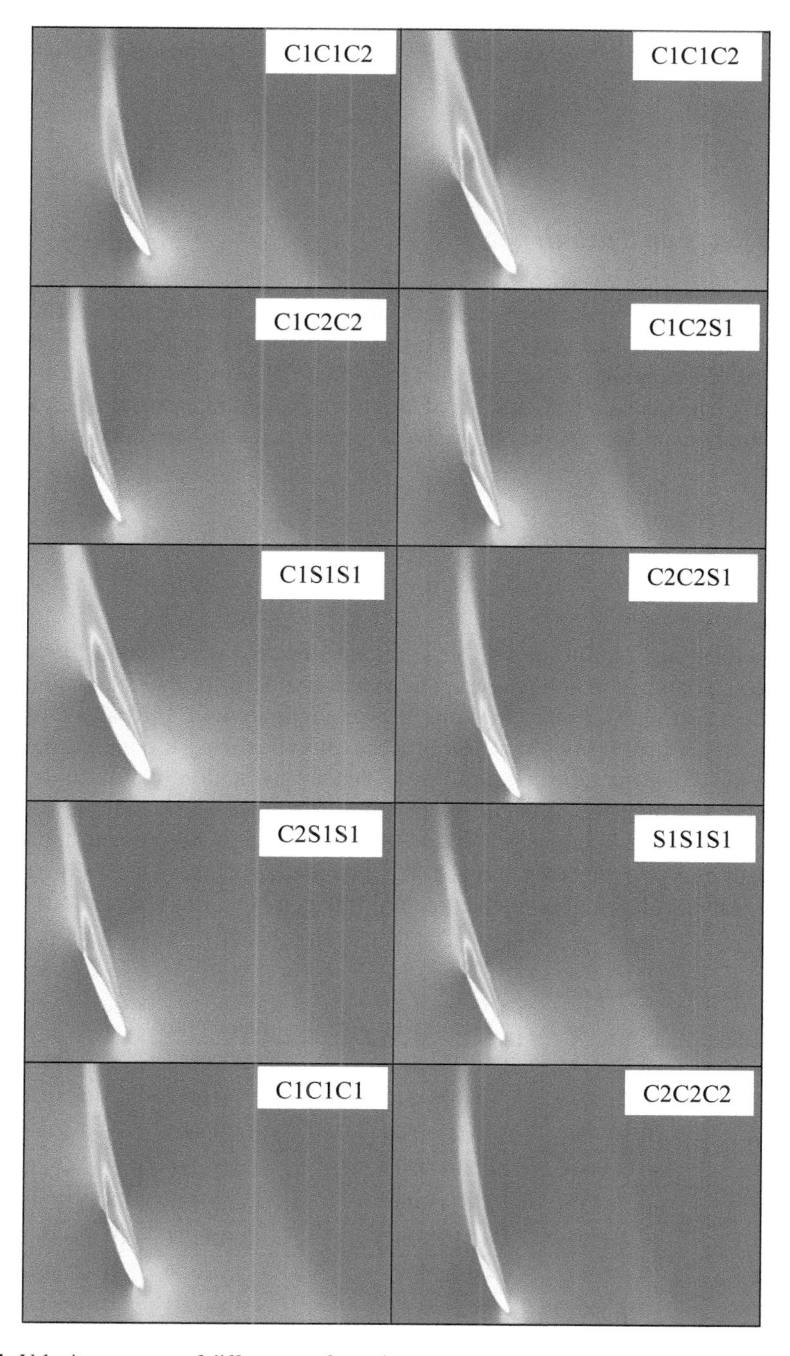

Fig. 7 Velocity contours of different configurations

Table 3 Optimum configuration values

Configuration	Static pressure (Pa)	Total pressure (Pa)	Velocity (m/s)
[C1C2C2]	−546	−515	40
[C2C2S1]	−843	−400	40

5.4 *Effects on Turbulence*

The turbulence contours define the disturbance created across the turbine. The issues related to the wake effect can be found out, further aiding in the identification of the optimized location for the successive turbines. This would help in the further research for designing an efficient wind farm for a particular location. The values of the required parameters for the optimum configurations have been tabulated in Table 3.

5.5 *Influence on T, C_p, C_L, C_d*

The simulated results of torque, C_m, C_L, C_d are tabulated which are further verified with the standardized combinations. The values are used for the calculation of C_p which decides the most efficient wind turbine configuration. The calculations and the contours aid in the optimization process, thus finalizing a combination to be the best fit. Using the information from Table 2, the equation relevant to torque is cited in Eq. (2), thus generating different values of torque. The simulated results for Torque are tabulated as shown in Table 4. The graph between net moment and various combinations is also projected in Fig. 8. We can infer from Table 4 that the magnitude of torque among all is highest in the 3rd, 6th, 10th combinations with 6.4, 6.0 and 6.8

Table 4 Simulated torque results

Simulations	C_m (average)	Net moment (Nm)	Torque (Nm)	Error torque
C1C1C2	−0.1285165	−5.3080789	−5.4720672	−0.2152845
C1C1S1	−0.1233788	−4.4784512	−5.2533122	−0.3103624
C1C2C2	−0.1504223	−4.2700546	−6.4047904	−0.4606697
C1C2S1	−0.1154360	−5.0322277	−4.9151160	−0.1717665
C1S1S1	−0.1092835	−4.2436484	−4.6531536	−0.2622346
C2C2S1	−0.1411275	−5.1726547	−6.0090311	−0.3036376
C2S1S1	−0.1209772	−4.9223381	−5.1510539	−0.2269606
S1S1S1	−0.0969050	−4.8946362	−4.1260909	−0.0403609
C1C1C1	−0.1306474	−4.2868604	−5.5628011	−0.3765924
C2C2C2	−0.1602907	−5.1173107	−6.8249749	−0.2502081

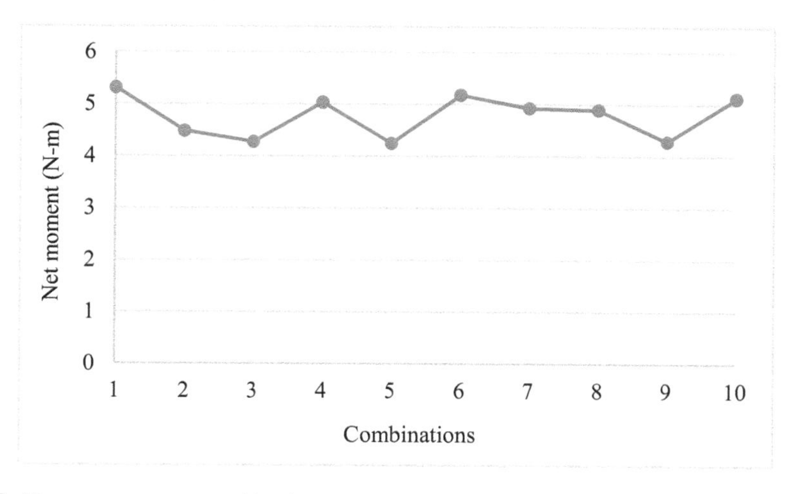

Fig. 8 Net moment versus combinations

Nm respectively. As the 10th combination, i.e. C2C2C2 is a cambered one, so it has its obvious preference for its optimistic lift and drag forces. Apart from that, the 3rd, i.e. C1C2C2 and the 6th, i.e. C2C2S1 combinations exhibit a decent performance in the torque simulations. These combinations can thus be considered on the basis of torque for maintaining a smooth rotation during the wind flow. Considering the 8th combination, i.e. S1S1S1 which consists of symmetrical configurations, though the torque is the least with around 4.1 Nm, but the performance exhibition is good because of the maintenance of stability during its circular movement. The comparison on the basis of $C_{p,t}$ is also shown in Table 5. The graph between $C_{p,t}$ and various combinations is plotted as shown in Fig. 9. The inference from Table 5 has the magnitude of $C_{p,t}$ highest in the 3rd, 6th, 10th combinations. The 10th combination, i.e. C2C2C2 being a cambered one has the highest $C_{p,t}$ (~0.67), which justifies its

Table 5 Simulated C_p results

Simulations	C_m (average)	$C_{p,t} = \lambda * C_m$(average)	$C_{p,t}$ (Theoretical)	Error $C_{p,t}$
C1C1C2	−0.1285165	−0.5371988	−0.5403119	−0.0017227
C1C1S1	−0.1233788	−0.5157234	−0.5362607	−0.0382972
C1C2C2	−0.1504223	−0.6287653	−0.5344376	0.1764990
C1C2S1	−0.1154360	−0.4825223	−0.5370213	−0.1014839
C1S1S1	−0.1092835	−0.4568052	−0.5371359	−0.1495538
C2C2S1	−0.1411275	−0.5899131	−0.5346631	0.1033362
C2S1S1	−0.1209772	−0.5056846	−0.5333925	−0.0519465
S1S1S1	−0.0969050	−0.4050628	−0.5315740	−0.2379935
C1C1C1	−0.1306474	−0.5461063	−0.5381258	0.0148300
C2C2C2	−0.1602907	−0.6700152	−0.5341726	0.2543048

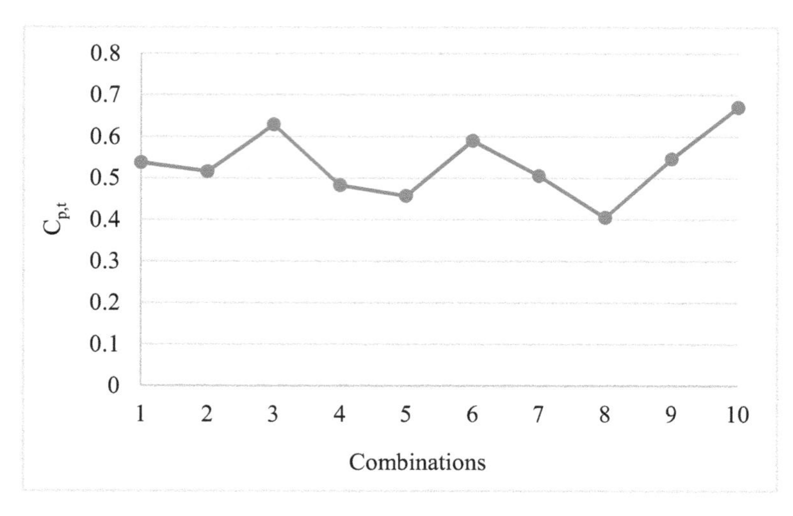

Fig. 9 $C_{p,t}$ versus combinations

priority. Apart from that, the 3rd, i.e. C1C2C2 and the 6th, i.e. C2C2S1 combinations exhibit quite good turbine efficiency. It can be clearly observed that the 3rd and 10th configurations have their efficiencies (~0.62 and ~ 0.67 respectively) much higher than the proposed Betz's Limit because of the consideration of uniform wind speed of 9 m/s, which is quite impractical. The reason behind such unrealistic values might be the blade tip losses or simulation issues that gave rise to such unexpected results. The comparison on the basis of C_L and C_d is shown in Table 6, which further becomes a mode of selection on the basis of C_L/C_d ratio. The graph between C_L/C_d and various combinations is represented in Fig. 10. From the above tabulation in Table 6, it can be easily noticed that the 7th combination, i.e. C2S1S1 would be preferable because of its high C_L/C_d ratio, i.e. 0.39, but the flaw is its high C_d value of 1.134 which acts as an impediment. The coefficients of lift, drag and moment are non-dimensional

Table 6 Simulated C_L and C_d results

Simulations	C_L	C_d	C_L/C_d
C1C1C2	−0.3722664	1.1672234	−0.3189333
C1C1S1	−0.4315963	1.1670209	−0.3698274
C1C2C2	−0.3753091	1.0993595	−0.3413889
C1C2S1	−0.3521510	1.1741136	−0.2999292
C1S1S1	−0.4112219	1.1756093	−0.3497947
C2C2S1	−0.3531262	1.0968312	−0.3219513
C2S1S1	−0.4464126	1.1346970	−0.3934201
S1S1S1	−0.3494801	1.1769194	−0.2969448
C1C1C1	−0.4467704	1.1637807	−0.3838957
C2C2C2	−0.3302124	1.0658855	−0.3098010

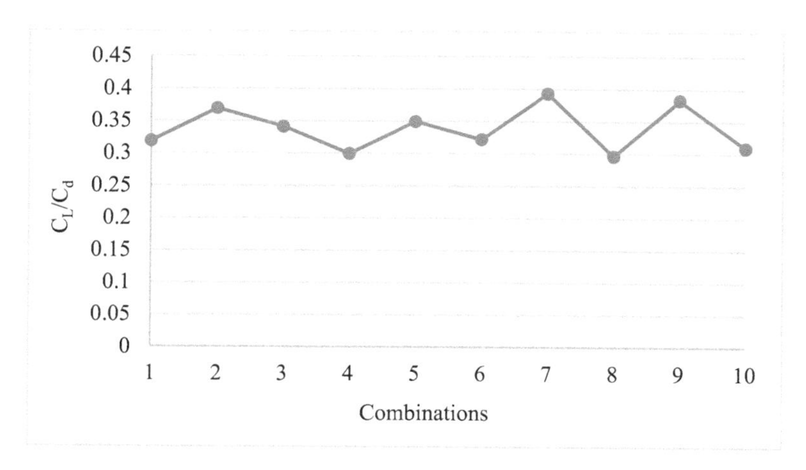

Fig. 10 C_L/C_d versus combinations

parameters representing the aerodynamic properties of a body. This helps in the identification of the most suitable configuration with sufficient amount of lift and minimum drag. A negative moment coefficient specifies the reduction in the angle of attack in the absence of a control input, which tends to resort in the linear lift region (stable) rather than the stall or post stall region (unstable). The reason behind the negative value of lift coefficient is the negative lift which directs the downward dropping of the aerofoil instead of the upward lifting.

Moreover, the 9th configuration also exhibits a similar high ratio of C_L/C_d (~0.38) and high lift coefficient (~0.44) due to its cambered shape. The 2nd combination of C1C1S1 has really high value of lift coefficient (around 0.43) which is an advantage but preferability would be less because of its low performance in Torque and $C_{p,t}$ simulations. An optimal lift to drag ratio is selected which depends on the angle of attack of the turbine blade, thus defining the aerofoil efficiency. The table also cites the low value of the 10th composition, i.e. C2C2C2 (about 0.30) which is an added advantage for its practical use. Along with that, the 3rd and the 6th combination (i.e. C1C2C2 and C2C2S1 respectively) have low values of drag coefficient of around 1.09 magnitude, which enhances the reliability to make a real-world use. The validation of these models have been done on the basis of coefficient of performance and the lift to drag ratio, suggesting a better optimized configuration.

6 Conclusion

This work has led to the conclusion that wind has a lot of potential in it and if properly harnessed, can solve the energy crises in the world. The study of the proposed H-rotor VAWT was done by configuring 10 different designs, comparing them with respect to the coefficients of moment, lift, drag, etc. The maximum output was realized by incorporating 3 different blade structures, i.e. DU 06-W-200 (at starting TSRs), NACA 0015, NACA 64-415 (TSR ~2–8) in a single H-rotor VAWT. From the current work, simulations of both cambered and symmetrical structured blades were executed which showed further scope of enhancement through static pressure, total pressure and velocity contours. The preferred optimized combination of aerofoils for the wind turbine was projected with C1C2C2 [DU 06-W-200, NACA 64-415, NACA 64-415] and C2C2S1 [NACA 64-415, NACA 64-415, NACA 0015]. Though installing different aerofoils in the same wind turbine would lead to stability issues, the problem might be mitigated by properly balancing the blades with the help of torque or moment. These stability issues in the proposed design can thus, create a scope for future work.

References

1. World Wind Energy Association, https://wwindea.org/blog/2020/09/29/wind-power-capacity-worldwide. Last accessed on 2020/12/06
2. Mohamed, M.H., Dessoky, A., Alqurashi, F.: Blade shape effect on the behaviour of the H-rotor Darrieus wind turbine: performance investigation and force analysis. Energy **179**, 1217–1234 (2019)
3. Zemamou, M., Aggour, M., Toumi. A.: Review of Savonius wind turbine design and performance. In: 4th International Conference on Power and Energy Systems Engineering, pp. 383–388. Energy Procedia, Germany (2017)
4. Bhutta, M.M.A., Hayat, N., Farooq, A.U., Ali, Z., Jamil, S.R., Hussain, Z.: Vertical axis wind turbine—a review of various configurations and design techniques. Renew. Sustain. Energy Rev. **16**, 1926–1939 (2012)
5. Batista, N.C., Melicio, R., Mendes, V.M.F., Calderon, M., Ramiro, A.: On a self-start Darrieus wind turbine: blade design and field tests. Renew. Sustain. Energy Rev. **52**, 508–522 (2015)
6. Battisti, L., Brighenti, A., Benini, E., Castelli, M.R.: Analysis of different blade architectures on small VAWT performance. J. Phys. **753**(6), 1–12 (2016)
7. Chen, J., Chen, L., Xu, H., Yang, H., Ye, C., Liu, D.: Performance improvement of a vertical axis wind turbine by comprehensive assessment of an airfoil family. Energy **114**, 318–331 (2016)
8. Qamar, S.B., Janajreh, I.: Investigation of effect of cambered blades on Darrieus VAWTs. In: 8th International Conference on Applied Energy, pp. 537–543. Energy Procedia, China (2017)
9. Sedighi, H., Akbarzadeh, P., Salavatipour, A.: Aerodynamic performance enhancement of horizontal axis wind turbines by dimples on blades: numerical investigation. Energy **195**, 1–9 (2020)
10. Kaushik, V., Shankar, R.N.: Innovative Design, Analysis and Development Practices in Aerospace and Automotive Engineering. Nicolas Gascoin, E. Balasubramanian, Springer, Singapore (2020)
11. Jin, X., Zhao, G., Gao, K., Ju, W.: Darrieus vertical axis wind turbine: basic research methods. Renew. Sustain. Energy Rev. **42**, 212–225 (2015)
12. Bedon, G., Betta, S.D., Benini, E.: Performance-optimized airfoil for Darrieus wind turbines. Renew. Energy **94**, 328–340 (2016)
13. Balduzzi, F., Drofelnik, J., Bianchini, A., Ferrara, G., Ferrari, L., Campobasso, M.: Darrieus wind turbine blade unsteady aerodynamics: a three-dimensional Navier-Stokes CFD assessment. Energy **128**, 550–563 (2017)
14. Hasan, M., Shahat, A.E., Rahman, M.: Performance investigation of three combined airfoils bladed small scale horizontal axis wind turbine by BEM and CFD analysis. J. Power Energy Eng. **5**, 14–27 (2017)

15. Ghasemian, M., Ashrafi, Z.N., Sedaghat, A.: A review on computational fluid dynamic simulation techniques for Darrieus vertical axis wind turbines. Energy Convers. Manage. **149**, 87–100 (2017)
16. Zhang, T., Elsakka, M., Huang, W., Wang, Z., Ingham, D.B., Ma, L., Pourkashanian, M.: Winglet design for vertical axis wind turbines based on a design of experiment and CFD approach. Energy Convers. Manage. **195**, 712–726 (2019)
17. Wang, Z., Wang, Y., Zhuang, M.: Improvement of the aerodynamic performance of vertical axis wind turbines with leading-edge serrations and helical blades using CFD and Taguchi method. Energy Convers. Manage. **177**, 107–121 (2018)
18. Airfoil Tools, https://airfoiltools.com/airfoil/naca4digit, last accessed on 2019/12/07
19. NASA, https://turbmodels.larc.nasa.gov/sst.html, last accessed on 2019/12/08

CFD Analysis of a Straight Bladed Darrieus Vertical Axis Wind Turbine Using NACA 0021 Aerofoil

Anand Raj, Sukanta Roy, Bhaskor Jyoti Bora, and Hakeem Niyas

Abstract This paper presents a 2D numerical simulation grid independence test (GIT) conducted on a straight bladed Darrieus vertical axis wind turbine. The objective is to analyse and understand the importance of performing a GIT during a numerical simulation process. Analysis of fluid flow is carried out through the velocity vectors before and after performing the GIT. NACA 0021 symmetrical aerofoil blade is chosen to carry out the numerical analysis. For performing the simulations, a 2D, SST k–ω turbulence model-based transient pressure-based solver is used. Wind velocity of 9 m/s is taken in the current study. It is shown that the grid independence test enhances the reliability of the results to a higher extent when the difference in coefficient of moment values between two identical computational domains comprised of different number of mesh elements is less than 1%.

Keywords Vertical axis wind turbine · Grid independence test · Rotating mesh

1 Introduction

The focus on renewable energy technologies has rapidly increased in recent years due to the diminishing reserves of fossil fuels, increase in energy demand, rise in fuel prices and growing environmental pollution. Various renewable energy sources include solar, biomass, geothermal, hydro, tidal, wave, ocean thermal and wind. Among all these, hydro, wind and solar energy accounts for most of the electricity production. Wind is one of the most sought-after sources of energy production owing to its clean nature, abundance availability and competitive cost per unit of electricity production. Based on the axis of rotation, wind turbines are of two types, i.e. horizontal axis wind turbine (HAWT) and vertical axis wind turbine (VAWT). In the past

A. Raj · B. J. Bora (✉) · H. Niyas
Energy Institute, Centre of Rajiv Gandhi Institute of Petroleum Technology, Bengaluru, Karnataka, India
e-mail: bhaskorb3@gmail.com

S. Roy
Curtin University, Miri, Sarawak, Malaysia

P. Muthukumar et al. (eds.), *Innovations in Sustainable Energy and Technology*, Advances in Sustainability Science and Technology,
https://doi.org/10.1007/978-981-16-1119-3_6

few decades, majority of research were carried out on HAWT, but recently there is an increase in attention towards VAWT. It is because of its inherent characteristics such as: (i) it can accept wind from any direction (omnidirectional), and (ii) it can deliver better performance even at high wind velocities and turbulent conditions [1]. Among the various available VAWT configurations, straight bladed Darrieus wind turbines are preferred over the others due to its simple and minimalistic design, which makes it cheaper to manufacture over other VAWTs.

Performance of a wind turbine majorly depends on the proper selection of aerofoil. Aerodynamic investigation for 20 different aerofoils was carried out by Mohamed [2] to improve the power output of H-rotor Darrieus turbine. He found that there was a relative increase of about 26.83% in power output for S-1046 aerofoil when compared to standard symmetric NACA aerofoils. Mohamed et al. [3] concluded that one of the most suitable aerofoils for straight bladed Darrieus VAWT is LS(1)-0413 aerofoil, which gives a coefficient of power (C_p) of 41.5% with an increase of 10% when compared to NACA 0018. They found that it also operated over a wide tip speed ratio (TSR) range, which means that stall can be delayed without any certain loss of efficiency. Hashem and Mohamed [4] found that VAWTs consisting of symmetric aerofoils have a relatively higher aerodynamic performance than the ones with asymmetric aerofoils. S-1046 symmetrical aerofoil was suggested by them for use in straight bladed Darrieus VAWT because it is having the highest C_p among all the other aerofoils considered in their investigation. Mohamed et al. [5] reported an absolute power output enhancement of 16% by using LS(1)-0413 aerofoil when compared to the conventional symmetric NACA 0021 aerofoil. The static torque coefficients for LS(1)-0413 were found to be higher than NACA 0021, which gives the added advantage of having better self-starting capability. Alsabri et al. [6] conducted several numerical simulations to increase the efficiency of H-rotor Darrieus turbine by improving the performance factor C_l/C_d ratio for different operating ranges of angle of attacks. They found that AH93W215 aerofoil has the highest efficiency among other aerofoils taken into consideration.

Usage of proper number of blades is crucial in the design of wind turbines to extract optimum power. Sabaeifard et al. [7] investigated the sensitivity of blade number in Darrieus VAWT. He concluded that more number of blades achieve maximum C_p for lower TSR, whereas a few number of blades achieve maximum C_p at higher TSR. Therefore, the best performance occurs when the rotor has 3 blades in their study. Li et al. [8] carried out a series of wind tunnel experiments starting with a single blade and increasing the blade numbers in consecutive experiments to determine the aerodynamic force characteristics. It was found that C_p reduces with the increase in number of blades. From the analysis carried, maximum C_p was found to be 0.410 and 0.326 for 2 and 5 blades, respectively.

Solidity of a wind turbine has a direct impact on the TSR range and C_p. Qamar and Janajreh [9] showed that the cambered blades with a lower solidity yield lower C_p, but over a wide range of TSRs. Similarly, they also found that higher solidity turbines, which have a solidity close to one, yield higher C_p, but at a lower TSR and for a shorter range of TSRs. Ghasemian et al. [10] deduced that higher solidity increases the C_p at lower TSR, whereas for high TSR, lower solidity turbines perform better.

Also, the operating range of TSR keeps on declining with the increase in solidity, due to the increase in wake effect. Mohamed [2] also recommended low solidity to be used for *H*-rotor Darrieus turbine to obtain wider operating range of TSR.

Distance between the wind turbines should be fixed in such a way that the overall performance of the wind farm is maximized. Shyu [11] conducted many wind tunnel experiments to analyse the performance gain by changing the distance between the VAWTs. It was found that when two VAWTs are moved towards each other (from 300 to 180 cm), both the wind turbines observed performance gain. When two VAWTs are kept at a distance of 1.5–2 times the diameter (*D*) of the turbine, the functioning of both the machines is amplified by about 11%. Ahmadi-Baloutaki et al. [12] performed wind tunnel experiments and reported that the optimal range of distance of the downstream wind turbine from a counter rotating pair of VAWT and the spacing between the pair was found to be three and one rotor diameters, respectively. Shaaban et al. [13] conducted numerical simulations with the aim of maximizing the output power density of VAWT farms. It was found that when the spacing between the wind turbines was kept small, the mutual interaction between the turbines reduced the average power coefficient of the wind turbines rather than enhancing it. Dabiri [14] performed full-scale field tests making use of counter rotating VAWT for maximizing the power density of a VAWT wind farm. Field tests indicated that the power densities of greater value can be achieved by positioning vertical axis wind turbines in layouts that allow them to extract energy from the neighbouring wakes. Power density for VAWT was found to be 18 W m^{-2}, whereas for HAWT, it is about 2–3 W m^{-2}.

In the development of numerical simulations for analysing the dynamic performance of wind turbines, the accuracy and relevance of the model for the chosen situation are of utmost importance. Daróczy et al. [15] conducted several URANS-based 2D numerical simulations using different turbulence models for comparison with the aid of four different configurations. They concluded that k–ω shear stress transport model with a cubic correction provides an exceptional prediction of the experimental results. The k–ω SST turbulence model is an amalgamation of superior elements of k–ω and k–ϵ turbulence model. Castelli et al. [16] conducted experiments and 2D numerical simulations for the evaluation of aerodynamic forces acting on a straight bladed VAWT along with its energy performance. A comparison between the numerical simulations and wind tunnel measurements was performed for certain performance parameters. They found that the numerical simulations accurately predicted the experimental results with the same pattern.

In view of the above literature review, in the field of VAWT, the present work focusses on the CFD analysis of a straight bladed VAWT using NACA 0021 as the aerofoil. The importance of mesh refinement and its consequences are elucidated.

2 Research Methodology

The detailed CFD analysis starts with the appropriate selection of wind turbine parameters followed by the relevant mathematical calculations for various parameters along

Table 1 Turbine features [16]

Turbine features	Values
Type of turbine	H-rotor Darrieus
Blade aerofoil	NACA 0021
Blade number (N)	3
Rotor radius (R)	515 mm
Height of rotor (H)	1 mm
Chord length (C)	85.8 mm
Rotor solidity (σ)	0.5
Inlet wind speed (V_w)	9 m/s
Tip speed ratio (TSR)	3.5

with mesh generation and grid independence test. For performing the simulations, a 2D, SST k–ω turbulence model-based transient pressure-based solver is used. Wind velocity of 9 m/s is taken in the current study. Pressure and velocity terms are coupled, and second-order upwind scheme is used for turbulent kinetic energy and dissipation rate. A computer with Intel Core i5 Processor, 8 GB DDR4 RAM and a dedicated 2 GB graphics card is employed for the numerical simulations.

2.1 Selection of Turbine Features

An appropriate turbine which aids in solving the objective of the current study should be chosen for carrying out the research task. Table 1 shows the turbine features that is employed for numerical simulations in the present work.

2.2 Performance Parameters

Various performance parameters meant for deducing the performance of the wind turbine are discussed in this subsection. Solidity is expressed as the ratio of the product of the number of blades and the chord length to the radius of the rotor. It expresses the percentage of material present in the circumference of the rotor and is given by Eq. (1). Tip speed ratio (λ) is defined as the ratio of the tangential velocity of the tip of the blade to the wind velocity. It signifies the relative rotational speed of the blade tip with respect to wind and is given by Eq. (2). Instantaneous torque is obtained in the wind turbine due to the lift/drag force exerted on the wind blades, and it is evaluated at the centre of turbine. Instantaneous torque is required to calculate the C_p of the turbine, and it is given by Eq. (3). C_p defines the amount of wind's total kinetic energy that is getting converted into mechanical power at the rotor shaft, and it is given by Eq. (4). Turbine power P_T is the power obtained at the turbine rotor shaft before the gear box. It is the power which is limited to 59.26% of wind power according to Betz's limit and can be calculated using Eq. (5).

$$\sigma = \frac{N.C}{R} \tag{1}$$

$$\lambda = \frac{\omega.R}{V_w} \tag{2}$$

$$T = \frac{1}{2}\rho.A.V_w^2.R.C_T \tag{3}$$

$$C_P = \lambda.C_T \tag{4}$$

$$P = C_P.P_{\text{input}} = C_P.\frac{1}{2}.\rho.A.V_w^3 \tag{5}$$

Two-dimensional CFD approach is employed for the purpose of numerical analysis. The approach predicts the fluid flow by solving the governing equations that represent the physical laws, using a numerical process. CFD is based on conservation of mass, momentum, energy and species. Navier–Stokes equations are the governing equations, which form the basis of CFD. Since the Navier–Stokes equations are nonlinear and coupled partial differential equations, they are difficult to solve, and therefore, there is a need for employing numerical methods for solving these equations.

CFD discretizes the fluid region of interest and then approximates the governing equations of fluid mechanics in that region. The governing partial differential equations are converted into a set of algebraic equations at each node, which are then solved numerically for the field variables. ANSYS Fluent is used for performing the numerical simulations to analyse the wind flow pattern around the VAWT. ANSYS Fluent is a robust software employed worldwide for the purpose of simulating fluid flow, heat transfer, mass transfer, etc.

2.3 Geometric Modelling

Figure 1 shows the 2D computer-aided design (CAD) model of the H-type Darrieus turbine with NACA 0021 aerofoil. Figure 2 represents a close-up view of the rotor rotating domain. The distance of the aerofoils from the axis of the rotor is kept at 515 mm (radius of the turbine), whereas the diameter of the blade domain and the rotating domain are 400 mm and 2000 mm, respectively.

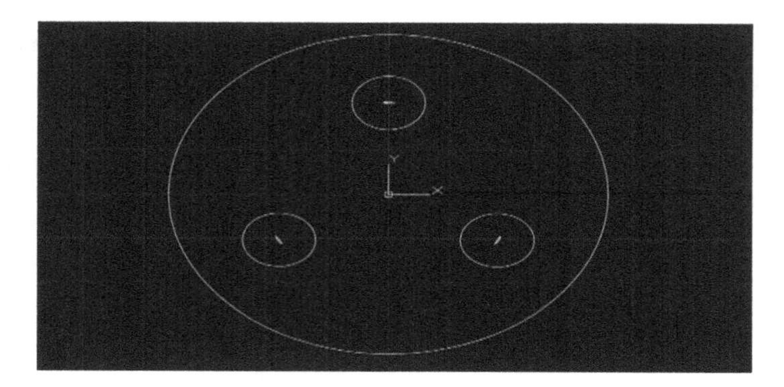

Fig. 1. 2D CAD geometric model

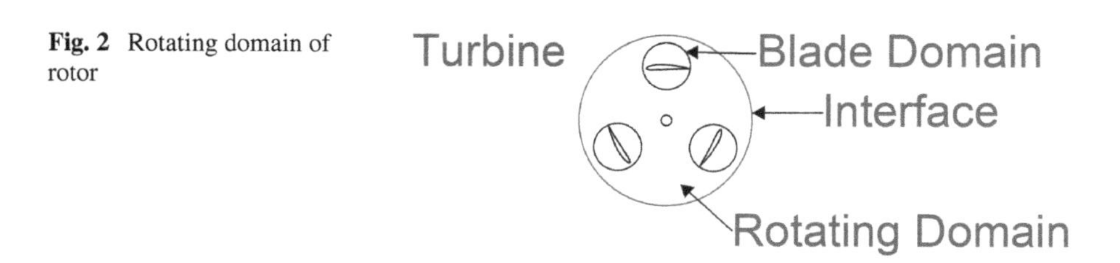

Fig. 2 Rotating domain of rotor

2.4 *Computational Domain and Boundary Conditions*

Figure 3 shows the *H*-Darrieus VAWT computational domain comprising of a 2D cross-sectional view of a three-blade wind turbine confined by an upstream path, far-field downstream path and symmetrical boundary. The domain resembles the actual top view of the VAWT. The symmetrical boundary condition is applied at the

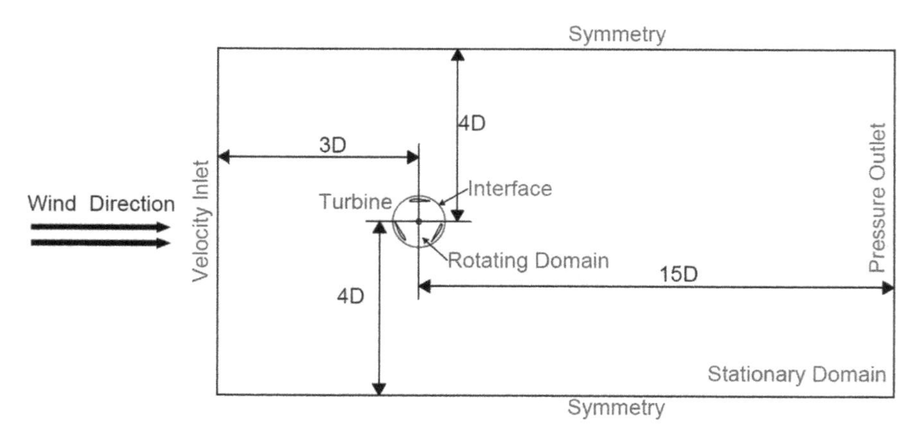

Fig. 3 Computational domain of wind turbine

horizontal side of the computational domain, whereas the boundary conditions of velocity inlet and pressure outlet are fixed at the inlet and outlet of the computational domain, respectively.

The computational domain consists of three sub-domains. The rotor rotating domain separates the stationary part from the rotating part. This rotating domain makes use of moving mesh which rotates about the axis of the turbine rotor. The portion from the outer interface of the rotor domain to the walls of the computational domain forms the stationary part. The mesh in this respective part is kept stationary and is refined to a high degree. The rotor rotating domain contains a blade domain (as seen from Fig. 2), which houses the chosen blade's aerofoil profile. This blade domain has a rotating mesh which rotates about their respective blade centres. Moving mesh technique is employed in the numerical simulation process.

Velocity inlet boundary condition is placed at a distance of 3 rotor diameters (3D) upstream from turbine. Pressure outlet boundary condition is given at 15 rotor diameters (15D) downstream from turbine to allow a full development of the wake. Both the top and bottom symmetrical boundary conditions are fixed at a distance 4 rotor diameters (4D) from the centre of the turbine.

2.5 Mesh Generation and Grid Independence Test

Mesh plays a very crucial role for accurate and reliable CFD simulation. Quality of mesh determines the rate of convergence, computational time, reliability of result, etc. Hence, finer mesh can capture even small changes in the flow field, thus proving beneficial, but it will also, at the same time, increase the computational time and demand high-end hardware. However, if coarser grid is used, although it will give the results in a shorter period of time. Mesh for the computational domain, rotor domain and blade domain is shown in Figs. 4, 5, 6 and 7. Hence, a good-quality grid size and/or element is required to obtain accurate results while also reducing the computational time. Hence, we go for the grid independent test (GIT). GIT is performed to ensure that the simulation solution is independent of the grid size and/or the number of elements. This is achieved by increasing the number of mesh elements for each simulation up to a point when the relative error in C_m or C_p becomes less than 1%. Then, the previous mesh is selected for further simulations. This also saves computational time as very fine mesh would give the same result (within 1% error) as the previous mesh once GIT is achieved.

Figure 4 shows the mesh of computational domain at which GIT was achieved for turbine with NACA 0021 aerofoil. Figure 5 gives an overlook of the mesh at the junction of the rotor domain and the refined mesh of stationary domain. Figure 6 presents a close-up view of the right blade domain showing the uniform shapes and sizes of mesh elements in the domain. The element size is 2.2 mm. Figure 7 shows the mesh for the right blade trailing edge at the blade–fluid interface which consists of a first layer thickness (inflation) of 0.5 mm with a maximum of six layers. Intense

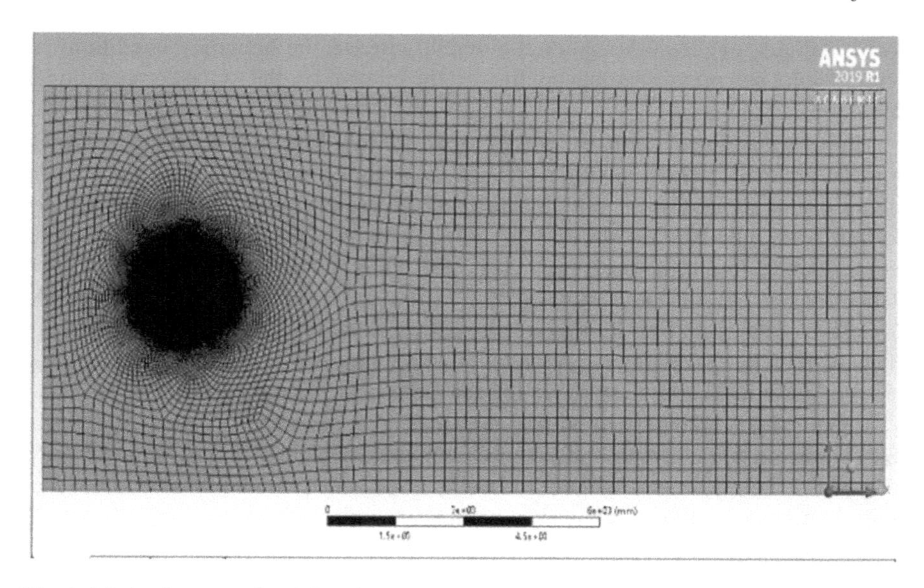

Fig. 4 Mesh of computational domain

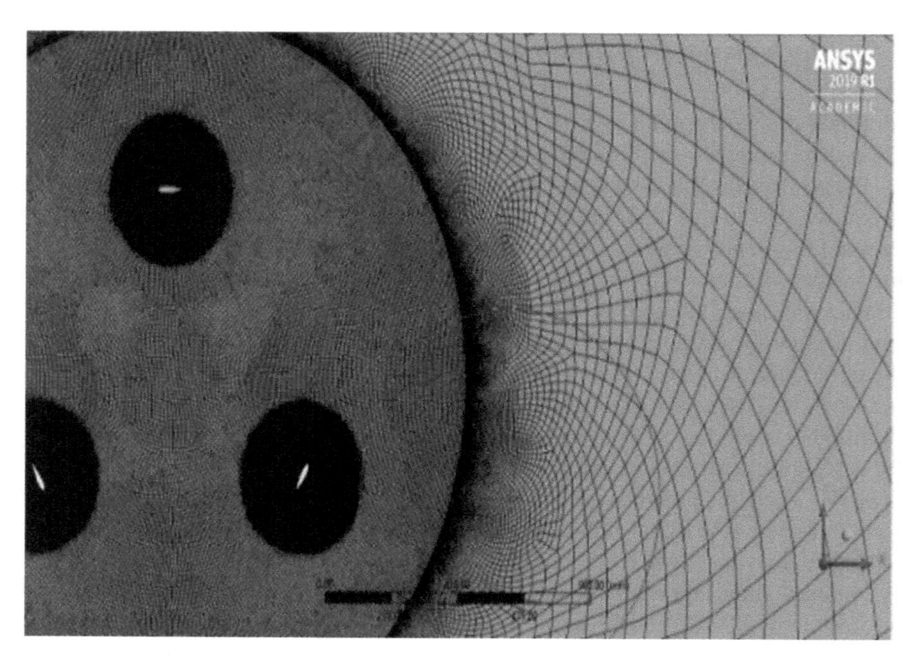

Fig. 5 Mesh of rotor domain

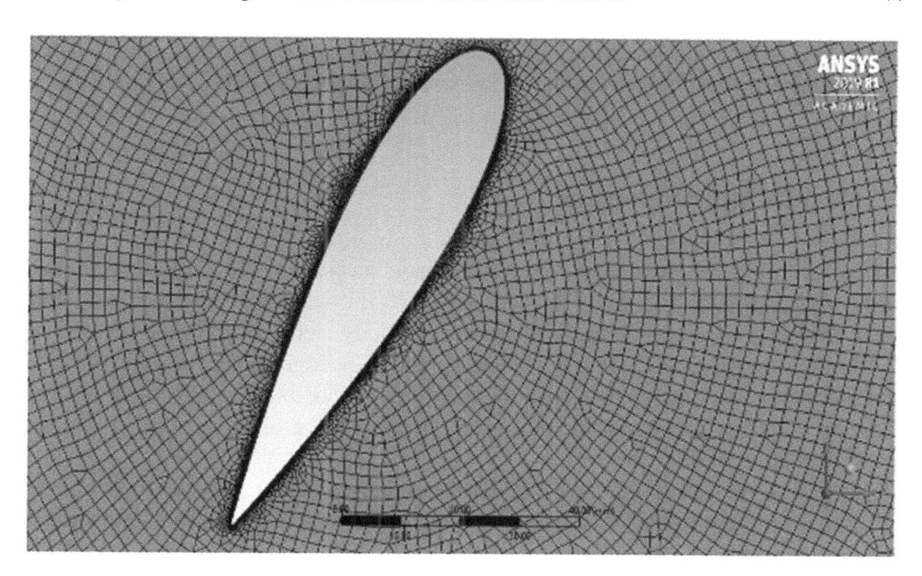

Fig. 6 Mesh of right blade

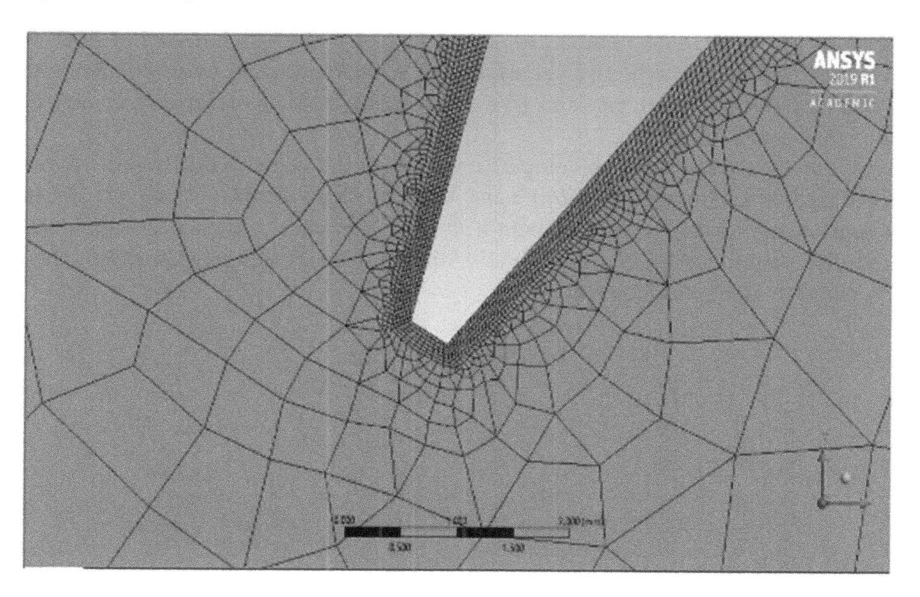

Fig. 7 Mesh of right blade trailing edge

meshing is performed to capture the minutest of interaction at the interface. This helps in arriving at reliable results. The growth rate was kept at 1.1 for the inflation.

3 Results and Discussion

A fine mesh must include inflation on the blade surfaces in the blade domain and refinement applied to far-field. The mesh at which GIT was achieved consists of 240 k mesh elements approximately with elements in body sizing of rotor sub-grid to be of 10 mm, whereas it is 2.2 mm for blade domain. The edge sizing for rotor sub-grid included 900 divisions, whereas for blade domain it was fixed to 600 divisions. The inflation height of first layer thickness is 0.5 mm with a growth rate of 1.1, and the maximum number of layers is 6.

It can be seen from Fig. 8 that the relative percentage error (between coefficients of moment values) reduces to below 1% at the mesh value of 240 k and further reduces to 0.03% for mesh elements of about 280 k. The relative error is calculated for the present mesh elements at a gap of 40 k mesh elements. The average values of coefficient of moment for one complete rotation of rotor with 240 k and 280 k mesh elements are 0.119285895 and 0.119326638, respectively, which results in an error of 0.034156% between the two values. Figure 9 shows the right blade domain with a total of 240 k mesh elements and gives an in-depth close-up of the leading edge of the right blade domain.

Figure 10 gives a close-up comparison between the velocity magnitude vectors of the right blade domain with 240 k and 160 k mesh elements, respectively. It can be seen clearly that right blade domain with total 240 k mesh elements gives a better fluid flow pattern and more detailed fluid velocity vectors and their direction when compared to 160 k mesh elements.

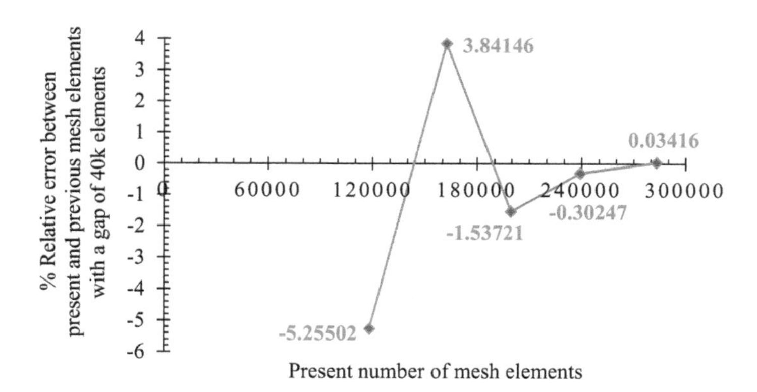

Fig. 8 Percentage relative error in grid independence test

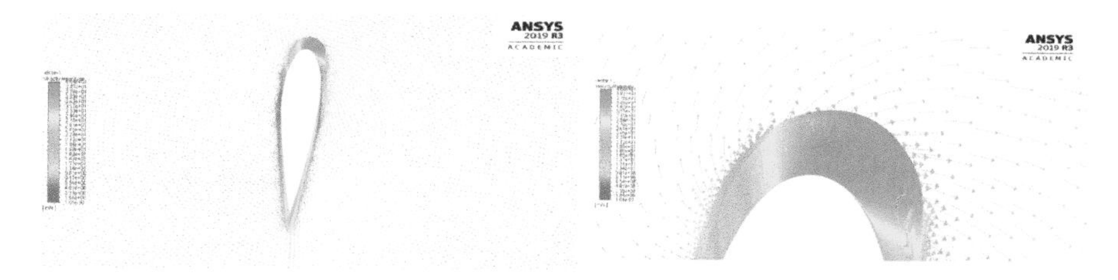

Fig. 9 Right blade (left) and its leading edge (right) with 240 k mesh elements

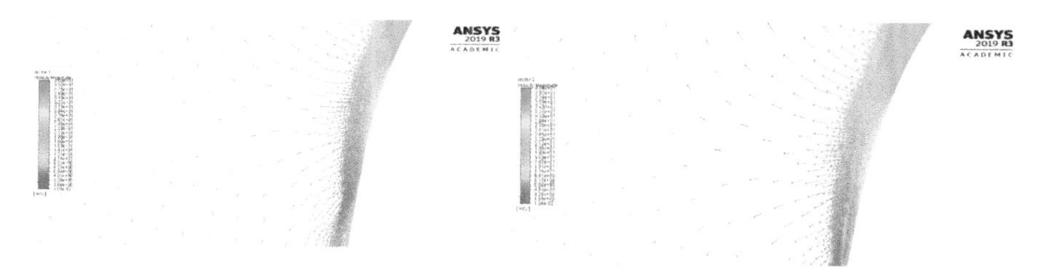

Fig. 10 Closeup of right blade domain with 240 k mesh elements (left) and 160 k mesh elements (right)

4 Conclusion

The following conclusions can be inferred from the present study:

- Variations in the fluid flow are better captured with a greater number of mesh elements primarily obtained after the grid independence test. This leads to reliable and accurate results, which could be further used for research purposes for different applications.
- Performing GIT gives the best compromise between the accuracy, reliability of the simulation and the computational power, and time requirements of the computing system.

Further research can be carried in continuation by replacing the existing airfoil by a different airfoil and analysing the flow dynamics of the wind around the turbine for aerodynamic performance enhancement of wind turbines.

References

1. Bhutta, M.M.A., Hayat, N., Farooq, A.U., Ali, Z., Jamil, S.R., Hussain, Z.: Vertical axis wind turbine—a review of various configurations and design techniques. Renew. Sustain. Energy Rev. **16**, 1926–1939 (2012)
2. Mohamed, M.H.: Performance investigation of *H*-rotor Darrieus turbine with new airfoil shapes. Energy **47**, 522–530 (2012)

3. Mohamed, M.H., Ali, A.M., Hafiz, A.A.: CFD analysis for H-rotor Darrieus turbine as a low speed wind energy converter. Eng. Sci. Technol. **18**, 1–13 (2015)
4. Hashem, I., Mohamed, M.H.: Aerodynamic performance enhancements of H-rotor Darrieus wind turbine. Energy **142**, 531–545 (2018)
5. Mohamed, M.H., Dessoky, A., Alqurashi, F.: Blade shape effect on the behaviour of the H-rotor Darrieus wind turbine: performance investigation and force analysis. Energy **179**, 1217–1234 (2019)
6. Alsabri, A., Oreijah, M., Mohamed, M.H.: A comparison study of aerofoils used in H-rotor Darrieus wind turbine. Int. J. Sci. Eng. Invest. **8**(90), 22–28 (2019)
7. Sabaeifard, P., Razzaghi, H., Forouzandeh, A.: Determination of vertical axis wind turbines optimal configuration through CFD simulations. In: International Conference on Future Environment and Energy 2012, IPCBEE, vol. 28, pp. 109–113. IACSIT Press, Singapore (2012)
8. Li, Q., Maeda, T., Kamada, Y., Murata, J., Furukawa, K., Yamamoto, M.: Effect of number of blades on aerodynamic forces on a straight-bladed vertical axis wind turbine. Energy **90**(1), 784–795 (2015)
9. Qamar, S.B., Janajreh, I.: A comprehensive analysis of solidity for cambered Darrieus VAWTs. Int. J. Hydrogen. Energy **42**(30), 19420–19431 (2017)
10. Ghasemian, M., Ashrafi, Z.N., Sedaghat, A.: A review on computational fluid dynamic simulation techniques for Darrieus vertical axis wind turbines. Energy Convers. Manage. **149**, 87–100 (2017)
11. Shyu, L.: A pilot study of vertical-axis turbine wind farm layout planning. In: Advanced Materials Research, vol. 953–954, pp. 395–399. Trans Tech Publications, Switzerland (2014)
12. Ahmadi-Baloutaki, M., Carriveau, R., Ting, D.S.-K.: A wind tunnel study on the aerodynamic interaction of vertical axis wind turbines in array configurations. Renew. Energy **96**, 904–913 (2016)
13. Shaaban, S., Albatal, A., Mohamed, M.H.: Optimization of H-rotor Darrieus turbines' mutual interaction in staggered arrangements. Renew. Energy **125**, 87–99 (2018)
14. Dabiri, J.O.: Potential order-of-magnitude enhancement of wind farm power density via counter-rotating vertical-axis wind turbine arrays. J. Renew. Sustain. Energy **3**, 043104 (2011)
15. Daróczy, L., Janiga, G., Petrasch, K., Webner, M., Thévenin, D.: Comparative analysis of turbulence models for the aerodynamic simulation of H-Darrieus rotors. Energy **90**(1), 680–690 (2015)
16. Castelli, M.R., Englaro, A., Benini, E.: The Darrieus wind turbine: proposal for a new performance prediction model based on CFD. Energy **36**, 4919–4934 (2011)

Fabrication of Dye-Sensitized Solar Cell Based on Natural Dye Sensitizer and ZnO Nanoflower Photoanode

Debasis De, M. Sreevidhya, and Chanchal Kumar De

Abstract Dye-sensitized solar cells (DSSCs) are solar cell of the third generation that transform visible light into electricity using the photoelectrochemical process. This one is the least inexpensive and more effective among all new photovoltaic technologies, and the material used in this is also plentiful and relatively nontoxic. This paper emphasizes the fabrication of photoanode based on ZnO nanoflowers and carotenoid natural dye sensitizer. Nanostructured ZnO is deposited on florin-doped tin oxide (FTO)-coated glass substrate using chemical bath deposition technique. A monolayer of dye sensitizer is adsorbed by dip coating on ZnO photoanode for absorbing near-UV and visible light region of the solar spectrum. The scanning electron microscopic (SEM) images show hexagonal patterned ZnO nanoflowers. The dye's function is similar to the role of the chlorophyll in plants by capturing solar light and transmitting energy by electron through ZnO photoanode for electricity generation. For preparing dye sensitizer, Yellow King Humbert (carotenoid) dye is used for this work. The developed cell's *I*–*V* and *P*–*V* characteristics are tested under stimulated light (100 mW/cm^2). The solar to electric conversion efficiency of 0.307% was achieved with short-circuit current (I_{sc}) of 1.42 mA/cm^2 and open-circuit voltage (V_{oc}) of 0.45 V for the fabricated cell with carotenoid natural dye.

Keywords Nanostructured ZnO · Chemical bath deposition method · Natural dye · DSSC

D. De (✉) · M. Sreevidhya
Energy Institute, Bengaluru (Centre of Rajiv Gandhi Institute of Petroleum Technology, Jais, Amethi), Bangalore, India
e-mail: debasisd@rgipt.ac.in

C. K. De
Department of Electronics and Communication Engineering, Haldia Institute of Technology, Haldia, India

P. Muthukumar et al. (eds.), *Innovations in Sustainable Energy and Technology*, Advances in Sustainability Science and Technology,
https://doi.org/10.1007/978-981-16-1119-3_7

1 Introduction

About 90% of the world's solar cell is first-generation solar cell that is made from wafers of crystalline silicon, and it shows efficiency up to 22–24%. Classic solar cells are generally of thin-film solar cell which are categorized as second-generation solar cell, and they are extremely thin, light and flexible, which can be laminated onto windows, skylights, roof tiles, but they lag in efficiency which is up to 11–15%. Third-generation solar cell is the combination of both first- and second-generation solar cell, and they promise relatively high efficiency. Ideally, that would make them cheaper, more competent and more functional than either first- or second-generation cells. Dye-sensitized solar cell (DSSC) is a third-generation solar cell with multiple layered structures [1–5]. DSSCs are cost-effective photochemical electric cell which can fabricate easily. A layer-by-layer stacking of conducting oxide substrate, photoanode, dye sensitizer, electrolyte and cathode are formed DSSC. A monolayer of dye sensitizer is sandwiched between anode and cathode in such cell. Each layer of the cell is important for achieving maximum efficiency from the cell. The photoanode consists of dye-absorbed semiconductor metal oxide (SMO) layer deposited on FTO which is an important part of the cell. Counter electrode based on conducting carbon-coated FTO and electrolyte are other important parts of the cell [4]. Under daylight condition, the electrons jump from highest occupied molecular orbital (HOMO) to lowest unoccupied molecular orbital (LUMO) of dye sensitizer. The electrons available in the LUMO level are inserted into the conduction band of SMO, and finally electrons are transported to the FTO. The electrons come back to dye through oxidation of electrolyte. In 1991, O. Regan and Michael Gratzel developed DSSC using TiO_2 as a photoanode and the cell achieved efficiency of 8% [6]. As reported in the literature, the maximum efficiency observed in TiO_2-based photoanode with synthetic dye sensitizer was >11%. This paper mainly focusses on ZnO nanoflowers as an anode for DSSC with natural dye sensitizer. Electron mobility in nanostructured ZnO is higher than TiO_2. Natural dye Yellow King Humbert (carotenoid) is used in this study because it is environmentally friendly, low cost, nontoxic in nature. There are many natural dyes available which are extracted from plants, and can be used as sensitizer [7–12]. These dyes contain different chemical groups, like anthocyanin, betacyanin, chlorophyll, etc., which can anchor on the surface of SMO. This can increase the electron transport from dye to anode. The paper is constructed as follows: synthesis of ZnO nanoflowers on conducting substrate, extraction of natural dye and cell fabrication are explored in Sect. 2. Performance of the ZnO as an anode and Yellow King Humbert as a dye sensitizer is assessed. The performance evaluation of DSSC is revealed in Sect. 3, and summary is specified at the end.

2 Experiment Procedure

2.1 Synthesis of ZnO

The fabrication of nanostructured ZnO as a photoanode material was coated on conducting FTO by chemical bath deposition (CBD) technique. The details of the synthesis process have been reported by De et al. [10, 13]. ZnO was grown on conducting FTO at 75 °C for 2 h on hot plate. During the process, it was observed that whitish ZnO was precipitated. The coated substrate was rinsed by deionized water properly followed by drying at an oven (at 70 °C). The ZnO-coated FTO was heated at 450 °C for 4 h for better anchoring and can be used further for fabrication of DSSC.

2.2 Extraction of Dye

Carotenoid dye was prepared from Yellow King Humbert flower as shown in Fig. 1a. For the extraction of the dye, fresh flowers' petals were washed properly using distilled water and then they were cut into small pieces and pounded with a mortar pestle into paste. Ethanol was used as solvent for extraction process. The samples were put into ultrasonicator for 10 min at 37 Hz frequency in 'degas' mode at room temperature. The colouring pigments of the samples were segregated by centrifuge at 2500 rpm for 20 min. The extracted dye is shown in Fig. 1b. The chemical structure of

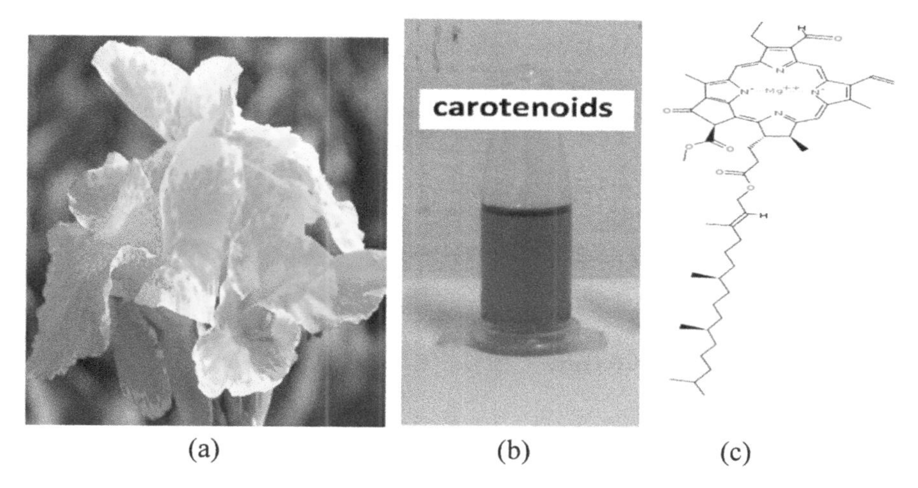

Fig. 1 Plant used as natural dye: **a** 'Yellow King Humbert' flower, **b** carotenoid dye and **c** chemical structure of carotenoid pigment

carotenoid dye is shown in Fig. 1c. Light absorption characteristics of the carotenoids were evaluated using UV–Vis spectroscopy.

2.3 Electrolyte Preparation

Electrolyte solution was prepared using potassium iodide and iodine salts. First, 0.26 M potassium iodide solution was prepared using acetonitrile solvent and then add slowly 0.033 M iodine in the solution. The final prepared solution was mixed properly on magnetic stirrer for 30 min and stored in dark container.

2.4 Fabrication of DSSC

The ZnO anode, extracted natural dye and electrolyte were used to make sandwiched-type DSSC. ZnO-coated FTO substrate was immersed into the extracted carotenoid dye for overnight. In this process, dye was properly anchoring on the surface on nanostructured ZnO. Prepared I^{-}/I^{-3} redox couple was used as electrolyte. One drop of electrolyte solution was put on the dye-absorbed ZnO-coated FTO. The counter electrode (cathode) was prepared using conducting carbon-coated FTO. The photo anode with electrolyte and counter electrode one over another as a layered structure to fabricate the solar cell and finally all the sides of the cell ware properly sealed by teflon tape. Silver paste was used to make external electrical contacts for the measurement of I–V characteristics. The performance of the DSSC using carotinoid dye was determined in terms of V_{oc}, I_{sc}, fill factor (FF) and efficiency (η). FF and efficiency of the fabricated cell were calculated using Eqs. (1) and (2), respectively

$$\text{FF} = \frac{V_{mp} X I_{mp}}{V_{\text{oc}} X I_{\text{sc}}} \tag{1}$$

$$\eta = \frac{V_{\text{OC}} I_{\text{SC}} \text{FF}}{P_{\text{in}}} \tag{2}$$

3 Results and Discussion

3.1 Microstructural Properties of Nanostructured ZnO

The X-ray diffraction (XRD) patterns of nanostructured ZnO were used to examine the crystal structure. The diffraction peaks can be readily indexed to a hexagonal wurtzite structure of ZnO, and the part of diffraction pattern is shown in Fig. 2

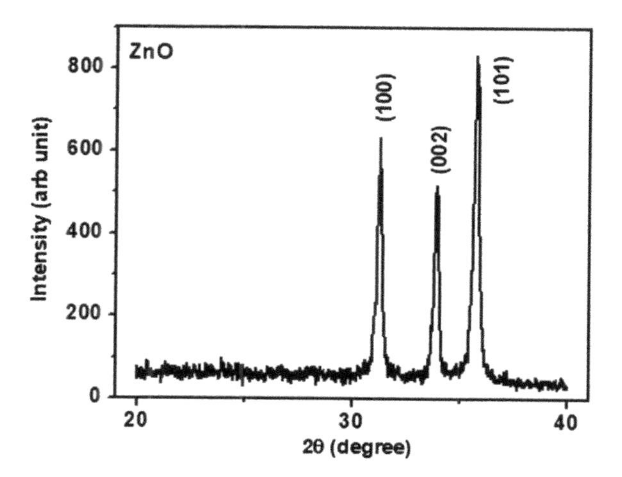

Fig. 2 XRD pattern of nanostructured ZnO flowers

[14]. Figure 3 shows the scanning electron microscopic (SEM) images of ZnO nanoflowers. It is observed from the image that all the flowers are similar in size and shape. Each flower has at least 6 petals in plane and 1 petal on the top of the flower along the *Z*-direction. The length of each petal is ~3 μm and the thickness ~40 nm. The magnified image of a ZnO nanoflower is shown in the top right corner of Fig. 3. The energy-dispersive X-ray spectroscopy (EDX) attached with SEM was conducted to classify the elements in the samples present in Fig. 4 which states that ZnO nanoflowers display distinct peaks of zinc and oxygen in the EDX range. This flower-like structure also provides a large surface area to anchor more amount of dye molecule on ZnO surface and improve the light harvesting.

Fig. 3 SEM image of ZnO nanoflowers with a magnified image of flower in the top right corner

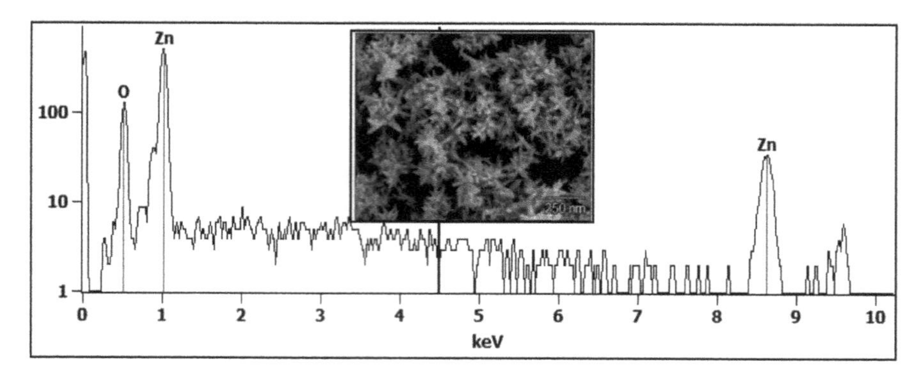

Fig. 4 EDX spectrum of nanostructured ZnO along with a SEM image

3.2 UV–Vis Spectroscopy of Carotenoid Dye

To understand the light absorbance/reflectance behaviour of the nanostructured ZnO, the photoanode was characterized using UV–visible diffused reflectance spectroscopy (DRS) and the spectrum is shown in Fig. 5. It is observed that the bare ZnO shows high reflectance of ~90% under visible light and such high reflectance is due to the scattering of light from the film surface. The caroteniod dye-absorbed ZnO shows reduced reflectance of ~40%. This is because the visible light is absorbed by the dye anchored on ZnO. The UV–visible absorption spectrum of caroteniod dye shows the maximum absorbance at ~510 nm in the wavelength range from 400 to 800 nm. A small absorption peak is also observed at 740 nm.

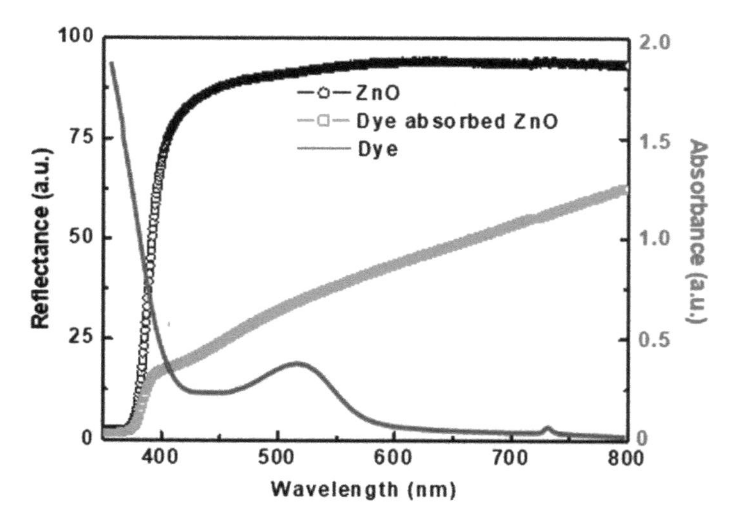

Fig. 5 UV–Vis absorbance spectrum of Yellow King Humbert dye and reflectance spectra of dye absorbed with and without nanostructured ZnO

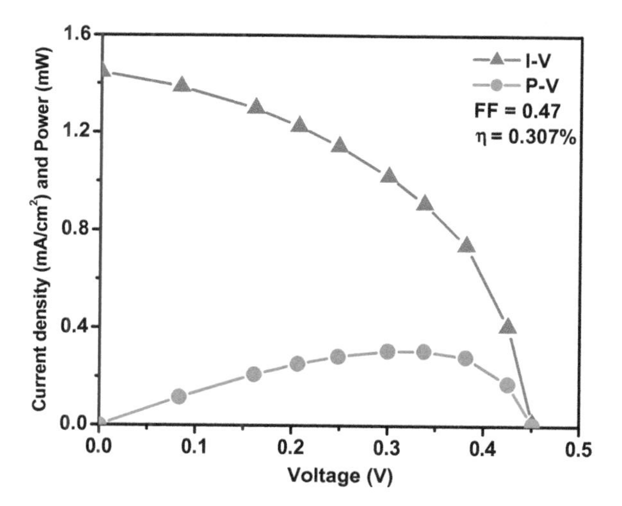

Fig. 6 *I–V* and *P–V* characteristics of DSSC with carotenoid dye photo sensitizer

3.3 Performance Analysis of DSSC

The output of DSSC depends primarily on natural dye used and their apparent absorption of energy. The *I–V* and *P–V* characteristics of the prepared DSSC are shown in Fig. 6. The DSSC fabricated using nanostructured ZnO photoanode and carotenoid dye sensitizer shows open-circuit voltage and short-circuit current density of 0.45 V and 1.42 mA/cm^2, respectively. The calculated value of current density and voltage at maximum power point are 0.38 mA/cm^2 and 0.34 V. Estimated value for solar to power conversion efficiency and fill factor is calculated as 0.307% and 0.47, respectively. The DSSC using natural carotenoid dye sensitizer shows better performance as compared to purple cabbage, spinach, beetroot-based dye sensitizer reported in our previous works [7, 9, 11].

4 Conclusions

The DSSC was fabricated using ZnO nanoflower photoanode and carotenoid dye photosensitizer. The ZnO nanoflowers were synthesized using chemical bath deposition technique on FTO substrate. Natural carotenoid dye was extracted from Yellow King Humbert flower petals. SEM images showed that the average length of ZnO nanoflowers petals was ~3 μm and the thickness of ~40 nm. The UV–Vis spectrum of the carotenoid dye showed the absorbance of 450–570 nm broadband in the visible range with a peak at 510 nm. The prepared solar cell showed the I_{sc} of 1.42 mA/cm^2, V_{oc} of 0.45 V with a FF of 0.47 and the cell efficiency of 0.307%.

References

1. Gong, J., Liang, J., Sumathy, K.: Review on dye sensitized solar cells (DSSCs): fundamental concepts and novel materials. Renew. Sustain. Energy Rev. **16**, 5848–5860 (2012)
2. Hoffmann, M.R., Martin, S.T., Choi, W., Bahnemann, D.W.: Environmental applications of semiconductor photocatalysis. Chem. Rev. **95**, 69–96 (1995)
3. Afifiand, K., Tabatabaei, A.M.: Efficiency investigation of dye-sensitized solar cells based on the zinc oxide nanowires. Orient. J. Chem. **30**, 155–160 (2014)
4. Longo, C., Paoli, M.-A.: Dye-sensitized solar cells: a successful combination of materials. J. Braz. Chem. Soc. **14**(6), 889–901 (2003)
5. Calogero, G., Yum, J., Sinopoli, A., Di Marco, Grätzel, G.M., Nazeeruddin, M.K.: Anthocyanins and betalains as light-harvesting pigments for dye-sensitized solar cells. Sol. Energy **86**, 1563–1575 (2012)
6. O'Regan, B., Gratzel, M.: A low-cost, high-efficiency solar cell based on dye-sensitized colloidal TiO_2 films. Nature **353**(6346), 737–740 (1991)
7. Goswami, D., Sinha, D., De, D.: Nanostructured ZnO and natural dye based DSSC for efficiency enhancement. In: Proceeding of IEEE 3rd International Conference on Science, Technology, Engineering and Management, pp. 556–560 (2017)
8. Sinha, D., De, D., Goswami, D., Ayaz, A.: Fabrication of DSSC with nanostructured ZnO photoanode and natural dye sensitizer. Proc. Mater. Today **5**, 2056–2063 (2018)
9. Sinha, D., De, D., Ayaz, A.: Performance and stability analysis of curcumin dye as a photo sensitizer used in nanostructured ZnO based DSSC. Spectrochim. Acta **193**, 467–474 (2018)
10. De, D., Sinha, D., Ayaz, A.: Performance evaluation of beet root sensitized solar cell device. Lect. Notes Electr. Eng. **602**, 223–228 (2019)
11. Sinha, D., De, D., Ayaz, A.: Photosensitizing and electrochemical performance analysis of mixed natural dye and nanostructured ZnO based DSSC. Sadhana **45**, 175–186 (2020)
12. Ayaz, A., Sinha, D., De, D.: Performance analysis of mixed natural dye and nanostructured TiO_2 based DSSC. Int. J. Innov. Technol. Explor. Eng. **9**, 453–456 (2020)
13. De, D., Sarkar, D.K.: Superhydrophobic ZnAl double hydroxide nanostructures and ZnO films on Al and glass substrates. Mater. Chem. Phys. **185**, 195–201 (2017)
14. JCPDS# 75-1526

Effect of Longitudinal Fin Configuration on the Charging and Discharging Characteristics of a Horizontal Cylindrical Latent Heat Storage System

Gurpreet Singh Sodhi, K. Vigneshwaran, Vishnu Kumar, and P. Muthukumar

Abstract In the present study, a three-dimensional numerical model for the heat transfer enhancement in a horizontal cylindrical latent heat storage system is presented. Sodium nitrate is used as the phase change material in the shell side, and air flowing through the tube side is used as the heat transfer fluid. Three different longitudinal fin configurations, namely straight fins, uniformly increasing fin height and uniformly decreasing fin height, are developed, and their melting/solidification behavior is compared with the system having no fins. The key objective is to distribute the fin surfaces effectively to achieve uniform heat transfer along the length of the system. The results show that the accurate positioning of fins is very critical for improving the heat transfer while designing a latent heat storage system. The fins with decreasing fin height are favorable for better heat transfer characteristics inside the PCM region.

Keywords Latent heat storage · Phase change material · Longitudinal fins · Melting · Solidification

1 Introduction

Latent heat thermal energy storage systems using phase change materials (PCMs) are an imperative solution for solar thermal applications requiring constant power characteristics. PCMs store the latent heat while undergoing the melting process,

G. S. Sodhi · V. Kumar · P. Muthukumar (✉)
Department of Mechanical Engineering, Indian Institute of Technology Guwahati, Guwahati 781039, India
e-mail: pmkumar@iitg.ac.in

G. S. Sodhi
e-mail: g.sodhi@iitg.ac.in

K. Vigneshwaran · P. Muthukumar
Centre for Energy, Indian Institute of Technology Guwahati, Guwahati 781039, India

P. Muthukumar et al. (eds.), *Innovations in Sustainable Energy and Technology*,
Advances in Sustainability Science and Technology,
https://doi.org/10.1007/978-981-16-1119-3_8

whereas this energy can be retrieved during their solidification. The thermal conductivity of PCMs is low [1], and hence the heat transfer rate is improved by external methods. Amongst all techniques used for heat transfer enhancement such as cascade PCM system, multi-tube heat exchanger, PCM encapsulation, the use of extended fin surfaces have been widely explored by various researchers [2].

Gasia et al. [3] employed a low-grade carbon-steel metal wool as an enhancement additive in replacement of fins. Although the performance of the system was inferior to fin-based system, an improvement of ~14% in charging rate was observed over a conventional heat exchanger. Deng et al. [4] discussed the effect of fin angle on the melting behavior inside the PCM using a numerical model. They observed that the heat transfer rate had improved by employing more fin surfaces in the bottom region of a shell and tube system. Hosseini et al. [5] conducted experimental studies on a tube-in-tube heat exchanger carrying longitudinal fins. They concluded that increasing the thickness of fins improved the melting rate, whereas increasing the fin height hindered the merging of convection vortices and reduced the energy density of the system. Abdulateef et al. [6, 7] developed a triplex tube heat storage system with HTF flowing in the center and outer tube, and different fin designs: external triangular, internal triangular and internal–external triangular fins in the intermediate tube. They found that the external triangular fin structure had 15 and 18% improvement in melting and solidification rate over an equivalent longitudinal fin design.

The major objective of discussing the literature is to underline the fact that using fins (especially longitudinal) has been a prime method for heat transfer improvement inside a cylindrical shell and tube latent heat storage system. Most of the studies have presented the effect due to shape, orientation and number of fins on the performance of the system [8]. However, most of the studies have developed two-dimensional models [4, 6, 7], and it is very important to study the effect of various configurations of fins along the system length. The major drawback of using fins is that they reduce the energy density of the system and also block the free PCM movement [9] during natural convection. The cost and fabrication of complex fin designs are again a challenge. Hence, it is very important to study the effect of fins using relatively simple fin shapes such as rectangular fins.

The current study discusses the performance enhancement of a horizontal tube-in-shell latent heat storage system using rectangular fins. A 3D numerical model validated with the experimental data is developed and presented. Further, a no-fin horizontal system (case A) is compared with three different fin configurations, namely: straight fins (case B), increasing fin height (case C) and decreasing fin height (case D). The study describes the importance of distributing the fin surfaces effectively for an economical design and improvement in the rate of heat transfer rate in a horizontal shell and tube latent heat storage system.

2 Numerical Modeling

2.1 Description of the System

The cylindrical tube in shell system is developed by fixing the storage capacity as 1 MJ. The PCM volume is determined by Eqs. (1) and (2).

$$Q_s = \rho V L_f \quad (1)$$

$$V = \frac{\pi}{4}\left(D_o^2 - d_o^2\right)L \quad (2)$$

where Q_s, ρ, V and L_f are the amount of latent heat stored (J), density (kg/m^3), volume (m^3) and latent heat (J/kg) of fusion of the PCM.

The thickness of inner tube, outer diameter (d_o) and length of the system (L) are 2 mm, 24.6 mm and 750 mm, respectively. The height, length and thickness of fins considered in all the cases are 10 mm, 750 mm and 2 mm, respectively. The physical models with different fin configurations are described in Fig. 1. The different configurations developed for comparison are case A having no fins, case B having four straight fins, case C having four increasing height fins and case D having four decreasing height fins. Sodium nitrate and air are chosen as the PCM and the heat transfer fluid (HTF), respectively. The thermo-physical properties of PCM are described in Table 1.

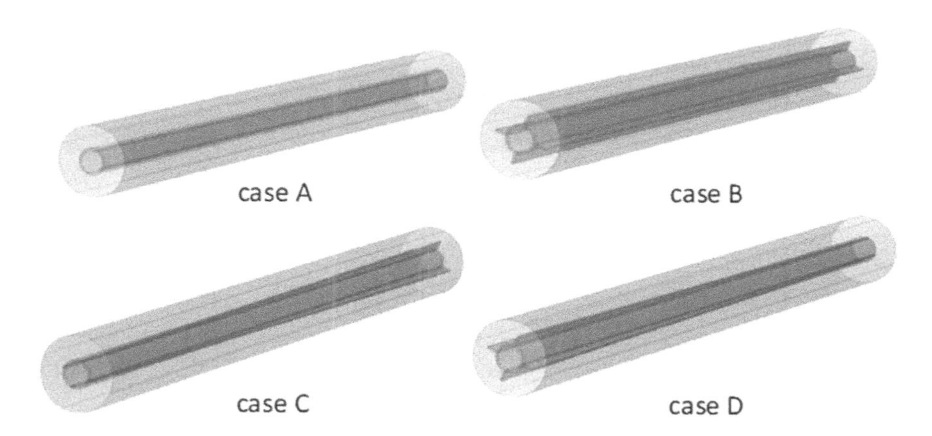

Fig. 1 Physical model with different fin configurations

Table 1 Thermo-physical properties of PCM [13]

Property	Values
Phase change enthalpy, L_f (kJ kg^{-1})	172
Heat conductivity, k (W m^{-1} K^{-1})	0.565
Density (kg m^{-3}), ρ	2261
Specific heat, c_s (J kg^{-1} K^{-1})	1823
Dynamic viscosity, μ (Pa s)	3.06×10^{-3}

2.2 Governing Equations

The governing equations of the PCM flow and energy are described below [10]:

Continuity Equation

$$\nabla.\vec{v} = 0 \tag{3}$$

Momentum Equation

$$\frac{\partial \vec{v}}{\partial t} + (v.\nabla)\vec{v} = \frac{1}{\rho}\left(-\nabla P + \mu\nabla^2\vec{v} + \vec{F} + \vec{S}\right) \tag{4}$$

where $\vec{F} = -\rho\vec{g}\beta(T - T_m)$ and $\vec{S} = \frac{-(1-LF)^2}{(LF^3+a)} A_m\,\vec{v}$.

Energy Equation

$$\rho c\frac{DT}{Dt} = k\nabla^2 T \tag{5}$$

Specific Heat of PCM (c)

$$c = \begin{cases} c_s & \text{for } T < T_s \\ c_{\text{eff}} & \text{for } T_s \le T \le T_l \\ c_l & \text{for } T > T_l \end{cases} \quad \text{and} \quad c_{\text{eff}} = \frac{c_s + c_l}{2} + \frac{L_f}{T_l - T_s} \tag{6}$$

Liquid Fraction (LF)

$$\text{LF} = \frac{T - T_s}{T_l - T_s} \tag{7}$$

Here, $\vec{F}$ is the body force term. The Boussinesq approximation is employed for density variation and natural convection during PCM melting. $\vec{S}$ is Darcy equation which describes dynamic velocity variations inside the PCM. The A_m is the mushy zone constant, and its value is assumed to be 10^4 in the present study [11]. The constant a has a value of 0.001 and prevents the division of Darcy term by 0. The effective heat capacity approach [12] is followed for the phase change problem, where c_{eff} is the effective specific heat (J/kg K) of the PCM.

Initial Conditions
Initially at $t = 0\,\text{s}$, the system temperature is assumed to be 286 °C and 326 °C during charging and discharging, respectively. The PCM melting temperature (T_m) is 306 °C, and the melting temperature range ($T_l - T_s$) is assumed to be 4 °C. T_s and T_l are the solid-state and liquid-state PCM temperatures, respectively.

Boundary Conditions

$$\vec{v} = 0 \quad \text{(no slip at walls)} \tag{8}$$

$$\frac{\partial T}{\partial z} = 0 \tag{9}$$

At time $t > 0\,\text{s}$, for charging, $v_{htf} = 6$ m/s and $T_{\text{inlet}} = 326$ °C, and for discharging, $v_{htf} = 6$ m/s and $T_{\text{inlet}} = 286$ °C where T_{inlet} and v_{htf} are the inlet temperature and velocity of HTF, respectively.

Assumptions The HTF flow is laminar, and the natural convection is symmetric about the vertical axes. Therefore, to reduce the computational effort, only half of the model is developed. Discharging is assumed to be purely conduction dominated, hence $\vec{F} = 0$.

The numerical simulations are performed using software tool COMSOL Multiphysics. Euler backward difference scheme is used for time-stepping, and the convergence criteria is taken as 10^{-3}.

2.3 *Experimental Validation*

The afore-mentioned modeling procedure is validated with the data from the experiments reported by Cao et al. [14]. They developed a circular annulus cavity system filled with lauric acid (melting point of 44.2 °C). A wall temperature of 80 °C was used as the input heat source, whereas the system was maintained at 25 °C initially. The position of thermocouple locations and the validated temperature profiles are shown in Fig. 2. It is observed that the present numerical simulation results match closely with the experimental data.

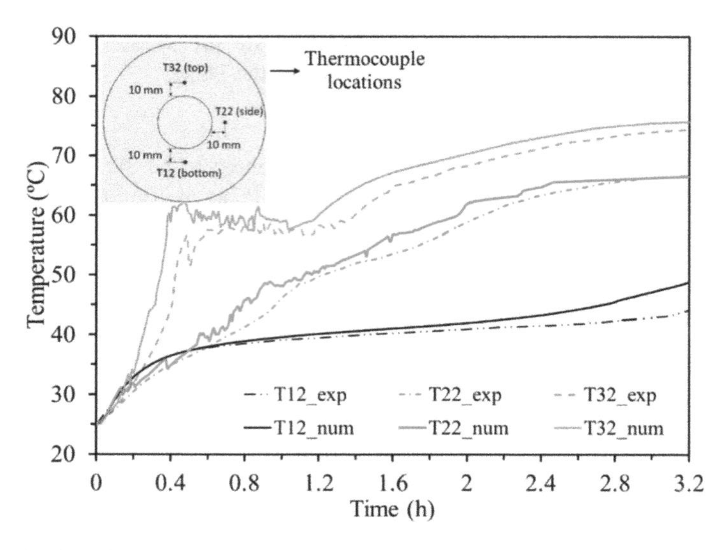

Fig. 2 Validation of the numerical simulation data with the experiments performed by Cao et al. [14]

3 Results and Discussion

3.1 Charging

The variation of liquid fraction (LF) during charging for all the cases is plotted in Fig. 3. The times taken by case A, case B, case C and case D for the completion of charging are 16.33, 12.9, 16.73 and 14.86 h, respectively. The heat transfer during melting is a combination of both the conduction and natural convection heat transfer. The reduction in times for case B and D over case A is 21 and 11.2%, whereas, in case C, the charging time increases by ~2.5%. Although it is evident that case B with straight fins performs the best due to higher fin surface area, it is important to note the performance of case C and case D. The case D performs significantly better than case A and case C. The conduction and natural convection heat transfer near front end is high due to a higher gradient between the HTF and PCM temperatures, and further accelerates in this region due to the presence of fins for case B and case D. In the case C, the fins have uniformly increasing height along the length of the system and do not yield any benefit for improving the rate of heat transfer. Instead, the charging performance deteriorates than no-fin system (case A). This indicates that using fins may not always be beneficial during the PCM melting. For case C, the heat transfer rate near the inlet section is high due to high natural convection rate caused by a high difference between the HTF and PCM temperatures, whereas at the outlet section, the natural convection rate is low due to low difference in the HTF and PCM temperatures. Even though heat transfer surface area is high in this region,

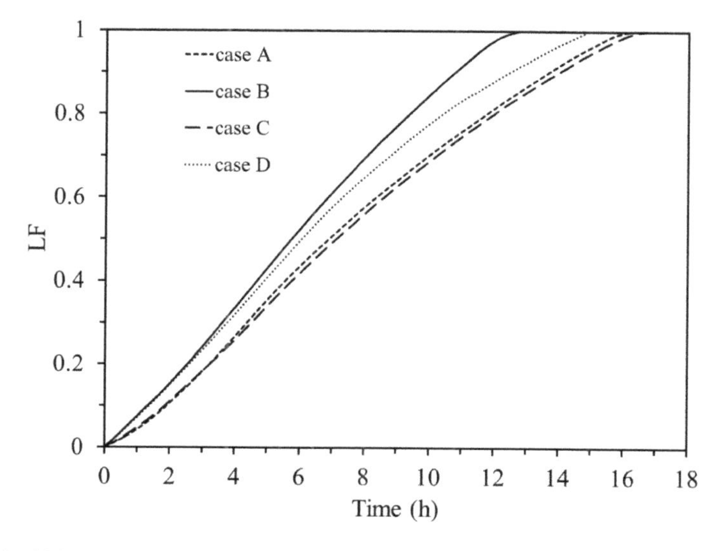

Fig. 3 Liquid fraction versus time for the charging process

there is a compromise between conduction and natural convection effect as the fins block the PCM flow. Hence, there is a reduction in the melting rate of the PCM.

Figure 4 shows the average temperature of the PCM varying with time for the charging process. The temperature profiles for all cases indicate an increase in temperature due to energy gained by the PCM. It can be inferred that the temperature rise

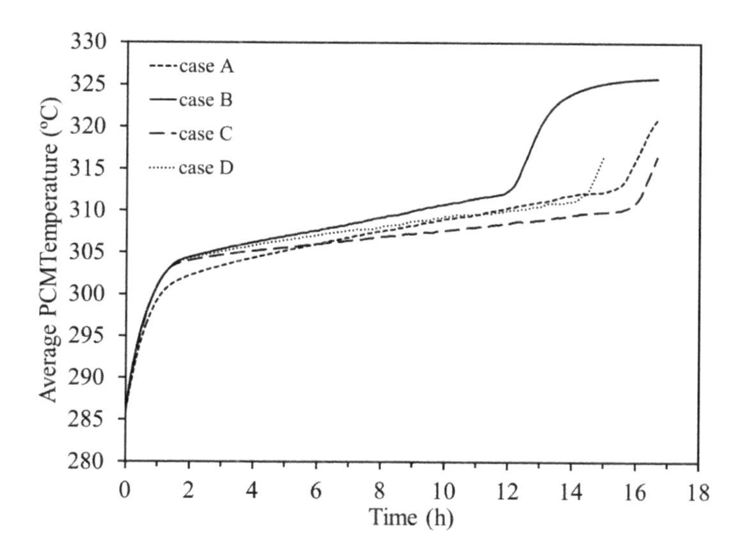

Fig. 4 Average PCM temperature variation versus time for the charging process

for case B is always higher than other cases, such that the rate of energy gained is high. The temperature rise in case A is higher than case C throughout the charging process, such that the rate of charging for case C does not improve by employing fins.

3.2 *Discharging*

During discharging, it is observed that case A, case B, case C and case D discharge completely in 19.93, 12.4, 13.76 and 14.56 h, respectively, as shown in Fig. 5. The heat transfer due to conduction dominated the discharging process. The reduction in time for case B, case C and case D over case A is 37.7%, 31% and 27%, respectively. Hence, having more fin surface area is always beneficial as in case B. By comparing case C and case D, it is found that case D discharges faster than case C till ~92% of the discharging process, which indicates that the heat transfer is high in the front end of the system having high difference between the HTF and PCM temperatures for most of the discharging process. The presence of fins near front end as in case B and case D accelerates the solidification even further, whereas near the rear end, the gradient between HTF and PCM temperatures is extremely low and the presence of fins as in case C helps to accelerate solidification only during the end of the discharging process.

Figure 6 shows the temperature profile variation for all the cases during the discharging process. There is a drop in the PCM temperature due to energy delivered

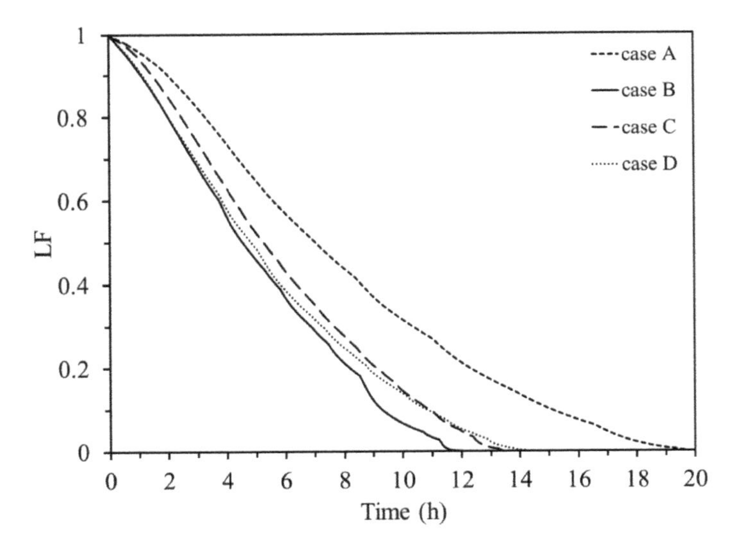

Fig. 5 Liquid fraction versus time for the discharging process

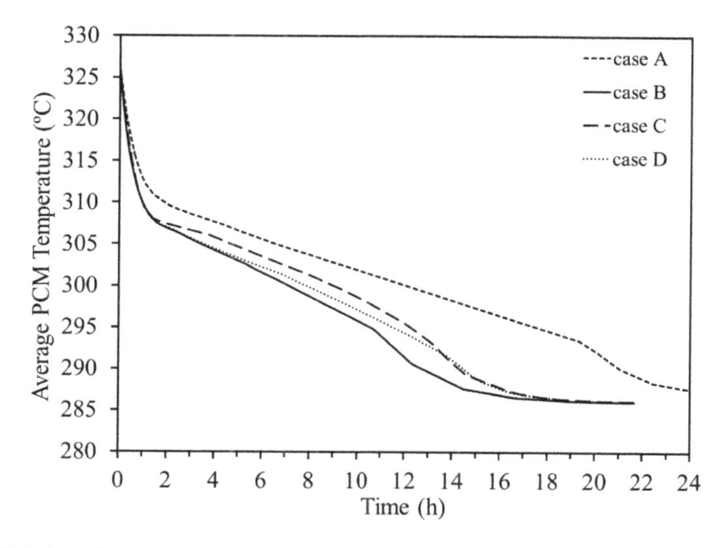

Fig. 6 Variation of average PCM temperature during the discharging process

to the HTF. The case B has the highest discharging rate; however, the case C and case D deliver similar performance throughout the discharging process.

It can be inferred that the charging and discharging performance of horizontal tube-in-shell latent heat storage system with longitudinal fins is improved comprehensively with the addition of fins. However, the optimal performance depends upon the distribution of the fin surfaces along the length of the system.

4 Conclusions

A 3D numerical model is developed to investigate the performance of a horizontal tube-in-shell system without fins and compared with different longitudinal fin configurations, namely: straight and increasing/decreasing fin height. The melting rate by adding straight fins (case B) improves by ~21%, whereas the performance of the system with increasing fin height (case C) reduces by 2.5% than system with no fins (case A). By using fins, the heat transfer rate during charging does not improve necessarily due to the addition of fins as there exists a trade-off between conduction and convection rate, whereas during discharging, the rate of heat transfer is always improved due to enhanced conduction of fins. The case B, case C and case D discharge faster than case A by 37.7%, 31% and 27%, respectively. Further, it is concluded that fins are beneficial to be used in front end of the system. Fins accelerate rate of heat transfer due to conduction near the front end during melting/solidification processes. At rear end, fins only enhance the rate of heat transfer due to conduction during the melting/solidification processes, but hinder the free PCM movement in melting case.

Acknowledgements The authors gratefully acknowledge the Department of Science and Technology (DST), Government of India, for their financial support (Project No.: DST/TMD/SERI/D12(C)).

References

1. Tao, Y.B., He, Y.L.: A review of phase change material and performance enhancement method for latent heat storage system. Renew. Sustain. Energy Rev. **93**, 245–259 (2018). https://doi.org/10.1016/j.rser.2018.05.028
2. Sivasamy, P., Devaraju, A., Harikrishnan, S.: Review on heat transfer enhancement of phase change materials (PCMs). Mater. Today Proc. **5**, 14423–14431 (2018). https://doi.org/10.1016/j.matpr.2018.03.028
3. Gasia, J., Maldonado, J.M., Galati, F., De Simone, M., Cabeza, L.F.: Experimental evaluation of the use of fins and metal wool as heat transfer enhancement techniques in a latent heat thermal energy storage system. Energy Convers. Manag. **184**, 530–538 (2019). https://doi.org/10.1016/j.enconman.2019.01.085
4. Deng, S., Nie, C., Wei, G., Ye, W.: Improving the melting performance of a horizontal shell-tube latent-heat thermal energy storage unit using local enhanced finned tube. Energy Build. **183**, 161–173 (2019). https://doi.org/10.1016/j.enbuild.2018.11.018
5. Hosseini, M.J., Ranjbar, A.A., Rahimi, M., Bahrampoury, R.: Experimental and numerical evaluation of longitudinally finned latent heat thermal storage systems. Energy Build. **99**, 263–272 (2015). https://doi.org/10.1016/j.enbuild.2015.04.045
6. Abdulateef, A.M., Mat, S., Sopian, K., Abdulateef, J., Gitan, A.A.: Experimental and computational study of melting phase-change material in a triplex tube heat exchanger with longitudinal/triangular fins. Sol. Energy. **155**, 142–153 (2017). https://doi.org/10.1016/j.solener.2017.06.024
7. Abdulateef, A.M., Abdulateef, J., Mat, S., Sopian, K., Elhub, B., Mussa, M.A.: Experimental and numerical study of solidifying phase-change material in a triplex-tube heat exchanger with longitudinal/triangular fins. Int. Commun. Heat Mass Transf. **90**, 73–84 (2018). https://doi.org/10.1016/j.icheatmasstransfer.2017.10.003
8. Pan, C., Vermaak, N., Romero, C., Neti, S., Hoenig, S., Chen, C.H.: Efficient optimization of a longitudinal finned heat pipe structure for a latent thermal energy storage system. Energy Convers. Manag. **153**, 93–105 (2017). https://doi.org/10.1016/j.enconman.2017.09.064
9. Sodhi, G.S., Jaiswal, A.K., Vigneshwaran, K., Muthukumar, P.: Investigation of charging and discharging characteristics of a horizontal conical shell and tube latent thermal energy storage device. Energy Convers. Manag. **188**, 381–397 (2019). https://doi.org/10.1016/j.enconman.2019.03.022
10. Clement, K.: Modern Fluid Dynamics. Springer (2018) www.springer.com/series/5980.
11. Kumar, M., Krishna, D.J.: Influence of mushy zone constant on thermohydraulics of a PCM. Energy Proc. **109**, 314–321 (2017). https://doi.org/10.1016/j.egypro.2017.03.074
12. Sodhi, G.S., Vigneshwaran, K., Jaiswal, A.K., Muthukumar, P.: Assessment of heat transfer characteristics of a latent heat thermal energy storage system: multi tube design. Energy Proc. **00**, 22–25 (2018). https://doi.org/10.1016/j.egypro.2019.01.737
13. Muhammad, M.D., Badr, O.: Performance of a finned, latent-heat storage system for high temperature applications. Appl. Therm. Eng. **116**, 799–810 (2017). https://doi.org/10.1016/j.applthermaleng.2017.02.006
14. Cao, X., Yuan, Y., Xiang, B., Highlight, F.: Effect of natural convection on melting performance of eccentric horizontal shell and tube latent heat storage unit. Sustain. Cities Soc. **38**, 571–581 (2018). https://doi.org/10.1016/j.scs.2018.01.025

Fabrication of Catalytic Sheet Filter of V_2O_5–WO_3/TiO_2-Supported SiC for Selective Catalytic Reduction of NO_x Emission from Combustion Engine

Ajit Dattatray Phule, Seongsoo Kim, and Joo Hong Choi

Abstract Nitrogen oxide (NO_x) emission is a global environmental problem, which arises mainly due to fuel used in diesel engines, power stations, industrial heaters and cogeneration plants. An effective standard technique for NO_x emission reduction is nothing but the selective catalytic reduction (SCR) using NH_3 over catalytic bed. Herein, we fabricated the SiC sheet filters with different sized (38, 38–53 and 53 μm) SiC powder. The thickness of the SiC filter had varied to see the change in V_2O_5–WO_3/TiO_2 catalyst loading and NO conversion performance at different reaction temperatures (240–340 °C). Increase in the thickness (double) of the SiC sheet filter increases the catalyst loading over the filter. As a result, NO conversion performance improved from 80% (BM25SF_SiC53) to more than 94% (BM25SF_SiC10mm_53) at reaction temperature range of 260–340 °C, where N_x slip concentration (outlet gas) <80 ppm observed in the reaction temperature range of 280–320 °C. The surface morphology of the SF with empty and buried (with catalyst) pores has been studied with the help of an optical image. In the presence of SO_2, the NO conversion performance of the BM25SiC10mm_53_6V9WT catalytic filter for the temperature window of 240–340 °C increases (i.e., >95%; temperature window of 260–340 °C) with N_x slip concentration <50 ppm. This catalytic sheet filter will be applicable to reduce the cost of the catalytic filter technology along with the environmental safety.

Keywords SiC sheet filter · V_2O_5–WO_3/TiO_2 catalyst · SCR of NO_x

1 Introduction

All in recent years, catalytic filter has made great attention due to its immense ability of simultaneous removal of NO_x and dust particles [1–5]. This provides

A. D. Phule (✉) · J. H. Choi
Department of Chemical Engineering/ERI, Gyeongsang National University, Jinju 52828, Korea
e-mail: phule.ajit@gmail.com

S. Kim
Department of Mechatronics Convergence, College of Engineering, Changwon National University, Bldg. 51, 20, Changwondaehak-ro, Changwon 641773, Korea

P. Muthukumar et al. (eds.), *Innovations in Sustainable Energy and Technology*,
Advances in Sustainability Science and Technology,
https://doi.org/10.1007/978-981-16-1119-3_9

immense benefits like compact process and space, and more economical equipments. In general, the external surface membrane of catalytic filter use to filtrate and separate the dust in flue gas and the selective catalytic reduction (SCR) with NH_3 over the catalyst coated on the inner part of filter subsequently eliminate the NO_x. The filter with narrow wall thickness along with the shorter reaction time constrains the NO conversion performance of the catalytic filter, which is different from the traditional honeycomb catalyst. To overcome this issue, high concentration of catalyst along with active components (>8 wt%) is needed to coat onto the filter. With this approach, the NO conversion performance of the catalytic filter can be enhanced; however, the increased catalyst content will increase the material cost in industrial uses. Also, the possibility of pressure drop through filter cannot be declined. So, the development of extremely efficient catalytic sheet filter with excellent activity is the current need of the filter world.

In recent years, versatile NH_3-SCR catalysts have been developed, such as transition metal supported on zeolites, e.g., Fe–Cu–ZSM-5 [6], Cu-SSZ-13 [7], Cu-SAPO-34 [8], and transition metal oxide catalyst, e.g., Fe-based [9, 10], Cu-based [11, 12], Mn-based [13, 14]. These catalysts displayed outstanding catalytic activity; however, the sulfate presented in the emission reduces their catalytic activity. Hence, the vanadium-based catalyst with strong sulfate tolerance becomes the appropriate candidate to reduce NO_x emissions. Due to low cost and high activity, V_2O_5–WO_3/TiO_2 catalyst is also widely used for commercial processes [15]. Other additives such as WO_3 [2, 4, 16] and Pt [17, 18] were also reported to enhance the NO conversion performance and lower down the SCR temperature for the V_2O_5–TiO_2-based catalytic filters. However, at high temperature above 300 °C, these catalytic filters show reduced NO conversion performance. This is due to high V_2O_5content, which promotes the severe oxidation of NH_3 over these catalysts. The catalysts are mostly commercially used in industrial field with the support on surface of honeycomb monolith filter. This filter occupies a little more space, so the concept of sheet-type filter has raised recently as they are more compact compared to candle type and other filters. Choi et al. [19] have used ceramic sheet-type filters coated with V_2O_5–WO_3/TiO_2 catalyst to achieve ~90% NO conversion with the reaction temperature range of 260–360 °C, which is not much satisfactory as a best for commercial use (which need >95% NO conversion). So, there is necessity to develop a catalytic SiC sheet-type filter with high NO conversion performance with the presence of sulfur.

In the present research work, we prepared the SiC sheet filter from different sized SiC powder and varied the thickness of the sheet filter to see the effect of thickness on the catalyst loading as well as on the NO conversion performance of catalytic sheet filter. These SiC sheet filters are coated with the V_2O_5–WO_3/TiO_2 catalyst with simple dip coating method, performed the NO conversion test at different temperature ranges 240–340 °C, at face velocity 2 cm/s, and discussed the results in comparison with ceramic catalytic sheet filter. The surface morphology of the SF with empty and buried (with catalyst) pores has been studied with the help of an optical image. The effect of SO_2 on the SiC catalytic sheet-type filter has been studied and discussed.

2 Experimental Section

2.1 Preparation of Catalyst Powder

V_2O_5–WO_3/TiO_2 powder catalysts were initially synthesized by using the wet impregnation method [20]. The composition 3 or 6 wt% V_2O_5 and 9 wt% WO_3 has been calculated on the basis of TiO_2. First, prepare the oxalic acid solution in distilled water at 60–70 °C temperature with magnetic stirring for 30 min. And then, add ammonium meta-tungstate hydrate [$(NH_4)_{10}W_{12}O_{41}$] (from Aldrich) slowly into the same beaker and allow it to mix with magnetic stirring for 30 min. Later, ammonium vanadate NH_4VO_3 (from Junsei Chemical Co. Ltd.) was added slowly to above solution in their desired compositions and continue stirring for 30 min, which turns to greenish-colored solution. TiO_2 (P25) powder (Degussa Ltd.) was added slowly to above-obtained greenish blue-colored solution. The sky bluish-colored VWT catalyst solution was heated at 60–70 °C with magnetic stirring with agitation and continued until the liquid phase disappeared. The paste formation has been observed with 120 °C temperature condition, which further continued to get dried cake. The catalyst dried cake was then grinded with mortar, after which it was calcined in electric furnace with an air stream at 450 °C for 5 h. The coating solution of different concentrations was prepared by using this calcined VWT catalyst powder.

2.2 Catalyst Coating Solution Preparation

The 25 wt% coating solution of the 6V9WT catalyst has been prepared by using high energy ball milling machine with the speed of 1000 rpm for the duration of 7 min. to obtain a particular sized VWT catalyst particle. A particle size and shape analyzer (1090LD Shape Analyzer from SCINCO) were used to measure the size of the catalyst particle coating solution during the ball milling process. These coating solutions have been further used to prepare the coated catalytic SF.

2.3 Preparation of SiC Sheet Filter

As purchased SiC powder (from Jay Chem. Co. Korea) has been sieved with varied size of 38, 38–53 and 53 μm, which further used to prepare the SiC sheet filter. Dry mixing of SiC powder with 1 wt% carboxymethyl cellulose sodium salt (CMC from Samchun, Korea) and 20 wt% borosilicate glass powder of 325 mesh (SiO_2 > 50, Al_2O_3 > 5, B_2O_3 > 15, Cao, NaO, K_2O, MgO, FeO_3) (BGP from Korea New Material Korea) have been done by stirred mill for 1 h. Then wet mixing of 0.2 wt% Calcium carbonate (from Wangpyo Chemical Korea), and 20 wt% distilled water have done by stirred mill for 1 h. The prepared sample aged for 4 h. The sheet filter shape formation is done by mold of 43 × 43 × 5 mm with 15.8 g/ea sample amount and 43 × 43 × 10 mm with 31.6 g/ea sample. The press force used during molding was 100 kg for each sample. Then, the molded sample dried at 80°C for 8 h in electric oven followed by sintering at 850 °C for 5 h in an electric furnace.

2.4 Preparation of Catalytic Sheet Filter

The ceramic SF (reference) and SiC sheet filter with 43 × 43 mm dimension and 5 mm/10 mm thickness have been coated with the VWT catalyst by using simple dip coating method. First, the sheet filter has been cleaned with the ultra-sonication process for 30 min at room temperature in distilled water and then dried at 120 °C for 2 h in an electric oven. The dried SF has been dipped into the coating solutions (25 wt% 6V9WT) for 5 min. and dried at 120 °C for 2 h to allow the solvent to evaporate followed by the calcination at 450 °C in an electric furnace for 5 h to obtain the 6V9WT coated catalytic SF. The coated SF with 5-mm thickness is named as BM25SF_6V9WT (ceramic sheet filter), BM25SF_SiC38, BM25SF_SiC38-53 and BM25SF_SiC53, and filter with thickness of 10 mm is named as BM25SF_SiC10mm_38, BM25SF_SiC10mm_38-53 and BM25SF_SiC10mm_53. The surface morphology of the VWT coated catalytic SiC SF has been studied by using optical microscope (HiMax Tech. Co. Ltd., Model HT004).

2.5 Catalytic Activity Measurement

A special experimental unit was used to conduct the catalytic activity of the catalytic filters. Experimental unit consists of a specially designed reactor to mount the sheet-type catalytic filter by keeping the commercial operation mode of the filtration in such a way that the reacting gas passes from outside into inside of the filter element. The typical reaction gas was composed of 700 ppm NO, 700 ppm NH_3, SO_2 1000 ppm, 5 vol% O_2, with balanced N_2. The total gas flow rate was controlled with a mass flow controller to adjust the actual face velocity of 2 cm/s. The catalytic activity was measured at a series of steady-state temperature condition.

To analyze simultaneously the concentrations of NO, N_2O, NO_2 and NH_3, we used the FT-IR spectrometer (MIDAC Prospect-IR) with a heated gas cell (permanently aligned 10-m gas cell from Gemini Scientific Instruments) and DTGS detectors. All gas line tubes were heated at 120 °C to avoid the condensation of gas in the system. Dry air was used to record the background spectra. Each component's concentration was calculated by integration of the specific absorption frequencies (cm^{-1}) based on ethylene 938–962: NO 1873–1881, N_2O 2188–2190, NO_2 1610–1614, NH_3 951–989 and H_2O 1987–1994. To estimate the catalytic performance of the filter, the following formula was used; N_x slip concentration (the summation of outlet concentrations of

$[NO]_{out} + [N_2O]_{out} + [NO_2]_{out} + [NH_3]_{out}$. This simple formula $([NO]_{in} - [NO]_{out})/[NO]_{in}$ was used to calculate NO, where $[X]_{in}$ and $[X]_{out}$ mean the concentration of component X (ppm) in the inlet and outlet, respectively.

3 Results and Discussion

SiC powder with different sizes impacts on the formation of SiC sheet filter as well as on catalyst coating process which ultimately shows the effect on the NO conversion performance of the catalytic sheet filter. Figure 1a–h shows an optical image of the virgin and coated SiC sheet filter with different thicknesses (5 and 10 mm). It revealed that the SiC sheet filter of 5-mm thickness prepared from SiC powder of 38 micron (SiC_38) and 53 μm (SiC_53) has less pore/gaps compared to SiC sheet filter prepared from 38 to 53 μm (SiC_38-53) sized SiC powder (see Fig. 1a, c, e). Figure 1b, d, f shows that the SiC sheet filter has been coated successfully by the catalyst and tried to burry the maximum pores/gaps as much as possible throughout the sheet filter. Further, we compared SiC_38-53 sheet filter with 10-mm thickness, which is devising less pores as compared to other sheet filters (see Fig. 1g, h), and the ball milled process helps the catalyst to reduce effectively to access the smaller pores which provides high surface area to improve better NO conversion performance of the catalytic sheet filter.

NO conversion, N_x slip concentration and N_2O formation of catalytic sheet filters tested with different sized SiC powder with varied thickness are compared with catalytic ceramic sheet filter with the VWT catalyst loading of same concentration (25 wt%) in Fig. 2a, b, and c, respectively. BM25SF_SiC10mm_53 catalytic filter displays the best NO conversion performance, i.e., ~ 92–94% for the window of 240–340 °C (broad reaction temperature range) among all the catalytic SiC sheet filters,

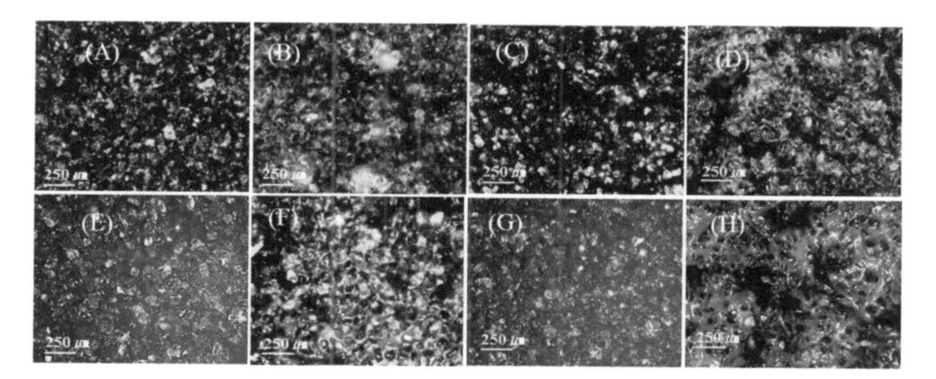

Fig. 1 Optical images of the virgin and coated SiC sheet filter of **a** SiC38 and **b** BM25SF_SiC38, **c** SiC38-53 and **d** BM25SF_SiC38-53, **e** SiC53 and **f** BM25SF_SiC53, **g** SiC10mm_38-53 and **h** BM25SF_SiC10mm_38-53; shows the surface with pores and catalyst coating (all images at scale bar 250 μm)

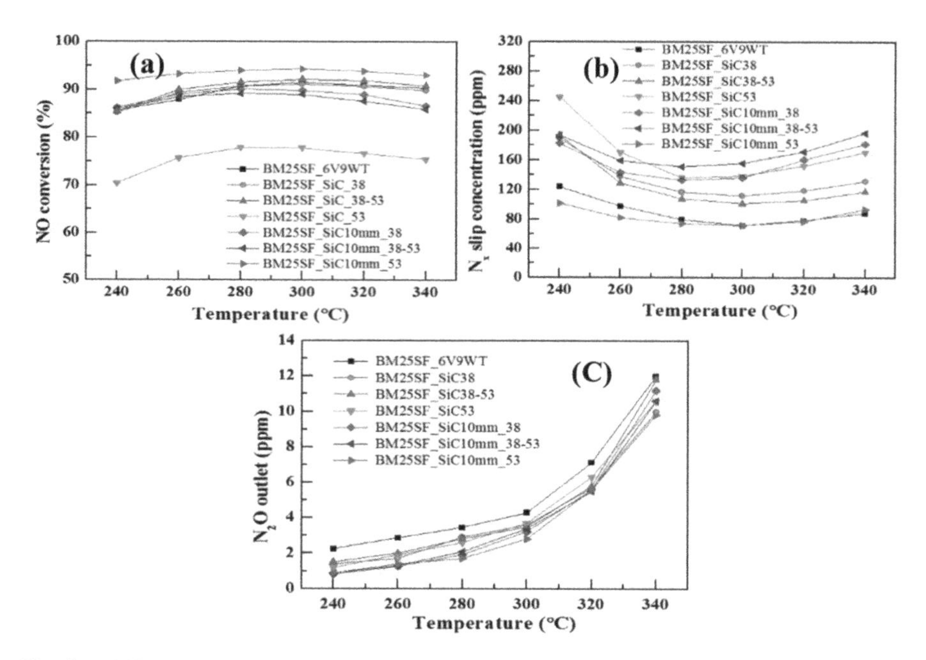

Fig. 2 **a** NO conversion, **b** N_x slip concentration and **c** N_2O outlet at different temperatures (240–340 °C) for BM25SF_6V9WT, BM25SF_SiC38, BM25SF_SiC38-53, BM25SF_SiC53, BM25SF_SiC10mm_38, BM25SF_SiC100mm_38-53 and BM25SF_SiC10mm_53 with reaction condition; NO and NH_3 = 700 ppm, O_2 = 5 vol%, face velocity = 2 cm/s

which is attributed to the well dispersion of the catalyst within the filter. The highest NO conversion, i.e., 94%, has been observed for the reaction temperature 300 °C, as shown in Fig. 2a. The reason for better NO performance is obviously the small pores in BM25SF_SiC10mm_53 compared to other sheet filters with less thickness. Increased thickness of the SiC sheet filter from 5 to 10 mm also provides more space to load more catalyst, which offers more reactive sites to perform better NO reduction. The reaction time also increased a little bit which also adds up the value to NO conversion performance. BM25SF_SiC10 mm_53 has highest loading of catalyst compared to BM25SF_SiC10mm_38 and BM25SF_SiC10mm_38-53 (see Table 1). The N_x slip concentration of BM25SF_SiC10mm_53 is <80 ppm in the temperature range from 260 to 320 °C (see Fig. 2b). N_2O formation concentration is below 10 ppm with broad temperature window at 240–340 °C (see Fig. 2c). However, in case of BM25SF_SiC53 it shows less than 80% NO conversion which is unexceptionally less in comparison with BM25SF_SiC38 and Bm25SF_SiC38-53 in 5-mm-thick sheet filter. It may be due to the small cracks present in sheet filter and less catalyst loading as compared to other SiC sheet filters (see Table 1).

The SO_2 effect on the NO conversion performance of the BM25SiC10mm_53_6V9WT catalytic filter has been carried out for the temperature window of 240–340 °C with 1000 ppm SO_2 concentration. It demonstrates the

Table 1 Catalyst loading on ceramic and SiC sheet filter with different thicknesses using 25 wt% catalyst coating solution concentration

S. No.	Catalytic sheet filter	Catalyst loading (g)
1	BM25SF_6V9WT	1.5
2	BM25SF_SiC38	1.42
3	BM25SF_SiC38-53	1.79
4	BM25SF_SiC53	1.5
5	BM25SF_SiC10mm_38	2.56
6	BM25SF_SiC10mm_38-53	2.66
7	BM25SF_SiC10mm_53	2.75

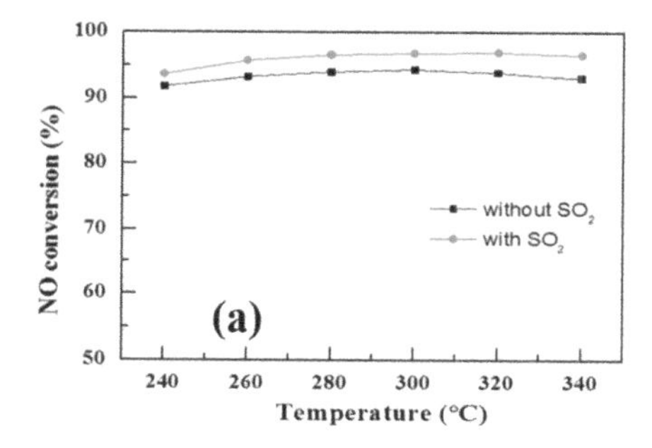

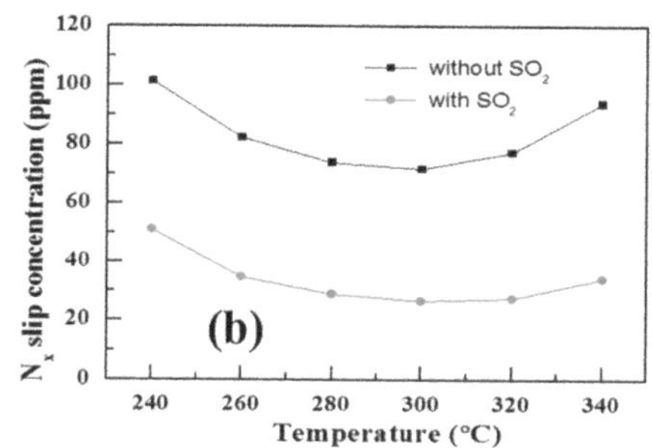

Fig. 3 SO_2 effect on **a** NO conversion and **b** N_x slip concentration at different temperatures (240–340 °C) for BM25SiC10mm_6V9WT catalytic sheet filter with and without SO_2 with reaction condition; NO and NH_3 = 700 ppm, SO_2 = 1000 ppm, O_2 = 5 vol%, V_f = 2.0 cm/s

increase in NO conversion performance of the catalytic filter with SO_2 (i.e., >95%) for the temperature window of 260–340 °C with N_x slip concentration less than 50 ppm (see Fig. 3a, b). The overall NO conversion performance of the developed SiC catalytic sheet filter is better than the ceramic sheet filter [19] with as well as without SO_2. It may be plausible that the adsorption of SO_2 improved the amount of Lewis acid sites, and thus the capacity of NH_3 improved which may help to little increase in the NO conversion performance of BM25SiC10mm_53_6V9WT. Further characterization is much more needed to establish the mechanism behind the increased NO conversion performance of the catalytic sheet filter of V_2O_5-WO_3/TiO_2-supported Sic in the presence of the sulfur content.

4 Conclusion

V_2O_5–WO_3/TiO_2-supported SiC catalytic sheet filter is prepared with varied thickness among which the high thickness (10 mm) and SiC powder size of 53 μm, i.e., BM25SF_SiC10mm_53, show the best NO conversion performance, i.e., 92–94% for the broad temperature window 240–340 °C with less N_2O formation, i.e.,

<10 ppm. Increased thickness (from 5 to 10 mm) of the SiC sheet filter increased the catalyst loading over the sheet filter as well as the reaction time which help to enhance the NO conversion performance of the filter. In the presence of the SO_2, BM25SF_SiC10mm_53 shows increased NO conversion performance, i.e., >95% for the temperature window of 260–340 °C with N_x slip concentration <50 ppm, which is very much satisfactory to use it in commercial applications.

References

1. Saracco, G., Specchia, S., Specchia, V.: Catalytically modified fly-ash filters for NO_x reduction with NH_3. Chem. Eng. Sci. **51**, 5289–5297 (1996)
2. Choi, J.H., Kim, S.K., Bak, Y.C.: The reactivity of V_2O_5–WO_3/TiO_2 catalyst supported on a ceramic filter candle for selective reduction of NO. Korean J. Chem. Eng. **18**, 719–724 (2001)
3. Zhang, K., Zhao, W., Dou, S., Wang, Q.: Catalytic performance of Ti^{+3} self-doped V_2O_5–TiO_2 catalystfor selective catalytic reduction with NH_3. J. Mater. Sci. Eng. **8**, 16–20 (2020)
4. Nacken, M., Heidenreich, S., Hackel, M., Schaub, G.: Catalytic activation of ceramic filter elements for combined particle separation, NO_x removal and VOC total oxidation. Appl. Catal. B **70**, 370–376 (2007)
5. Zuercher, S., Pabst, K., Schaub, G.: Ceramic foams as structured catalyst inserts in gas-particle filters for gas reactions-effect of backmixing. Appl. Catal. A **357**, 85–92 (2009)
6. Zhang, T., Liu, J., Wang, D.X., Zhao, Z., Wei, Y.C., Cheng, K., Jiang, G.Y., Duan, A.J.: Selective catalytic reduction of NO with NH_3 over HZSM-5-supported Fe–Cu nanocomposite catalysts: the Fe–Cu bimetallic effect. Appl. Catal. B: Environ. **148–149**, 520–531 (2014)
7. Fickel, D.W., D'Addio, E., Lauterbach, J.A., Lobo, R.F.: The ammonia selective catalytic reduction activity of copper-exchanged small-pore zeolites. Appl. Catal. B: Environ. **102**, 441–448 (2011)
8. Martinez-Franco, R., Moliner, M., Franch, C., Kustov, A., Corma, A.: Rational direct synthesis methodology of very active and hydrothermally stable Cu-SAPO-34 molecular sieves for the SCR of NO_x. Appl. Catal. B: Environ. **127**, 273–280 (2012)
9. Liu, F.D., He, H., Zhang, C.B.: Novel iron titanate catalyst for the selective catalytic reduction of NO with NH_3 in the medium temperature range. Chem. Commun. **17**, 2043–2045 (2008)
10. Yang, S.J., Li, J.H., Wang, C.Z., Chen, J.H., Ma, L., Chang, H.Z., Chen, L., Peng, Y., Yan, N.Q.: Fe–Ti spinel for the selective catalytic reduction of NO with NH_3: mechanism and structure–activity relationship. Appl. Catal. B: Environ. **117–118**, 73–80 (2012)
11. Si, Z.C., Weng, D., Wu, X.D., Li, J., Li, G.: Structure, acidity and activity of CuO_x/WO_x–ZrO_2 catalyst for selective catalytic reduction of NO by NH_3. J. Catal. **271**, 43–51 (2010)
12. Liu, J., Li, X.Y., Zhao, Q.D., Hao, C., Wang, S.B., Tade, M.: Combined spectroscopic and theoretical approach to sulfur-poisoning on Cu supported Ti–Zr mixed oxide catalyst in the selective catalytic reduction of NO_x. ACS Catal. **4**, 2426–2436 (2014)
13. Song, Z.X., Ning, P., Li, H., Zhang, Q.L., Zhang, J.H., Zhang, T.F., Huang, Z.Z.: Effect of Ce/Mn molar ratios on the low-temperature catalytic activity of CeO_2–MnO_x catalyst for selective catalytic reduction of NO by NH_3. J. Mol. Cat. (China) **29**, 422–430 (2015)
14. Putluru, S.S.R., Schill, L., Jensen, A.D., Siret, B., Tabaries, F., Fehrmann, R.: Mn/TiO_2 and Mn–Fe/TiO_2 catalysts synthesized by deposition precipitation—promising for selective catalytic reduction of NO with NH_3 at low temperature. Appl. Catal. B: Environ. **165**, 628–635 (2015)
15. Parvulescu, V., Grange, P., Delmon, B.: Catalytic removal of NO. Catal. Today **46**, 233–316 (1998)
16. Heidenreich, S., Nacken, M., Hackel, M., Schaub, G.: Catalytic filter elements for combined particle separation and nitrogen oxides removal from gas streams. Powder Technol. **180**, 86–90 (2008)

17. Choi, J.H., Kim, J.H., Bak, Y.C., Amal, R., Scott, J.: Pt–V_2O_5–WO_3/TiO_2 catalysts supported on SiC filter for NO reduction at low temperature. Korean J Chem. Eng. **22**, 844–851 (2005)
18. Kim, Y., Choi, J.H., Yu, L., Bak, Y.: Modification of V_2O_5–WO_3/TiO_2 catalysts supported on SiC filter for NO reduction at low temperature. Solid State Phenom. **124**, 1713–1716 (2007)
19. Phule, A.D., Choi, J.H., Kim, J.H.: Improvement of V_2O_5–WO_3/TiO_2 supported catalytic sheet filter for simultaneous reduction of NO_x and particulates. Int. J. Environ. Sci. Dev. **9**, 244–249 (2018)
20. Kim, J.H., Choi, J.H., Phule, A.D.: Development of high performance catalytic filter of V_2O_5–WO_3/TiO_2 supported–SiC for NO_x reduction. Powder Technol. **327**, 282–290 (2018)

CEPSO-Based Load Frequency Control of Isolated Power System with Security Constraints

Santigopal Pain, Dilip Dey, Kamalika Tiwari, and Parimal Acharjee

Abstract Considering physical limitations like generation rate constraint (GRC), governor dead band (GDB) and time delay (TD), a unique chaotic exponential particle swarm optimization (CEPSO) algorithm is proposed to design the control parameters of PID controller for an isolated realistic power system which consists of thermal and hydro-generating units. An exclusive cost function is framed by taking both transient and steady-state response specifications providing proper weighting coefficients. For avoiding the local optima and to obtain faster and sure convergence, the tuning parameters of CEPSO algorithm like inertia weight, constriction factors and chaotic variables are properly designed. Because of exponential inertia weight and newly developed chaotic variables, optimal solutions are obtained. The simulation outcomes establish the superiority of the proposed CEPSO algorithm compare to genetic algorithm (GA), particle swarm optimization (PSO) and exponential particle swarm optimization (EPSO) algorithms.

Keywords Load frequency control · PID controller · Physical constraints · Chaotic exponential particle swarm optimization

1 Introduction

The load frequency control (LFC) is an important factor for quality and consistent electric power supply. The objective of the LFC is to supply electricity with nominal system frequency and keeping inter area tie-line power exchange at the schedule values. In the power system, frequency is controlled by active power balance. As frequency is same all over the system, any deviation in active power demand at a

---

S. Pain (✉) · D. Dey
Department of Electrical Engineering, Haldia Institute of Technology, Haldia, India
e-mail: pain.santigopal@gmail.com

K. Tiwari
Department of AEIE, Dr. B. C. Roy Engineering College, Durgapur, India

P. Acharjee
Department of Electrical Engineering, National Institute of Technology, Durgapur, India

P. Muthukumar et al. (eds.), *Innovations in Sustainable Energy and Technology*, Advances in Sustainability Science and Technology,
https://doi.org/10.1007/978-981-16-1119-3_10

point will be reflected in the whole system. The LFC provides an effective control mechanism in each area to minimize the frequency deviation and regulate tie-line power flow by controlling the generation for safe and reliable operation [1].

A lot of research papers had been published over the years for better design of LFC. From the literature, it is evident that many control areas still adopt the concept of fixed gain proportional integral (PI) controller in the design of LFC [1]. Though this PI control technique is simple to implement but its performance is not satisfactory for large complex power system. Therefore, many control strategies based on advance control theory such as quantitative feedback theorem (QFT), robust LFC method and sequential quadratic programming technique [2–4] had been suggested for the design of LFC and complex power system. The control design methods based on robust performance index and linear matrix inequality [5, 6] were proposed for the power system with communication delay. The advance control methodologies used so far for the design of LFC are model specific, increase the complexity by enhancing the order of the system and unable to handle the nonlinarites present in the system. The performance of the LFC is generally affected by complexity due to large structure, presence of physical nonlinear security restrictions (GDB, GRC and TD), load variation, parameter uncertainty, changing operating condition, etc. Therefore, the direct/analytical conventional tools or methods are unable to solve the problem. For this practical scenario, to obtain the optimal tuning parameters (K_p, K_i, K_d) of controllers, soft computing (SC) techniques are essential because of their ability to handle the complex nonlinear systems. Different types of SC techniques such as fuzzy logic [7], genetic algorithm [8, 9], differential evolution (DE) [10], bacteria foraging optimization (BFO) [11] and modified bat inspired algorithm [12] were efficaciously used to solve many LFC problems. Nowadays, popularly PSO [13] and its different forms such as craziness-based PSO [14], exponential PSO (EPSO) [15] and hybrid PSO [16] were used to tune the control variables of linear and nonlinear models. To overcome the local optima and for faster and sure convergence, chaos was also used with the evolutionary techniques [17, 18]. A large number of research papers had been published in the last few years in which the LFC designs of isolated power system were executed by considering linear simple model with one power generating units or more identical power generating units. Moreover, in most of the literature, the similar values of parameters for all the connected generating units were considered for simplicity and easier implementation of the algorithms. The important physical security constraints were also neglected in most cases. This makes the design and analysis impractical which may not be useful for practical implementation. To get precise design of the LFC model, it is obligatory to consider all the constraints (GRC, GDB and TD) imposed by the physical system dynamics. It is evident from the literature that in most cases, the primary concern of LFC design was the minimization of the deviation of frequency and tie-line power deviation. In the restructured power system, the steady-state performance (steady-state error SSE) is not the only concern. But the transient performances such as maximum overshoot (M_p) and settling time (t_s) are also highly important which are overlooked in most cases. In this paper, both transient and steady-state performances are examined thoroughly by developing a unique and logical cost function.

Simple PSO sometimes fails to reach the global minimum and stuck to local minima. Exponential inertia weight-based exponential PSO (EPSO) algorithm can be effectively used to overcome this problem because of the adaptive nature of inertia weight. Chaos can be introduced to achieve optimum performances, to overcome stagnation and to obtain more useful solutions. The combination of EPSO and chaos, i.e., chaotic exponential PSO (CEPSO) is developed and effectively used in this paper. An effort has been given for the optimal design of CEPSO-based PID controller for LFC of single area practical power system. In this work, thermal and renewable power generation units like reheat, non-reheat and hydro-generation units with dissimilar and real-time practical parameter values are taken to create the practical model of the LFC system. The physical security limitations (GRC, GDB and TD) are also incorporated to achieve the realistic design. The aim of this study is to achieve not only zero SSE but also best transient performances of the realistic single area power system. In this strategy, the problem is framed as a minimization problem. A unique cost function with proper weighting factors is developed to enhance both the steady-state and transient performances of the LFC system. Simulation results show the good transient and steady-state response provided by the controller and established the superiority of CEPSO over GA, PSO and EPSO algorithms.

2 System Model

In this study, a single area isolated power system with diverse type of power generating unit is considered. The system is built with most widely used sources—thermal (reheat and non-reheat) and hydro-generation units to make the analysis useful one. Three sources of power generation units are taken together to make the power system model more practical. In most of the literatures, LFC design is carried out by overlooking vital physical limitations as a whole or by part. This makes the analysis non-realistic. The important physical limitations/constraints which prominently affect the performance of power system are GRC, GDB and TD [1]. These practical limitations are inflicted by turbine, governor, crossover elements of thermal system, penstock dynamics of hydraulic unit and communication channels. The constraints must be considered to make the analysis practical; however, these constraints make the system highly a nonlinear system. Consideration of all the limitations in LFC design may be difficult task but it is very much useful to get accurate dynamic behavior of the power system. The physical limitations govern the transient as well as steady-state output of the power system by enhancing M_p, t_s, SSE and also degrade the stability of the system. So, designing the controller without considering these constraints is impractical. The GRC of the reheat thermal unit is quite low and it is in the order of 3% p.u. per minute. For non-reheat units, it is in the order of 5–10% p.u. per minute, and for hydro-unit, it is in the order of 100% maximum continuous rating (MCR) per minute. Another important constraint is the GDB which is produced by backlash effects and coulomb friction in various governors' linkages and overlapping of valve in the hydraulic relays. GDB is expressed in percent of the frequency or rated

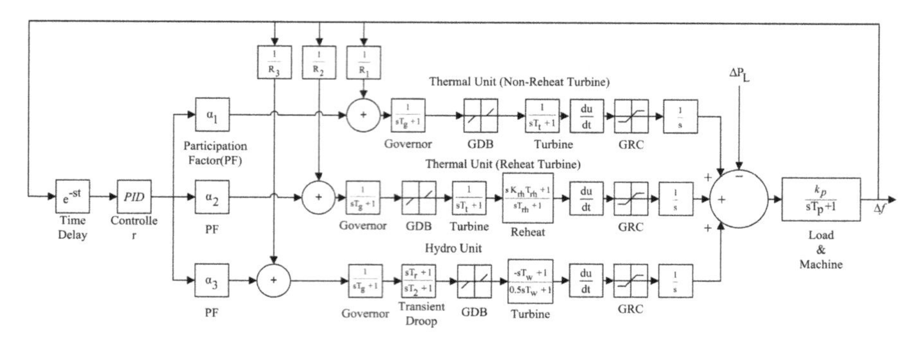

Fig. 1 System model of single area power system

speed. For governor of steam turbine, the maximum dead band is 0.06% of rated speed (frequency), i.e., 0.036 Hz, and for hydraulic turbines, the maximum speed dead band is 0.02% of rated speed. In the standard LFC arrangements, the transmission of measurements from remote terminal units (RTU) to the control center and control output from the control center to the generation unit is made through specific communication channels. There is a constant delay in the communication channel which has a great impact on the LFC design. Increasing the decentralization, control of the power system highly needs the support of open communication infrastructure in deregulated market environment. Future use of open communication network will also incorporate another type of delay, i.e., time varying delay. Those delays, if does not cater in proper way, would worsen the dynamic performance of the LFC. In the extreme case, it will destabilize the whole system. In single area isolated power system, frequency bias has no effect in control actions. The transfer function model of the realistic single area power system considering all the physical constraints is represented in Fig. 1, and the power system parameters are taken from reference [19]. The time delay is taken as 0.1 s. Following a load disturbance, the participating generating units minimize or generate the power in accordance with their participation factors to match the load demand.

3 Cost Function

In LFC system, a particular control strategy is applied to get the true optimal solution. The true optimal solution means fast response, zero overshoot, minimum settling time, zero SSE and better damping performance of the power system. In order to achieve the true optimal solution and fast convergence, a unique cost function is framed considering specific transient and steady-state performance specifications with proper weighting factors. The proper selection of specifications for formation of cost function is highly necessary; otherwise, simulation time may unnecessarily increase and there would be a possibility of getting erroneous result. In this section, a cost function is obtained by adding the different objectives (response specifications) with proper weighting factors. The weighting factors for each objective are selected in such a way that each and every objective gets equal importance during the optimization process. In this work, the unique cost function (J) is developed by finding out the best combination (trial and error approach) of response specifications and weighting factors

$$J = 0.1 \times E + M_p + \text{SSE} + 0.0001 \times t_s \tag{1}$$

where E = (Integral of frequency deviation)^2, SSE = steady-state error, M_p = Maximum overshoot, t_s = Settling time.

4 Proposed Method

The SC techniques can effectively handle the complex and nonlinear control problems compared to conventional control approaches. The SC techniques are robust, not model specific and are useful when the system is functioning over an indeterminate operating range. In this paper, an evolutionary computation-based algorithm is developed for LFC design of isolated realistic power system.

4.1 Exponential PSO Method

The PSO [13] method is useful to optimize complex problems but it has some drawbacks as well. This method is dependent on proper parameter settings; otherwise, there will be a chance that the convergence process may trap in local optima. In PSO, the velocity and position of ith particle are updated as

$$v_i^k = w_i \times v_i^k + c1 \times \text{rand1} \times (p_i^k - x_i^k) + c2 \times \text{rand2} \times (g_i - x_i^k) \tag{2}$$

$$x_i^k = x_i^k + v_i^{k+1} \tag{3}$$

where $c1$ and $c2$ are the constriction factors, and rand1, rand2 are the random numbers lies in between 0 and 1. The best position of the particle i is $p_i{}^k$ in the kth iteration and it is called *pbest* position and the best position of the ith particle up to the kth iteration is g_i which is represented as the *gbest* position. The inertia weight 'w_i' is for the velocity of ith particle. The PSO may not converge at all, if $c1$ and $c2$ are not appropriately selected according to the problem. Inertia weight (w) plays very important role in PSO optimization process. 'w' must not be constant for better optimization. To produce better result, w should vary with the iteration in proper way.

Exponential Inertia Weight. If w is dependent on random number, it is not adaptive. To make it adaptive, a new equation where inertia weight w varies exponentially with the iteration is developed as follows:

$$w_k = w_0 e^{-m(k/K)^n} \tag{4}$$

where k = current iteration, w_0 = initial inertia weight, m = local search attractor, n = global search attractor and K = maximum no. of iteration. The adaptive weight facilitates w high at starting and w low at the end to stop premature convergence and enhance local search, respectively. In adaptive case, w is reliant on both current and maximum no of iteration. It does not depend on cost function value or convergence pattern. 'w' is the factor which governs the effect of the preceding velocity on its present one. In PSO, better solution is achieved by local exploitation and global exploration. The equilibrium between local exploitation and global exploration is achieved by manipulating the value of m and n. So, the value of m and n must be selected carefully. In EPSO algorithm, w is designed in such a manner that it upholds the correct population diversity and correct convergence proficiency.

4.2 Chaotic Exponential Particle Swarm Optimization (CEPSO) Method

When practical limitations are considered for the complex nonlinear problems, sometimes the EPSO method may also suffer from stagnation problem and it may not provide the useful solutions. To overcome this phenomenon, chaos can be introduced in the EPSO method.

In recent studies, it is well established that the chaos theory is a powerful tool which can be applied to many engineering optimization problems [17, 18]. Chaos is a common natural nonlinear phenomenon which is stochastic in nature. A chaotic system exhibits a nonlinear complex behavior when input initial condition changes. It has unique properties of periodic oscillation, bifurcation, regularity and periodicity. To optimize a complex problem using chaotic optimization, chaotic variables are presented as disturbance variables to the decision variables. Due to periodicity property of the chaos variables, the population diversity greatly improves and the algorithm starts to search for the global solution. For avoiding proposed algorithm to trap in local minima or stagnation, chaotic local search (CLS) is incorporated during evolution using CEPSO. CLS is done to execute local exploitation for finding out global solution (*gbest*). CLS is based on chaotic maps of chaos theory represented by a logistic function. The logistic function is very sensitive to initial conditions.

$$y_i^{k+1} = \begin{cases} 2y_i^k & 0 < y_i^k \le 0.5 \\ 2(1-2y_i^k) & 0.5 < y_i^k < 1 \end{cases} \tag{5}$$

The logistic equation-based CLS procedure is illustrated bellow for this minimization problem:

(i) Identify best_obj (minimum cost function value) and the corresponding ith particle ($x_{\min,\,i}^k$) among the whole particles at kth iteration.

(ii) Identify worst_obj (maximum cost function value) and the corresponding ith particle ($x^k_{\max,\,i}$) among the whole particle at kth iteration.

(iii) The chaotic variable $y_i^k \in (0, 1)$ for kth iteration will be calculated as

$$y_i^k = \frac{\left|x_i^k - x^k_{\max,\,i}\right|}{\left|x^k_{\min,\,i} - x^k_{\max,\,i}\right|} \tag{6}$$

(iv) For the next iteration, obtain the chaotic variable y_i^{k+1} using the logistic Eq. (5) and y_i^k.

(v) Obtain the decision variables using EPSO position variables and chaotic variables from the equation represented bellow for the next iteration.

$$x_i^{k+1} = x_i^k + y_i^{k+1} \tag{7}$$

(vi) Find out the new solution with the obtained decision variables x_i^{k+1}.

5 LFC Using Proposed CEPSO

The practical isolated power system model is taken in this study where security limitations GDB, GRC and TD are considered. The cost function is formulated in such a way that it could produce the desired steady-state and transient performances. In LFC, SSE is the primary objective. If the SSE becomes less than 10^{-4}, it can be concluded that the desired solution is achieved. If the algorithm reaches its maximum iteration, but SSE is greater than 10^{-4}, then the *gbest* solution is not the desired solution and the problem is not converged. The tuning parameters of the LFC problem are K_p, K_i and K_d which are randomly generated within a definite real range. Range determination of control parameters is a vital issue which regulate the convergence time. Wrong selection of range will prolong the convergence time and aggravate the performance of the optimization process. In this problem, by trial and error approach, the practical range of K_p, K_i and K_d is obtained as -2 to 2. According to Eq. (1), the cost function values are calculated for the whole population set and the local solution (*pbest*) is obtained. In the next step, the particles are updated using Eqs. (2)–(7). To achieve quick convergence, elitism is incorporated in this optimization process. After getting the updated solutions from the CEPSO method, fitness values are calculated for the updated solutions and compare with the old solutions. According to the cost function values (fitness), best solution set is obtained keeping the population size constant. If the fitness value of *pbest* is less than that of *gbest*, then the *pbest* will be the *gbest* solution. The iteration stops, if termination condition is fulfilled. Otherwise, the process again starts from calculation of cost function values and continues in the same manner to obtain the desired result. The detail flowchart of the developed technique is given in Fig. 2.

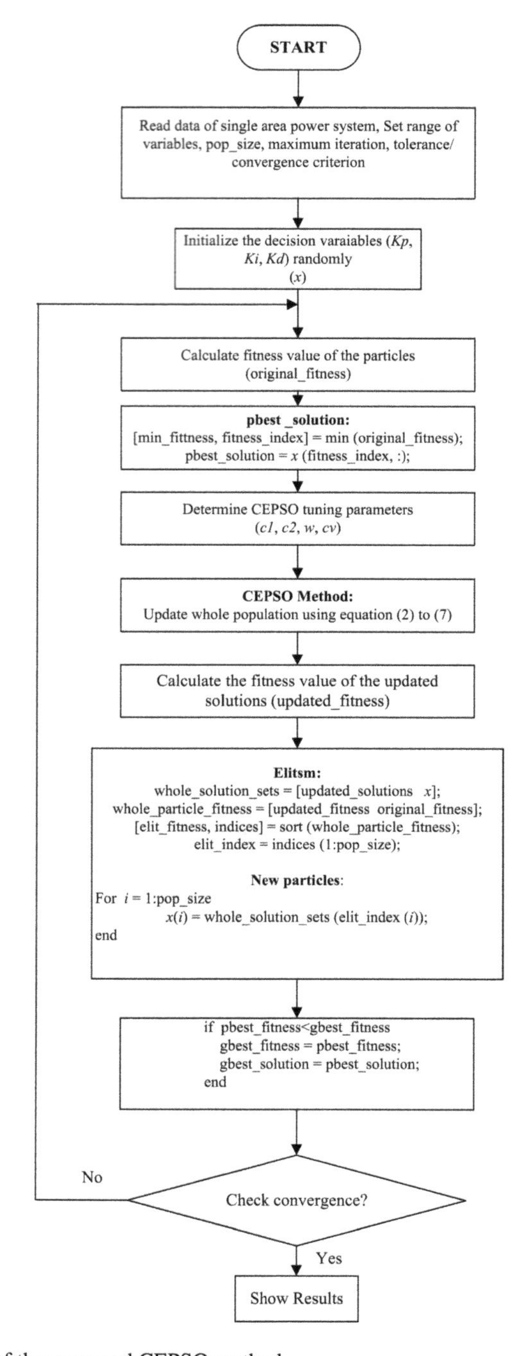

Fig. 2 Flowchart of the proposed CEPSO method

6 Result and Analysis

6.1 System Performances

The control variables of the PID controller are optimized using MATLAB 7.9 with a core i5 PC of 4 GB RAM and 2.5 GHz clock speed. For the global search, 100 particles are taken and 50 iterations are considered for the optimization process. Comparative study has been carried out with different techniques (GA, PSO and EPSO) to show the effectiveness of the proposed CEPSO method. The simulations are realized by step load change of 0.01 ($\Delta P_L = 0.01$). In this problem, the simulation procedure is complex because the power system model is large and highly nonlinear. There are two sections, program file (.m file) and Simulink model (Fig. 1) of the LFC system. At first, PID parameters (K_p, K_i and K_d) are randomly generated through program file. The generated PID parameters are sent to PID controller of the Simulink model. From the Simulink model, E and output response with time are obtained. By programming, frequency deviation (SSE), M_p and t_s are calculated from the output response of Simulink model. The cost function is now determined using E, M_p, t_s and SSE. Using CEPSO method, the PID parameters are updated. The updated parameters are again sent to the PID controller of the Simulink model. This iteration continues until the convergence criterion is reached. Same step-by-step process is applied for other methods also. The optimal control gains of the PID are obtained by running the simulation 50 times (trial run) for each method. Table 1 shows the tuned parameters (K_p, K_i and K_d) of the control system, steady-state performance (SSE) and time response specifications for GA, PSO, EPSO and CEPSO methods. Figure 3 demonstrates the time response behavior of the LFC scheme using the proposed CEPSO method and other (GA, PSO and EPSO) methods.

It is noted from Table 1 and response characteristics (Fig. 3) that the developed control configuration with uniquely designed cost function achieves good dynamic performance for the realistic power system. Comparing GA and PSO methods, it is clear that both are unable to produce the satisfactory control parameters and desired steady-state and transient performances. PSO method gives low t_s and M_p but high

Table 1 Control parameters of different methods

Method	$[K_p\ K_i\ K_d]$	Settling time (t_s) (s)	Max. overshoot (M_p)	Steady-state error (SSE)
GA	[−0.0064 −1.0317 −1.4783]	17.4557	0.0326	1.3026e−005
PSO	[−1.6308 −1.1188 −2.8318]	16.6386	0.0134	6.4335e−004
EPSO	[−1.7467 −0.3258 −1.0732]	15.0243	0.0037	6.5281e−006
CEPSO	[−1.7127 −0.3245 −1.2348]	14.1456	0	4.9867e−08

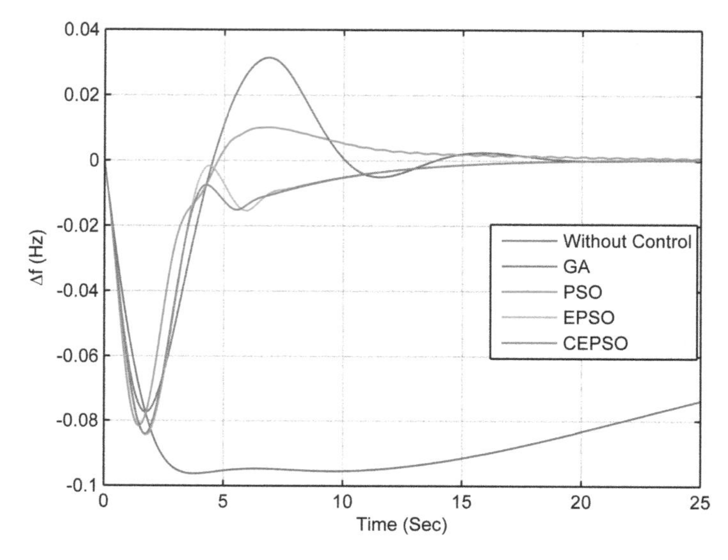

Fig. 3 Response characteristics

SSE compare to GA method. EPSO method produces better solution compare to GA and PSO methods. The newly developed CEPSO method gives best results compare to GA, PSO and EPSO methods for both transient and steady-state performances.

6.2 Comparative Performance Index

How the performance of one algorithm is better than the other algorithm—can be determined by the comparative performance index (CPI). The CPI is expressed in percentage. The calculation of CPI is given in Appendix 2. The CPI in percentage for the algorithms is given in Table 2. From Table 1 and Table 2, it is clear that the CEPSO algorithm yields minimum SSE, zero overshoot and lowest t_s. The

Table 2 CPI in percentage (%) for the algorithms

	Comparative performance index (%)		
	t_s	M_p	SSE
CEPSO versus GA	18.96	100	99.62
CEPSO versus PSO	14.98	100	99.99
CEPSO versus EPSO	5.85	100	99.24
EPSO versus GA	13.93	88.65	49.88
EPSO versus PSO	9.7	72.39	98.99
PSO versus GA	4.68	58.9	−97.97

percentage improvement of these quantities is significantly high in case of CEPSO method compare to others. Negative (−) CPI of the last column indicates percentage improvement of SSE of GA compare to PSO. Hence, compared to other methods, CEPSO gives true optimal solutions, increase the stability of the system and improve the system damping characteristic.

7 Conclusion

In this study, the LFC control problem of an isolated power system with thermal and hydro-generating (renewable) units is solved using newly developed CEPSO method. The authors' contribution in this problem can be considered in three aspects. First, a realistic power system configuration is used considering the practical limitations like GDB, GRC, and TD. Second, an exclusive cost function is introduced. Providing proper selection of weight factors, the cost function is logically designed by taking appropriate time response specifications. Third, a new method, i.e., CEPSO algorithm is proposed to attain the optimum PID control gains for achieving desired control performances. A comparison study between CEPSO-based PID tuning method with other tuning methods is being carried out. The comparison results show that the CEPSO -based controller design methodology provides superior steady-state and transient performances by generating true optimal gains compared to GA, PSO and EPSO methods.

Appendix 1: Parameters of GA, PSO, EPSO and CEPSO

Probability of Crossover = 0.8; Rate of Mutation = 0.03; Constriction Factors ($c1 = c2$) = 1.49455, $w_0 = 0.95$, $m = 1$; $n = 1$.

Appendix 2: CPI Calculation

As an example, the procedure of CPI calculation for t_s is given below. The CPI for M_p and SSE can be calculated in the same way.

Here, CPI (t_s) = comparative performance index for settling time (t_s), GA (t_s) = settling time of GA method, CEPSO (t_s) = settling time of CEPSO method.

Now, $\mathrm{CPI}(t_s) = \frac{\mathrm{GA}(t_s) - \mathrm{CEPSO}(t_s)}{\mathrm{GA}(t_s)} \times 100\% = \frac{17.4557 - 14.1456}{17.4557} \times 100 = 18.96\%$.

References

1. Kundur, P.: Power System Stability and Control. McGraw-Hill, New York (1994)
2. Stanković, A.M., Tadmor, G., Sakharuk, T.A.: On robust control analysis and design for load frequency regulation. IEEE Trans. Power Syst. **13**, 449–455 (1998)
3. Tan, W., Xu, Z.: Robust analysis and design of load frequency controller for power systems. Electr. Power Syst. Res. **79**, 846–853 (2009)
4. Khodabakhshian, A., Pour, M.E., Hooshmand, R.: Design of a robust load frequency control using sequential quadratic programming technique. Int. J. Electr. Power Energy Syst. **40**, 1–8 (2012)
5. Zhang, C.K., Jiang, L., Wu, Q.H., He, Y., Wu, M.: Delay-dependent robust load frequency control for time delay power systems. IEEE Trans. Power Syst. **28**, 2192–2201 (2013)
6. Yu, X., Tomsovic, K.: Application of linear matrix inequalities for load frequency control with communication delays. IEEE Trans. Power Syst. **19**, 1508–1515 (2004)
7. Ho, J.L., Jin, B.P., Young, H.J.: Robust load frequency control for uncertain nonlinear power systems: a fuzzy logic approach. Inf. Sci. **176**, 3520–3537 (2006)
8. Daneshfar, F., Bevrani, H.: Multiobjective design of load frequency control using genetic algorithms. Int. J. Electr. Power Energy Syst. **42**, 257–263 (2012)
9. Golpîra, H., Bevrani, H., Golpîra, H.: Application of GA optimization for automatic generation control design in an interconnected power system. Energy Convers. Manage. **52**, 2247–2255 (2011)
10. Sahu, R.K., Panda, S., Rout, U.K.: DE optimized parallel 2-DOF PID controller for load frequency control of power system with governor dead-band nonlinearity. Int. J. Electr. Power Energy Syst. **49**, 19–33 (2013)
11. Ali, E.S., Abd-Elazim, S.M.: Bacteria foraging optimization algorithm based load frequency controller for interconnected power system. Int. J. Electr. Power Energy Syst. **33**, 633–638 (2011)
12. Ali, S., Yang, G., Huang, C.: Performance optimization of linear active disturbance rejection control approach by modified bat inspired algorithm for single area load frequency control concerning high wind power penetration. ISA Trans. **81**, 163–176 (2018)
13. Abdel-Magid, Y.L., Abido, M.A.: AGC tuning of interconnected reheat thermal systems with particle swarm optimization. In: IEE Proceedings—International Conference on Electronics, Circuits and System, pp. 376–379 (2003)
14. Gozde, H., Taplamacioglu, M.C.: Automatic generation control application with craziness based particle swarm optimization in a thermal power system. Int. J. Electr. Power Energy Syst. **33**, 8–16 (2011)
15. Wu, J.X., Liu, W.Z., Zhao, W.G., Li, Q.: Exponential type adaptive inertia weighted particle swarm optimization algorithm. In: 2nd International Conference on Genetic and Evolutionary Computing, pp. 79–82 (2008)
16. Panda, S., Mohanty, B., Hota, P.K.: Hybrid BFOA–PSO algorithm for automatic generation control of linear and nonlinear interconnected power systems. Appl. Soft Comput. **13**, 4718–4730 (2013)
17. Jing, H., Kwong, C.K., Chen, Z., Yisim, Y.C.: Chaos particle swarm optimization and T-S fuzzy modelling approaches to constrained predictive control. Exp. Syst. Appl. **39**, 194–201 (2012)
18. He, Y., Yang, S., Xu, Q.: Short-term cascade hydroelectric system scheduling based on chaotic particle swarm optimization using improve logistic map. Nonlinear Sci. Numer. Simulat. **18**, 1746–1756 (2013)
19. Pain, S., Acharjee, P.: AGC of practical power system using backtracking search optimization algorithm. In: 2016 International Conference and Exposition on Electrical and Power Engineering (EPE 2016), pp. 687–692 (2016)

Biogas Cook Stove with a Novel Porous Radiant Burner—An Alternate for LPG Cook Stoves in Rural and Semi-urban Indian Households

M. Arun Kumar, Lav K. Kaushik, Sangjukta Devi, and P. Muthukumar

Abstract Biogas is one of the abundantly available renewable resources in hot and tropical regions. Biogas is used as a fuel for cooking and power generation. Biogas is one of the traditional cooking fuels used in India for a long time. In India, several attempts, such as the national biogas and manure management program and off-grid biogas power generation program, have been made since 1970s to promote biogas as an alternate fuel for cooking and lighting. Conventional cook stoves work on free flame combustion and are less efficient and emit high concentrations of pollutants. To minimize environmental pollution and consumption of fossil fuels, there is an urgent need for research to develop energy-efficient and less polluting cook stoves. The present article introduces a novel porous radiant burner (PRB) applicable in domestic biogas cook stove. The novel PRB is designed to work on porous media combustion (PMC) technology. Because of improved combustion, the burner is found to be highly efficient and resulted in almost clean burning. In the operational biogas flow rate range of 360–480 l/h, the thermal efficiency (obtained using water boiling test as per IS: 8749:2002 (Bureau of Indian Standard (BIS) in Biogas Stove-Specification. New Delhi, 2002 [1]) of the PRB varies 47.5–60%, whereas same is only 44.5–48% for burners available in the market. In PRB, the maximum CO and NO_x emissions are found to be 52 ppm and 3 ppm, respectively. Compared to its conventional counterpart, a maximum reduction of ~84 and ~90% in CO and NO_x emission levels is achieved using PRB. The overall performance showed that the novel PRB could be a potential alternative to the market available biogas burner for domestic cooking. Also, it has the potential to empower millions of rural and semi-urban Indian households by giving them access to cleaner cooking.

Keywords Biogas · Conventional burner · Porous radiant burner · Thermal efficiency · Emissions

M. Arun Kumar
Centre for Energy, Indian Institute of Technology Guwahati, Guwahati, India

L. K. Kaushik · S. Devi · P. Muthukumar (✉)
Department of Mechanical Engineering, Indian Institute of Technology Guwahati, Guwahati, India
e-mail: pmkumar@iitg.ac.in

P. Muthukumar et al. (eds.), *Innovations in Sustainable Energy and Technology*,
Advances in Sustainability Science and Technology,
https://doi.org/10.1007/978-981-16-1119-3_11

1 Introduction

Ingress to energy is a primary requisite for the development of society and livelihood. Energy poverty in developing countries negatively affects the well-being of its population as it highlights the low power consumption, the use of highly polluting fuels, and spending of excess time for collecting fuels. Fossil fuel is the main contributor to energy in the present scenario. Every person needs food to sustain their life. The majority of staple foods need to be cooked before eating, and most people cook 2–3 times every day. Thus, the need for clean cooking fuel and an efficient cook stove with less emission is of great concern. The most commonly used clean cooking fuels in India are liquefied petroleum gas (LPG), piped natural gas (PNG), and kerosene. The government of India provides a massive subsidy for LPG and kerosene to reduce the burden on consumers. But this, in turn, affects the economic growth of the country. In 2016, the Pradhan Mantri Ujjwala Yojana (PMUY) scheme was launched in India to promote the use of LPG, which reduced the exposure of women and children to unhealthy smoke levels. However, recent studies show that Indian households, depending on circumstances and needs, use more than one technology or one source of energy for cooking. To decrease the import of LPG and kerosene, usage of locally available alternative cooking fuels like biogas, ethanol, methanol, plant oil, and blends of these fuels with high calorific value fuels has been encouraged. The government of India is promoting the usage of alternative cooking fuels by implementing various schemes like the "Methanol Economy Program," "National Biogas and Manure Management Program (NBMMP)," "Unnat Chulha Abhiyan Program," etc. Biogas is one of the environmentally friendly and clean cooking fuels [2] and its applications are shown in Fig. 1.

Air quality can be improved by replacing the fossil fuel or traditional biomass with biogas as the emission from biogas combustion is comparatively less [3]. The

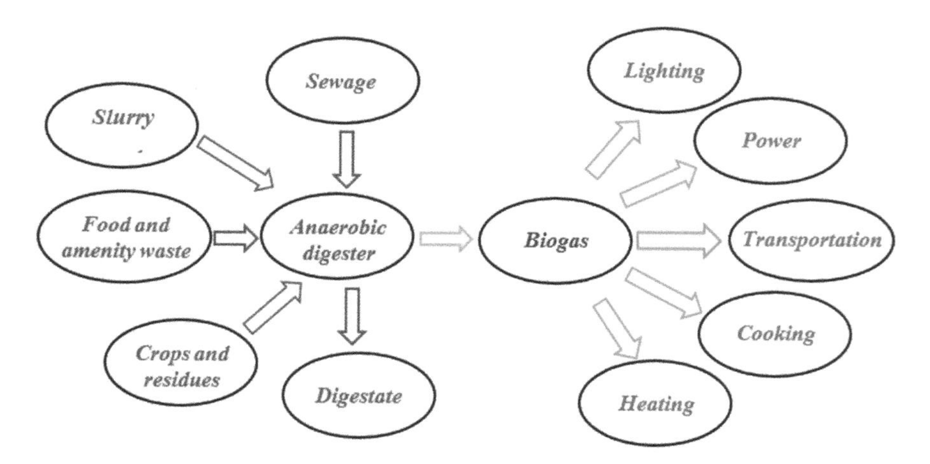

Fig. 1 Biogas sources and applications

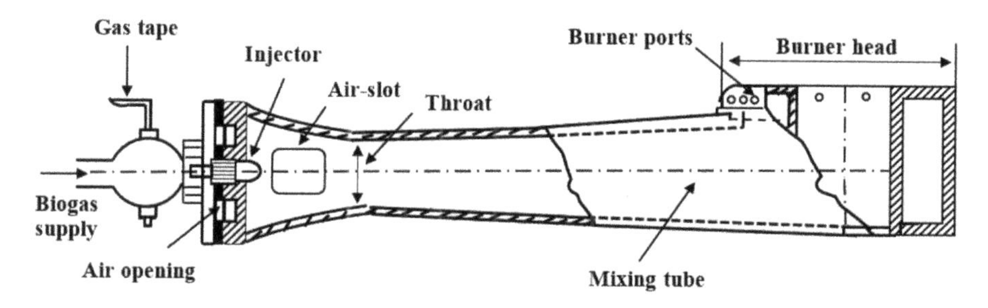

Fig. 2 Assembly of a typical biogas burner working on FFC

utilization of biogas also helps in waste management. It allows communities to be energy independent, as the quantity and quality of the output are entirely dependent on the use and maintenance practices of the plant users. Furthermore, the Ministry of New and Renewable Energy (MNRE) is promoting family biogas digesters through its National Biogas and Manure Management Program (NBMMP). There were about five million individual and community-level biogas plants in operation as of 2016, against an estimated potential of 12.3 million [4]. Many of them were installed as part of NBMMP launched in 2002–03, earlier known as National Project on Biogas Development. In 2018–19, the New National Biogas and Organic Manure Program has succeeded the NBMMP as the primary biogas scheme in the country, albeit with a scaled-down target of installing at least 0.25 million plants by 2019–20 [5]. Despite having government support by way of central government schemes, the entire value chain of biogas faces several critical challenges. One among them is the availability of inefficient burners for biogas applications.

The development of biogas burners begun in the late 1970s. The assembly of the typical conventional biogas burner working on free flame combustion (FFC) is shown in Fig. 2. Chandra et al. [6] tested different biogas cooking stoves and found that there was inconstancy in design and performance (thermal efficiency (η_{th}): 38–54%). Similar η_{th} range was also reported by Mahin [7]. A biogas burner design guide with a 1.5 kW 'DCS' biogas stove with 55% η_{th} was reported by Fulford [8]. Kurchania et al. [9, 10] developed biogas cook stoves for domestic (input power (P_i) = 2.3 kW, η_{th} = 60.1%) and community cooking (P_i = 5.3 kW, η_{th} = 43.96%) applications. Kebede and Kiflu [11] developed a burner for injera baking application with gas consumption rating, P_i and η_{th} as 0.93 m^3/h, 5.7 kW and 25%, respectively. Obada et al. [12] developed a domestic cooking burner with a biogas consumption rate of 0.47 m^3/h and found η_{th} as 21%. Tumwesige et al. [13] studied eight different biogas stoves available in Sub-Saharan Africa and found that stoves were poorly designed and less efficient. Reported η_{th} ranged between 20.2 and 28% in the P_i range of 3.9–6.8 kW.

From the above literature, a significant variation in η_{th} of biogas burner has been found. This is because of the inherent demerits of the FFC. A new technology, popularly known as porous medium combustion (PMC) finds its traction in the development of efficient and eco-friendly burners for cooking. In PMC technology, the combustion of the fuel happens in a highly radiating and conducting porous matrix. Because of the enhanced heat transfer within the porous matrix, η_{th} is high and concentrations of CO and NO_x in the exhaust gas are low. PMC works on the principle of excess enthalpy combustion in which the preheating of the fuel–air mixture enables the PRB to combust even fuel with low energy density and lean mixtures. Also, the reaction zone (RZ) in the PRB is wide and the temperature on the burner surface is almost uniform. Because of these advantages, higher thermal efficiency and reduced emissions can be achieved from PRB.

Mishra and Muthukumar [14] developed LPG cook stove with PRB for domestic cooking (1–3 kW) which yielded a maximum η_{th} of 75.1%, whereas the conventional burner (CB) of the same power range yield 65%. Similarly, Mishra et al. [15] developed a PRB for medium-scale LPG cooking stove. Compared to its conventional counterpart, the developed PRB showed ~22–28% higher η_{th} and reported a maximum reduction of ~84% and 90% in CO and NO_x emissions, respectively. Mujeebu et al. [16] developed LPG operated PRB for domestic cooking and reported a maximum η_{th} of 71% at burner P_i of 0.62 kW. Gao et al. [17] studied the combustion of biogas in PRB with a two-layer spherical alumina bed and found that increment in CO_2 content, moved upper and lower stability limit to higher values. Keramiotis and Founti [18] reported the variation of radiation efficiency for firing rate range of 200–800 kW/m^2 in biogas operated PRB and also found that for equivalence ratio of 0.8, efficiency varied between 18.5 and 26%. It is concluded from the above reported works that no researchers explored the development of self-aspirated PRB for biogas cook stoves.

In an experimental investigation by Kaushik et al. [19], a biogas operated PRB has been explored for its feasibility in domestic cooking. The work primarily highlighted the PRB-based cook stove as a potential alternative to CB. However, its application in self-aspirated mode was not examined. Intrigued by improvement in η_{th} and low emission level of pollutants from PRB, present study aims to extend the research work and present a detailed performance assessment of self-aspirated PRB-based domestic biogas cook stove. Also, based on market potential in India, economic and societal benefits have been projected.

2 Description of PRB and CB

The schematic of the developed PRB is shown in Fig. 3. The PRB consists of a disk of silicon carbide (SiC) foam and alumina (Al_2O_3) ceramic matrix. The SiC foam and Al_2O_3 ceramic act as combustion zone (CZ) and preheat zone (PZ), respectively. The high-temperature resistant castable cement is used for the fabrication of burner casing. Figure 4 illustrates the pictorial view of the CB applicable in domestic-scale biogas cook stoves. Here, combustion takes place over the surface of the burner, which is precisely above the burner head.

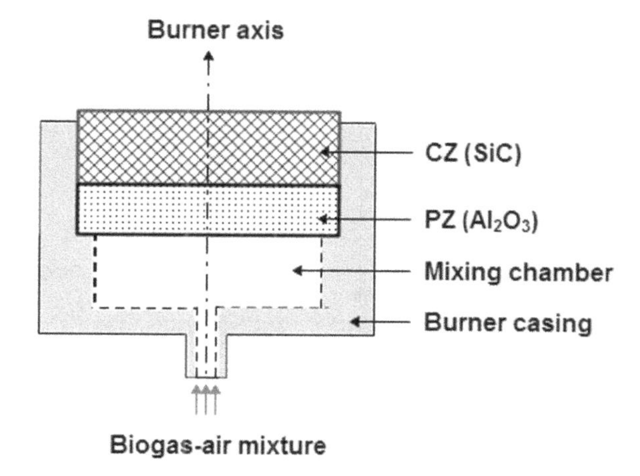

Fig. 3 Schematic diagram of PRB

Fig. 4 Different types of domestic-scale biogas conventional cooking stove's burner head

3 Experimental Setup

The schematic of the experimental setup used for performance assessment is shown in Fig. 5. The biogas is supplied from a balloon through a pressure regulator and a control valve. The flow rate of biogas is monitored through a mass flow meter (MFM). At a pressure of 1.2 bar, biogas reaches the burner through an orifice in the mixing tube. Due to the venturi effect, the high-velocity biogas jet creates a low static pressure near the mixing tube, which causes primary air to entrain through the two air slots. The air and biogas move via a mixing tube and reaches the burner casing. Mixing of biogas and air takes place in the mixing tube and mixing chamber provided at the bottom of the PRB. The combustion is initiated by using an igniter. After some time, the PRB becomes entirely red hot, indicating the stable operation of the burner. The measured composition and thermophysical properties of the biogas used in the present experimental study are presented in Table 1.

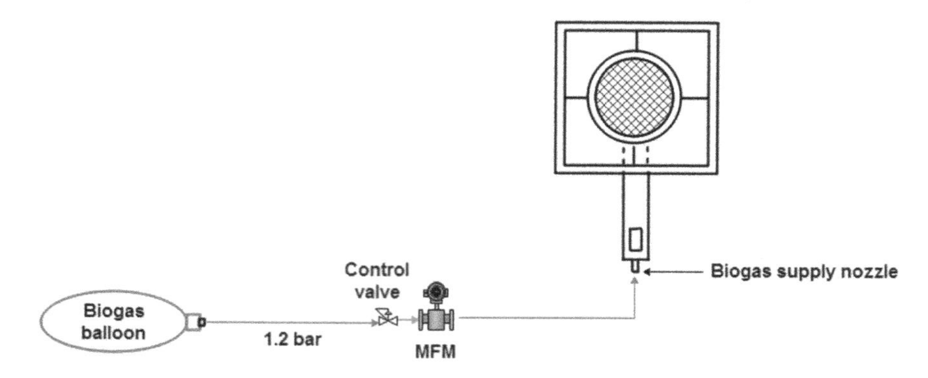

Fig. 5 Schematic diagram of the experimental setup

Table 1 Composition and thermophysical properties of biogas

Parameters	Text
CH_4	43–56%
CO_2	34–38%
Lower calorific value	17 MJ/kg
Density at 1 atm and 15 °C	1.2 kg/m^3
Autoignition temperature	650 °C

4 Experimental Procedure

In this section, the procedures adopted for thermal efficiency (η_{th}) estimation and emission measurement are presented. Schematic of the experimental setup used for η_{th} and emission measurement is shown in Fig. 6.

4.1 Thermal Efficiency

Thermal efficiency (η_{th}) of the burners is measured by conducting the standard water boiling test (WBT) as per IS: 8749:2002 [1]. Based on biogas flow rate, aluminum pan and mass of water used during WBT were selected. The aluminum pan was filled with a known amount of water and the temperature (T_1) was measured using a mercury-in-glass thermometer. Once the burner reached the stable condition, the pan with a stirrer was kept above pan stand. Stirring was started once the temperature of water in pan reached 80 °C and it was continued until the temperature (T_2) of water reached 90 ± 0.5 °C. At this stage, the burner was turned off. Throughout the measurement, the flow rate of biogas was kept constant, and the amount of biogas consumed during the test was noted. The test was repeated three times, and the average of these three results was considered for estimating the η_{th} of the PRB. The

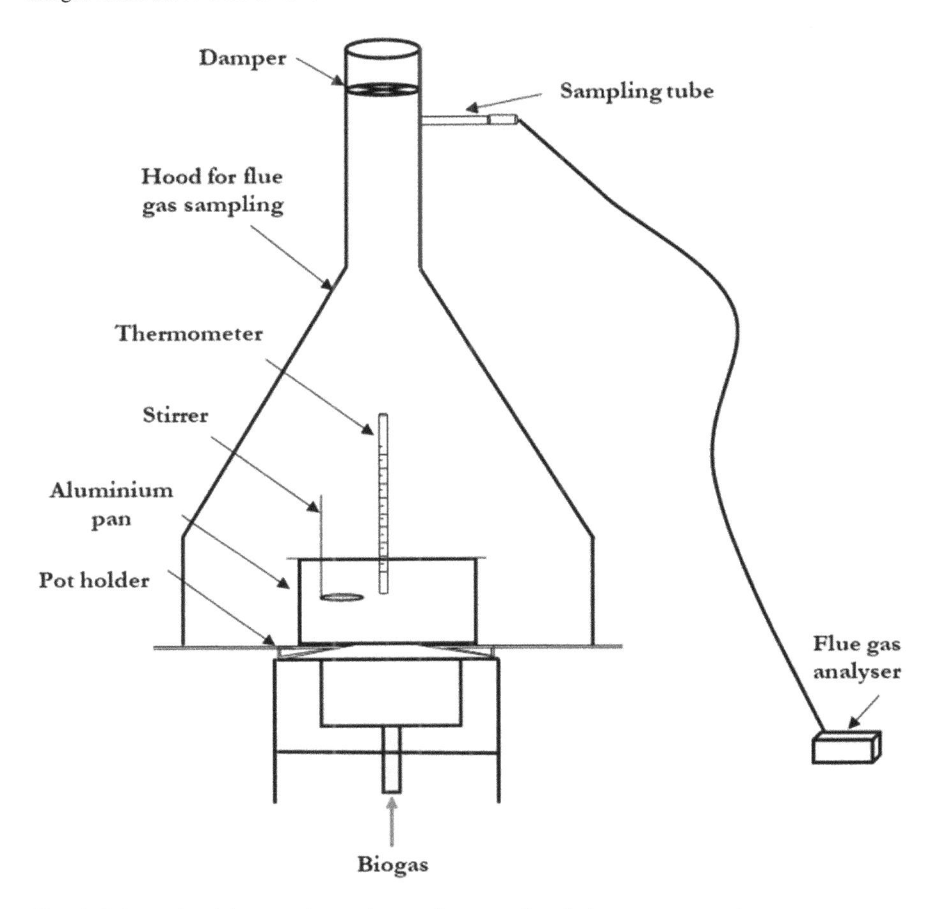

Fig. 6 Schematic of the experimental setup for η_{th} and emission measurement

equation given below is used for the calculation of η_{th} .

$$\text{Thermal efficiecncy, } (\eta_{th}) = \frac{\text{Heat utilized}}{\text{Heat produced}}$$

$$\eta_{th} = \frac{\left(M \times C_p + m \times C_w\right) \times (T_2 - T_1)}{\dot{m}_{\text{Biogas}} \times t \times \text{LCV}_{\text{Biogas}}}$$

where M is mass of pan along with stirrer (kg), m is mass of water in the pan (kg), C is specific heat (kJ/kg-k, p: pan and w: water), T_2 and T_1 are final and initial temperatures of water (°C), $\dot{m}_{\text{Biogas}}$ is flow rate of biogas (kg/s), t is time of experiment (s) and $\text{LCV}_{\text{Biogas}}$ is lower calorific value of biogas (kJ/kg).

Table 2 Accuracy of the measuring instruments

Instruments	Uncertainty
Weighing balance (SES15TH)	±0.1 g
Portable Flue gas analyser (Testo 340)	±2 ppm
Mass flow meter	0.001 g
Thermometer	±0.5 °C

4.2 CO and NO_x Emissions

For the emission measurements, the flue gas sampling is done as per IS: 8749:2002 [1]. Hood showed in Fig. 6 was used to isolate the flue gases from the atmosphere, and then CO and NO_x emission concentrations were recorded in the portable flue gas analyser (Testo 340). The hood was placed above the burner along with the vessel, and the portable flue gas analyzer probe was placed in the sampling hole. The reported emission values are taken on dry basis, with correction to a 3% oxygen level.

4.3 Uncertainty Calculation

The accuracies of the instruments used during the experiments are given in Table 2. The error associated with the estimation of η_{th} is calculated using the expression given below.

$$\delta\eta_{th} = \sqrt{\left(\frac{\delta\eta_{th}}{\delta M}\Delta M\right)^2 + \left(\frac{\delta\eta_{th}}{\delta m}\Delta m\right)^2 + \left(\frac{\delta\eta_{th}}{\delta(\Delta T)}\Delta(\Delta T)\right)^2 + \left(\frac{\delta\eta_{th}}{\delta\dot{m}_{fuel}}\Delta\dot{m}_{fuel}\right)^2}$$

The maximum relative uncertainty $\left(\frac{\delta\eta_{th}}{\eta_{th}}\right)$ of η_{th} during the experiment was found to be around ±1%. To check the repeatability, the experiments were performed three times for each performance parameter and the average values are presented.

5 Results and Discussion

Thermal efficiency (η_{th}) variations of PRB and CB with biogas flow rate (360–480 l/h) are shown in Fig. 7. The η_{th} shows a decreasing trend with increase in fuel flow rate. Within the biogas flow rate range, the lowest flow rate yields the maximum efficiency. The maximum efficiency of 60% was obtained for the biogas flow rate of 360 l/h. Similarly, the minimum value of 47.5% was achieved for 480 l/h. Previous works by Mishra et al. [15] also reported similar trends in η_{th}. Within the biogas flow

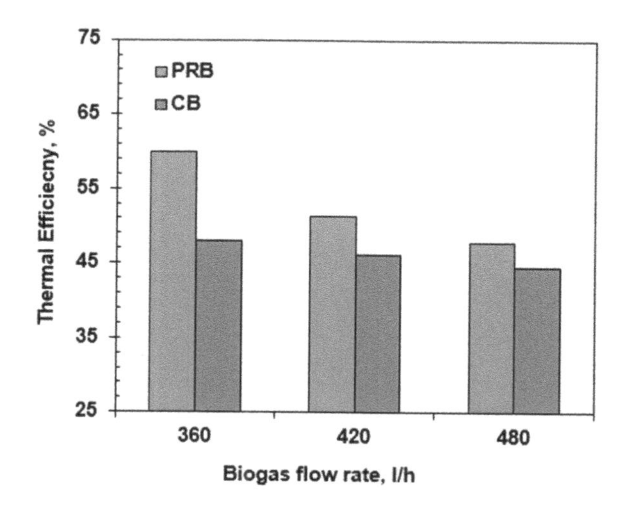

Fig. 7 Effect of biogas flow rate on thermal efficiency of PRB and CB

rate, η_{th} for CB ranged between 48 and 44.5%. The reason behind decreasing η_{th} of CB with biogas flow rate is due to increasing flame height with increase in flow rate, which results in more convective heat loss. From efficiency data, it is clear that the percentage improvement in η_{th} of PRB is in the range of ~7–24%. For all the cases, PRB shows higher η_{th} than CB because of the combined effect of radiative and convective heat transfer within the porous matrix and also better combustion.

Since the developed PRB is projected for domestic cooking, the measurement of CO and NO_x emissions is critical due to direct contact of the burner flue gases with the user. A comparison of CO and NO_x values between PRB and CB is shown in Fig. 8. In PRB, measured CO and NO_x were found in the range of 38–52 ppm and 1–3 ppm, respectively, in the whole range of biogas flow rate. Whereas, in the case of CB, the same were 235–263 ppm and 10.7–13.4 ppm, respectively. Compared to its conventional counterpart, a maximum of ~84% and ~91% reduction in CO and NO_x, respectively was achieved in the novel PRB. Measured CO emissions of PRB are lower than that of CB, because of better combustion. In the PRB, the lower surface temperature of the burner is the reason behind less NO_x emission. Similar emission patterns from biogas combustion in PRB were also presented in previous work by Devi et al. [20, 21].

6 Economic and Societal Benefits of the PRB

Over the past five years, India has seen huge progress in scaling up of clean cooking reach to its diverse population. Following the implementation of the Pradhan Mantri Ujjwala Yojana, the number of LPG connections has been increased considerably. But, concerns about the cost, as well as the adequacy of LPG supply, are still a challenge. The government has earmarked Rs. 34,085 crores as subsidy for LPG

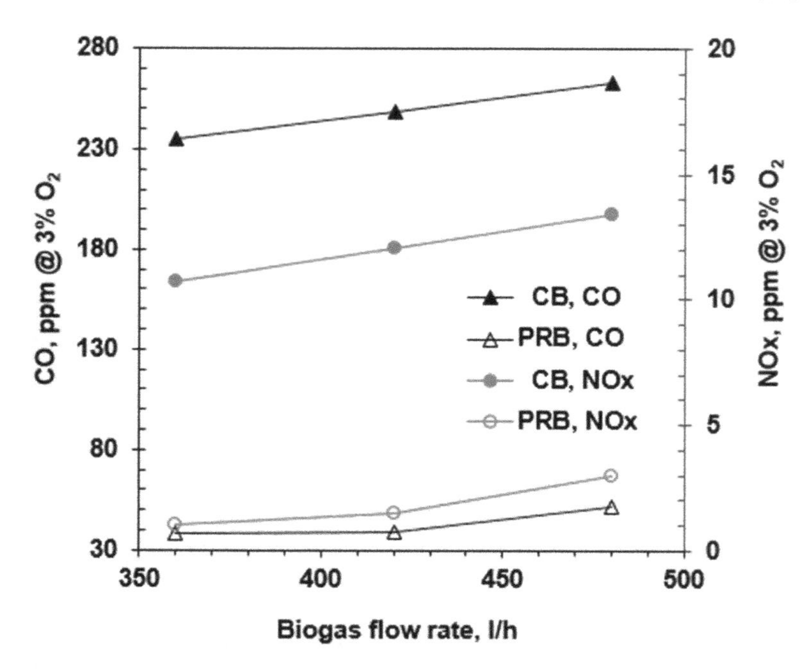

Fig. 8 CO and NO_x emissions reported in PRB and CB

in the financial year 2019–2020 (India budget, 2019). Such a substantial economic burden affects the financial health of the country, and this emphasizes the need for exploring locally available clean cooking fuel alternatives. During 2018–19, LPG consumption was 24.9 million metric tons (MMT) out of which approximately 12.1 MMT was imported [22]. Massive LPG import highlights that domestic production is incapable of fulfilling the demand. Biogas, as a local alternative, can serve to offset a significant portion of the LPG import requirement. Mittal et al. [23] estimated the yearly biogas production potential in India as 239 billion m^3 (raw biogas), which is equivalent to 74.09 MMT of LPG per year (1 m^3 of biogas is equivalent to 0.31 kg of LPG; present case *LCV* of biogas: 17 MJ/kg; LPG: 45.6056 MJ/kg). In 2016, the annual production of biogas in India was only ~2 million m^3 [24] for cooking. If India's huge biogas potential is taped, this can considerably reduce India's LPG import and generate huge revenue for the country. Earlier, because of less efficient conventional biogas burners, biogas was not considered as a choice, as the LPG deal was attractive. With newly developed PRB, which results in maximum efficiency of ~60%, there is a probability that it can compete with the conventional LPG burners whose average thermal efficiencies are in the range of 60–65% [14]. Biogas use can lead to the empowerment of women by giving them access to clean fuel similar to LPG and enabling them to have time-saving for other livelihood activities. Another benefit of biogas application is nutrient-rich organic manure as a by-product, which enhances crop yield and maintains soil health and fertility by reducing the use of chemical fertilizers.

7 Conclusion

The impact of traditional cooking practices on the environment, climate and human health has been in the limelight of national and international development organizations for a long time. Access to clean cooking fuels with efficient burner has the transformative potential to curb the health risks posed by traditional cook stoves. In India, the government has planned to install a total of ~5.255 million biogas plants in rural and semi-urban households, which can be a local alternative to LPG as clean cooking fuel. In this article, a feasibility study on a newly developed commercially viable self-aspirated PRB, applicable for biogas cooking stove has been presented. For PRB, the thermal efficiency in the biogas flow rate range of 360–480 l/h was found as 60–47.5%, whereas the same was only 48–44.5% in the case of conventional burner available in the market. Employing the newly developed novel PRB, the highest values of CO and NO_x emissions were limited to 52 ppm and 3 ppm, respectively. These values signify that almost clean combustion can be achieved with the newly developed novel PRB as its counterpart delivered much higher amounts of CO (263 ppm) and NO_x (13.4 ppm) emissions. The overall performance of PRB shows that it is suitable for domestic cooking applications. The new PRB has the potential to replace its conventional counterpart, which can lead to the empowerment of rural and semi-urban households by giving them access to clean fuel like LPG and enabling them to save time for other livelihood activities.

Acknowledgements The authors are thankful to the MHRD for their financial help vide IMPRINT project No. 6727.

References

1. Bureau of Indian Standard (BIS): Biogas Stove-Specification, 2nd rev. IS 8749: 2002, New Delhi
2. Putti, V.R., Tsan, M., Mehta, S., Kammila, S.: The State of the Global Clean and Improved Cooking Sector. ESMAP (Energy Sector Management Assistance Program), Technical Paper No. 007/15. World Bank, Washington, DC (2015)
3. Lukehurst, C., Bywater, A.: Exploring the viability of small scale anaerobic digesters in livestock farming (2015). www.iea-biogas.net/files/daten-redaktion/download/Technical%20Brochures/Small_Scale_RZ_web1.pdf. Last accessed on 14 Mar 2020
4. Ministry of New and Renewable Energy (MNRE): Annual Report 2016–2017 (2017)
5. Ministry of New and Renewable Energy (MNRE): Guidelines for implementation of the Central Sector Scheme, New National Biogas and Organic Manure Programme during the Period 2017–18 to 2019–20 (2018)
6. Chandra, A., Tiwari, G.N., Yadav, Y.P.: Hydrodynamical modelling of a biogas burner. Energy Convers. Manage. **32**(4), 395–401 (1991)
7. Mahin, B.: Biogas in developing countries. In: Bioenergy System Report to USAID, Washington, (1992)
8. Fulford, D.: Biogas stove design—a short course. University of Reading, United Kingdom (1996)

9. Kurchania, A.K., Panwar, N.L., Pagar, S.D.: Design and performance evaluation of biogas stove for community cooking application. Int. J. Sustain. Energ. **29**(2), 116–123 (2010)
10. Kurchania, A.K., Panwar, N.L., Pagar, S.D.: Development of domestic biogas stove. Biomass Convers. Biorefin. **1**, 99–103 (2011)
11. Kebede, A., Kiflu, A.: Design of Biogas stove for injera baking application. Int. J. Novel Res. Eng. Sci. **1**(1), 6–21 (2014)
12. Obada, D.O., Obi, A.I., Dauda, M., Anafi, F.O.: Design and construction of a biogas burner. Palest. Techn. Univ. Res. J. **2**(2), 35–42 (2014)
13. Tumwesige, V., Fulford, D., Davidson, G.C.: Biogas appliances in Sub-Sahara Africa. Biomass Bioenerg. **70**, 40–50 (2014)
14. Mishra, N.K., Muthukumar, P.: Development and testing of energy efficient and environment friendly porous radiant burner operating on liquefied petroleum gas. Appl. Therm. Eng. **129**, 482–489 (2018)
15. Mishra, N.K., Mishra, S.C., Muthukumar, P.: Performance characterization of a medium-scale liquefied petroleum gas cooking stove with a two-layer porous radiant burner. Appl. Therm. Eng. **89**, 44–50 (2015)
16. Mujeebu, M.A., Abdullah, M.Z., Mohamad, A.A.: Development of energy efficient porous medium burners on surface and submerged combustion modes. Energy **36**, 5132–5139 (2011)
17. Gao, H.B., Qu, Z.G., Tao, W.Q., He, Y.L.: Experimental investigation of methane/(Ar, N_2, CO_2)–air mixture combustion in a two-layer packed bed burner. Exp. Therm. Fluid Sci. **44**, 599–606 (2013)
18. Keramiotis, C., Founti, M.A.: An experimental investigation of stability and operation of a biogas fueled porous burner. Fuel **103**, 278–284 (2013)
19. Kaushik, L.K., Mahalingam, A.K. Palanisamy, M.: Performance analysis of a biogas operated porous radiant burner for domestic cooking application. Environmental Science and Pollution Research (2020)
20. Devi, S., Sahoo, N., Muthukmar, P.: Combustion of biogas in Porous Radiant Burner: low emission combustion. Energy Proc. **158**, 1116–1121 (2019)
21. Devi, S., Sahoo, N., Muthukumar, P.: Experimental studies on biogas combustion in a novel double layer inert Porous Radiant Burner. Renew. Energy **149**, 1040–1052 (2020)
22. Petroleum Planning and Analysis Cell (PPAC): Ministry of Petroleum & Natural Gas, Snapshot of India's Oil & Gas data, January, 2020
23. Mittal, S., Ahlgren, E.O., Shukla, P.R.: Future biogas resource potential in India: a bottom-up analysis. Renew. Energy **141**, 379–389 (2019)
24. Jain, S., Newman, D., Nizhou, A., Dekker, H., Le Feuvre, P., Richter, H., Gobe, F., Morton, C., Thompson, R.: A report on global potential of biogas. World Biogas Association, June 2019

Numerical Investigation of a Multi-tube Conical Shell and Tube-Based Latent Heat Energy Storage System

Vishnu Kumar, Gurpreet Singh Sodhi, Suraj Arun Tat, and P. Muthukumar

Abstract In the present study, the performance of the high temperature tube-in-shell latent heat thermal energy storage system (LHTESS) is evaluated using a three-dimensional numerical model. The cylindrical shell model has been optimized to conical shell model based on the solidification process. Single and multi-tube conical shell models are developed where sodium nitrate is used as the phase change material filled in the shell side. Air used as the heat transfer fluid flows in the tube side. Solidification contours have been plotted to evaluate the behavior of heat transfer along the length of the system to analyze the effect of radial distance between tubes. From the simulation study, the optimized cone angles obtained for single and multi-tube conical shell model are 4.5° and 2.3°, respectively. The rate of heat transfer for both the conical single and multi-tube system is better than the corresponding system with cylindrical shell due to better PCM distribution. However, for a fixed ratio of mass of the PCM to the surface area of the tube, the multi-tube conical shell model leads to ~22% reduction in discharging time than single tube conical shell model.

Keywords Phase change material · Latent heat storage · Solidification

1 Introduction

In the present scenario, energy is the most important prerequisite for the sustainable human development. Solar energy-based renewable technologies are a futuristic approach to overcome the disparity between the supply and demand. However, due to the intermittent nature of solar energy, thermal energy storage is a viable option to match the required demand [1]. Latent heat thermal energy storage systems (LHTESS) are more prominent due to advantages such as high energy storage density and near isothermal nature [2]. However, the low thermal conductivity of phase change materials (PCMs) (~0.1–1 W/m/K) is the major challenge to be addressed while developing LHTESS [3]. Therefore, various techniques to enhance the rate of

V. Kumar · G. S. Sodhi · S. A. Tat · P. Muthukumar (✉)
Department of Mechanical Engineering, Indian Institute of Technology Guwahati, Guwahati, India
e-mail: pmkumar@iitg.ac.in

P. Muthukumar et al. (eds.), *Innovations in Sustainable Energy and Technology*,
Advances in Sustainability Science and Technology,
https://doi.org/10.1007/978-981-16-1119-3_12

heat transfer such as multi-PCM, fins, varying the LHS geometry, and using nano composite PCMs have been explored in the literature [4, 5].

Shell and tube LHTESS geometry has been of key interest to many researchers. Mahdi et al. [6] designed a vertical shell and tube-based LHS system for discharging at medium temperature (~200 °C) and found that the discharging efficiency improves by increasing the HTF flow rate and decreasing inlet HTF temperature. Khalifa et al. [7] reported an 84% increase in energy extraction rate and 24% increase in the effectiveness of heat pipes by employing axial fins. Ma et al. [8] found uniform distribution of temperature within the PCM by addition of annular fins, which increases the charging and discharging power by 6.8 and 9.1%, respectively. Esapour et al. [9] studied the melting behavior of shell and tube-based LHTESS and observed that the melting time reduced by 29% when the number of tubes were increased from one to four due to increased natural convection rate. Joybari et al. [1] conducted an experimental study to compare single and multi-tube shell and tube-based LHTESS unit at low temperature range below 100 °C. In a complete charging and discharging cycle, multi-tube heat exchanger outperformed the single tube heat exchanger by a reduction of 73% in cycle time and also found that maximum average temperature in multi-tube was higher than the single tube LHS unit. Kousha et al. [10] studied the behavior of shell and tube type LHS models and found that increasing the number of tubes from one to four showed a reduction in melting and solidification time by 43% and 50%, respectively. Seddegh et al. [11] compared cylindrical and conical shell based vertical LHTESSs and found that the energy storage rate enhanced by 12% using the conical shell design. Sodhi et al. [12] numerically investigated the performance of a horizontal conical shell and tube LHTESS with optimized shell cone angle and observed a reduction in charging and discharging time by 17% and 28% than cylindrical shell system.

The literature study reveals that most of the reported multi-tube LHTESSs have cylindrical shell-based heat exchanger designs. Unlike previous studies, the present work involves the development of an optimized conical shell multi-tube system and its comparison with single tube conical shell system. A 3D numerical model is developed to find the optimized cone angle (α) for single and multi-tube models having energy storage capacities 1 MJ and 4 MJ, respectively. The two models are compared for a fixed PCM volume to heat transfer surface area ratio. Further, the analysis is extended to study the effect of radial distance between tubes on the discharging performance.

2 Thermal Modeling

2.1 Model Design

Two different LHS modules, single tube and multi-tube models of storage capacity 1MJ and 4 MJ are designed where the PCM shell is changed from cylindrical (0°) to

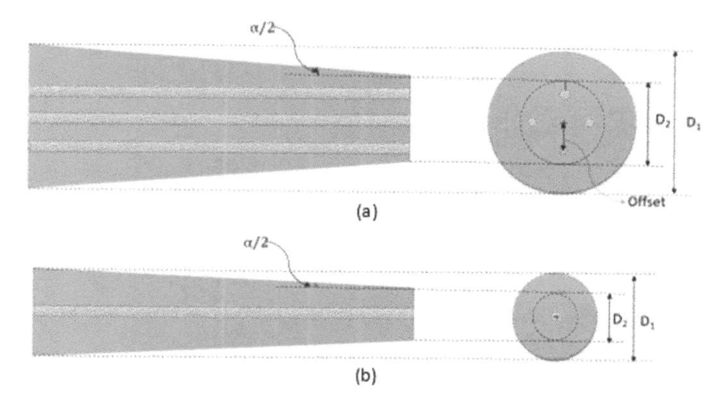

Fig. 1 Physical model with dimensions. **a** Multi-tube module. **b** single tube module

Table 1 Dimensions and discharging characteristics of the single and multi-tube models

Cone angle	Single tube module			Multi-tube module		
	D_1 (mm)	D_2 (mm)	Discharging time (min)	D_1 (mm)	D_2 (mm)	Discharging time (min)
0°	92.52	92.5	1353	154.7	154.7	977
1.1°	97.19	87.6	1167	161.7	147.4	970
2.3°	102.3	82.3	1147.5	169.4	139.3	862.5
3.4°	106.9	77.3	1141.5	176.3	131.8	988.5
4.5°	111.5	72.2	1102.5	183.1	124.2	1098
5.6°	115.8	66.9	1123.5	189.8	116.5	1200

conical shape (5.6°). The HTF tubes have a thickness of 2 mm and inner diameter 24.6 mm. The developed models are shown in Fig. 1, and their dimensions are described in Table 1. The PCM material selected is sodium nitrate ($NaNO_3$), and the thermo-physical properties of which are adopted from Sodhi et al. [12].

2.2 Governing Equations

The governing equations (Eqs. 1–7) described below include mass, momentum and energy balance equation along with Boussinesq approximation to consider the density-driven buoyancy effects. Effective heat capacity (EHC) method is used to model the phase change problem, where effective specific heat is given by Eq. (7). Equation (5) is a source term which is used to characterize the mushy zone.

$$\text{Continuity Equation: } \nabla \vec{v} = 0 \tag{1}$$

$$\text{Momentum Equation: } \frac{\partial \vec{v}}{\partial t} + (\vec{v} \cdot \nabla)\vec{v} = \frac{1}{\rho}\left(-\nabla P + \mu \nabla^2 \vec{v} + \vec{F} + \vec{S}\right) \quad (2)$$

$$\text{Energy Equation: } \rho C_P \frac{DT}{Dt} = k \nabla^2 T \quad (3)$$

$$\text{Boussinesq Approximation: } \vec{F} = \rho \vec{g} \beta (T - T_M) \quad (4)$$

$$\text{Source term: } \vec{S} = \frac{(1-\theta)^2}{(\theta^3 + \varepsilon)} A_{\text{MUSH}} \vec{v} \quad (5)$$

$$\text{Specific heat of PCM: } C_{\text{P}} = \begin{cases} C_{\text{P,S}} & \text{for } T < T_{\text{S}} \\ C_{\text{P,EFF}} & \text{for } T_{\text{S}} \le T \le T_{\text{L}} \\ C_{\text{P,L}} & \text{for } T > T_{\text{L}} \end{cases} \quad (6)$$

$$\text{Effective specific heat: } C_{\text{P,EFF}} = \frac{C_{\text{P,S}} + C_{\text{P,L}}}{2} + \frac{L_{\text{F}}}{T_{\text{L}} - T_{\text{S}}} \quad (7)$$

$$\text{Melt fraction: } \theta = \frac{T - T_{\text{S}}}{T_{\text{L}} - T_{\text{S}}} = \begin{cases} 0 & \text{for } T < T_{\text{S}} \\ 0-1 & \text{for } T_{\text{S}} \le T \le T_{\text{L}} \\ 1 & \text{for } T > T_{\text{L}} \end{cases} \quad (8)$$

2.3 Assumption and Boundary Condition

To formulate the flow and heat transfer problem, some assumptions are made; (i) phase change happens over a small temperature interval, (ii) the effect of radiation heat transfer is neglected, (iii) heat loss to the ambient is neglected, and (iv) HTF flow is assumed to Newtonian.

Boundary conditions: adiabatic outer walls and no slip at the HTF tube walls. For the discharging model, the initial PCM temperature is assumed to be 599 K, and the HTF inlet temperature and velocity is given a value of 559 K and 6 m/s.

2.4 Numerical Procedure and Validation

COMSOL 4.3a Multiphysics is used to perform the numerical simulation with finite element approach. For meshing, all the boundaries and domains, free tetrahedral and triangular elements are used. 3D simulations for the discharging process were performed using conjugate heat transfer model, and PARDISO solver is used to solve the system of nonlinear equations. Convergence criterion value is set as 10^{-3} for temperature and velocities. Grid independency test is performed to ensure the

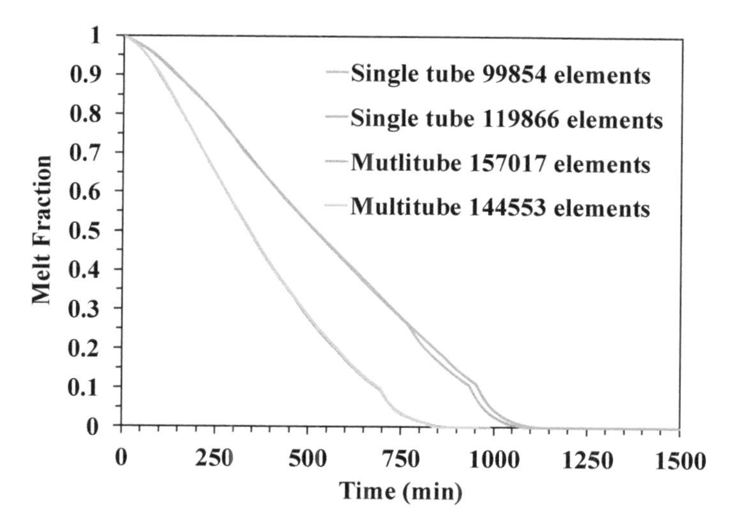

Fig. 2 Grid independency test results for single tube and multi-tube models

quality of the solution at an optimum grid size of the developed numerical model. Figure 2 shows the result of grid independency test performed for both single and multi-tube models. It can be seen that the models are grid independent over a wide element range, and hence, single tube and multi-tube model with 99,854 and 144,553 elements are selected. The model has been validated with the experimental data in our previous study [12]. Further, the model is also verified with the CFD model developed (see Fig. 3) based on enthalpy-porosity technique used by Kumar and Krishna [13].

3 Results and Discussion

3.1 Cone Angle Optimization

The heat transfer characteristics across the PCM are governed mainly by distribution of PCM across the HTF. The HTF temperature varies along the length of HTF tube from inlet to outlet as it exchanges heat with the PCM. The temperature gradient between the PCM and HTF is high at inlet side, but toward the outlet, the temperature gradient is minimal which results in slow heat transfer rate. With the variation of the cone angle, the distribution of the PCM varies across the length. For the optimized cone angle, there is enhanced heat transfer near the inlet section caused by a high temperature difference between the HTF and PCM. However, the heat transfer is enhanced at the outlet section due to reduced PCM mass in this section. Table 1

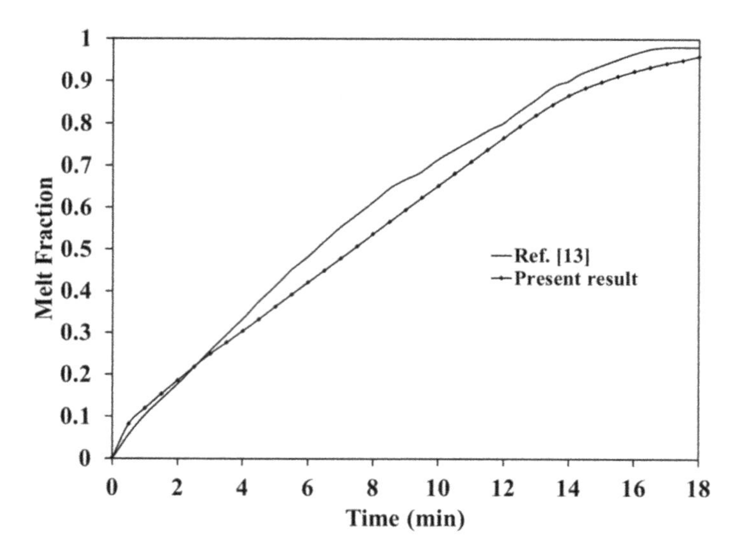

Fig. 3 Validation of the model with the CFD model reported by Kumar et al. [13]

enlists the discharging times for both single and multi-tube models at various cone angles.

Figure 4 shows the variation of melt fraction with time for models with different cone angles. From this result, it can be concluded that the shell with cone angle 4.5° gives the best performance for single tube model, whereas cone angle of 2.3° is optimized value for multi-tube. By converting the cylindrical shell into the conical and keeping the PCM mass to heat transfer surface area ratio the same, the shell diameter at the HTF inlet and outlet starts increasing and decreasing, respectively, with the increase of cone angle. Thus, more mass of PCM gets accumulated near the inlet section where we have a high temperature gradient which results in better heat transfer. It is also evident from melt fraction contours (see Fig. 5) that due to intensification of heat at outlet side, a new solidification front is developed specially in case of single tube model which travels axially in reverse direction and meets the solidification front started from inlet at almost the middle section leading to faster and uniform melting.

3.2 *Performance Comparison Between Optimized Single and Multi-tube Models*

Figure 6 shows the variation of average PCM melt fraction with time during discharging process for the single and multi-tube design for both cylindrical and optimized conical models. It shows that the discharging time in optimized multi-tube model is reduced by 21.76% than single tube model. In case of single tube

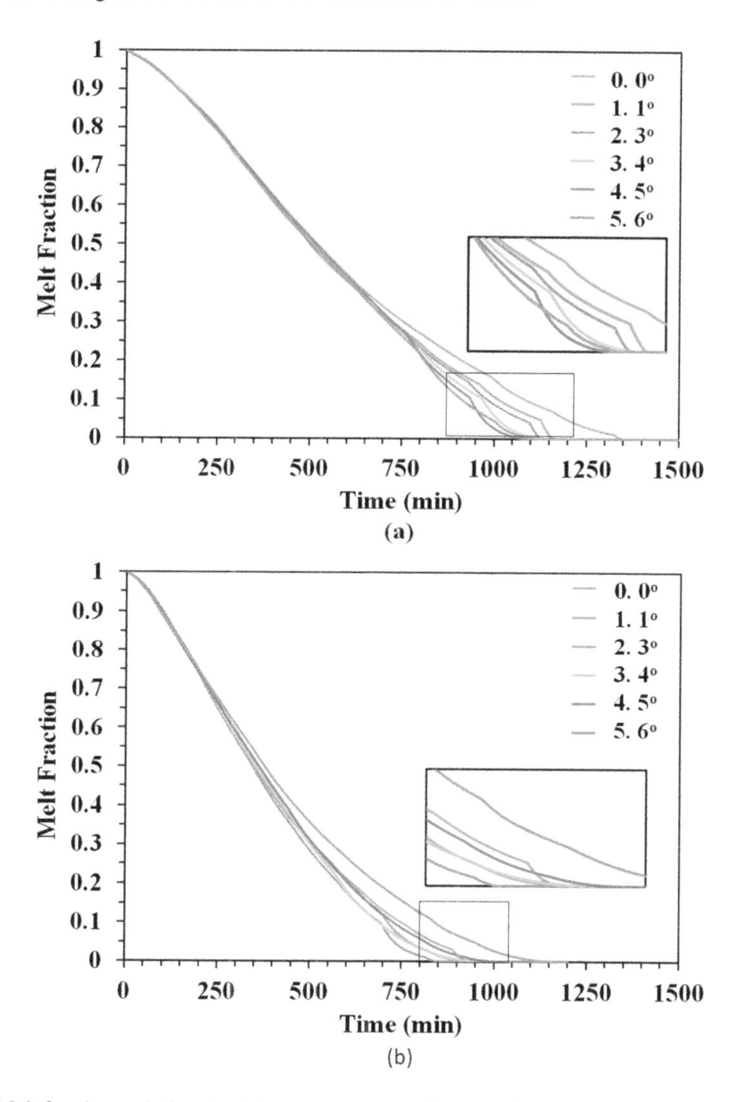

Fig. 4 Melt fraction variation for different cone angles for **a** single tube model **b** multi-tube model

model, only one solidification front progresses radially outward and along the system length; however, in case of multi-tube model, solidification fronts originate from each of the tubes as shown in Fig. 5. Thus, the thermal resistance for the movement of solidification fronts decreases, and the heat transfer rate improves. This leads to higher discharging rate. Comparing the performance of cylindrical and conical shell designs, it was found that there was a reduction of 18.51 and 11.7% in discharging time in single and multi-tube conical shell systems than corresponding cylindrical shell systems.

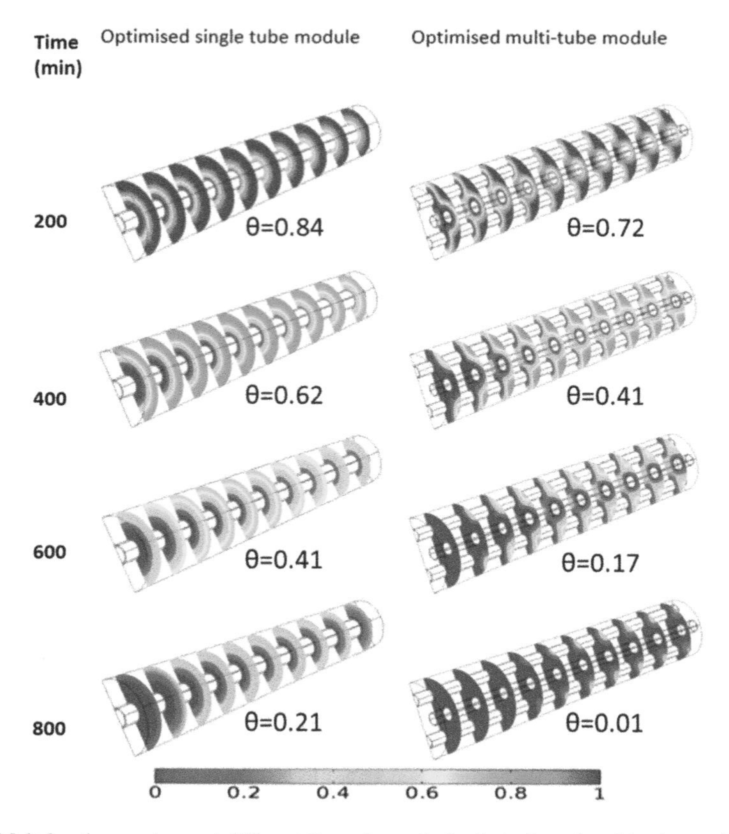

Fig. 5 Melt fraction contours at different times for optimized single and multi-tube models

3.3 *Effect of Radial Distance of Tubes in Multi-tube Conical LHS System*

Radial distance of tubes from the shell center also affects the heat transfer inside the PCM shell. Three different values of offset (41.6, 36.6 and 31.6 mm) from the center of PCM shell as shown in Fig. 1 are taken, and simulations are performed for the conical multi-tube model with a cone angle 2.3°. The discharging times for different offset values are reported in Table 2. Figure 7 shows the variation of average melt fraction of PCM with time for various offset values. It can be inferred that as we decrease the radial distance, the discharging time increases. This happens because by decreasing the offset, the HTF tubes come closer to the center of the PCM shell, and thus, the heat transfer to the PCM distributed toward the circumferential side weakens.

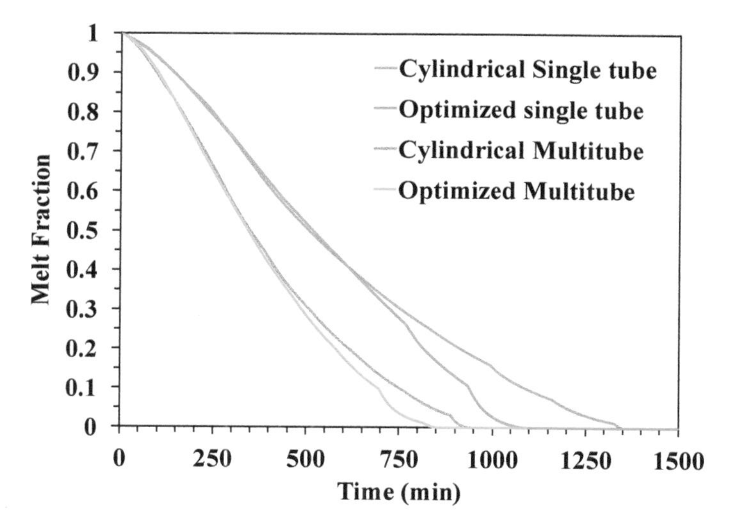

Fig. 6 Average melt fraction comparison between single and multi-tube for cylindrical and optimized conical shell models

Table 2 Discharging times for the optimized conical shell model for different offset values

Offset (mm)	Discharging time (Hours)
41.6	14.375
36.6	14.55
31.6	16.6

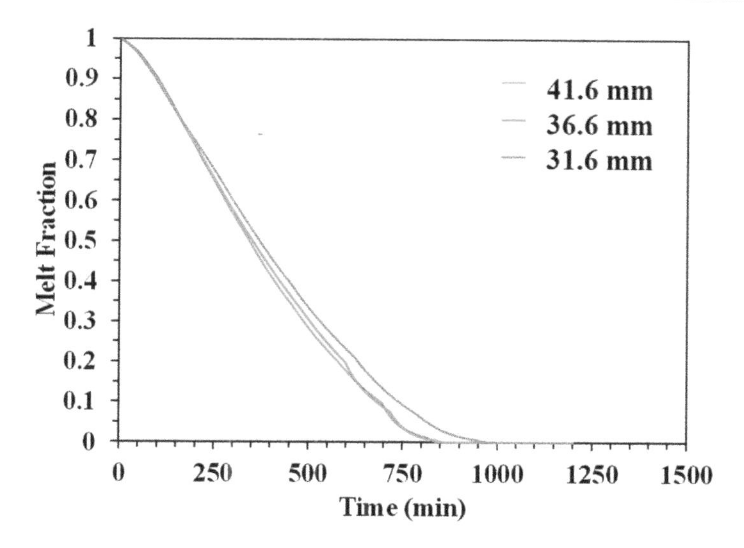

Fig. 7 Average melt fraction variation with time for different offset values for the optimized conical shell model

4 Conclusion

The current study focuses on studying the discharging behavior of a LHTESS and heat transfer enhancement by modifying the geometrical shape of the shell. A comparative study has been performed between the conical shell single and multi-tube models. 3D numerical simulations are performed, and the following conclusions are drawn:

- As the cone angle increases, the discharging rate increases up to the optimum value, whereas the discharging rate decreases thereafter. For the optimized conical models with cone angle 4.5° for single tube model and 2.3° for the multi-tube model, the discharging times are reduced by 18.51 and 11.7%, than the respective cylindrical models.
- The discharging rate for the optimized multi-tube model improved by 21.76% than the optimized single tube conical shell model.
- As the offset between the HTF tubes decreases, the heat transfer rate and hence the discharging rate decrease.

References

1. Joybari, M.M., Seddegh, S., Wang, X., Haghighat, F.: Experimental investigation of multiple tube heat transfer enhancement in a vertical cylindrical latent heat thermal energy storage system. Renew. Energy **140**, 234–244 (2019). https://doi.org/10.1016/j.renene.2019.03.037
2. Ahmed, N., Elfeky, K.E., Lu, L., Wang, Q.W.: Thermal and economic evaluation of thermocline combined sensible-latent heat thermal energy storage system for medium temperature applications. Energy Convers. Manag. **189**(March), 14–23 (2019). https://doi.org/10.1016/j.enconman.2019.03.040
3. Sodhi, G.S., Vigneshwaran, K., Jaiswal, A.K., Muthukumar, P.: Assessment of heat transfer characteristics of a latent heat thermal energy storage system: multi tube design. Energy Procedia **158**(2018), 4677–4683 (2019). https://doi.org/10.1016/j.egypro.2019.01.737
4. Cárdenas, B., León, N.: High temperature latent heat thermal energy storage: Phase change materials, design considerations and performance enhancement techniques. Renew. Sustain. Energy Rev. **27**, 724–737 (2013). https://doi.org/10.1016/j.rser.2013.07.028
5. Tao, Y.B., He, Y.L.: A review of phase change material and performance enhancement method for latent heat storage system. Renew. Sustain. Energy Rev. **93**(May), 245–259 (2018). https://doi.org/10.1016/j.rser.2018.05.028
6. Mahdi, M.S., et al.: Numerical study and experimental validation of the effects of orientation and configuration on melting in a latent heat thermal storage unit. J. Energy Storage **23**, 456–468 (2019). https://doi.org/10.1016/j.est.2019.04.013
7. Khalifa, A., Tan, L., Date, A., Akbarzadeh, A.: A numerical and experimental study of solidification around axially finned heat pipes for high temperature latent heat thermal energy storage units. Appl. Therm. Eng. **70**(1), 609–619 (2014). https://doi.org/10.1016/j.applthermaleng.2014.05.080
8. Ma, Z., Yang, W.W., Yuan, F., Jin, B., He, Y.L.: Investigation on the thermal performance of a high-temperature latent heat storage system. Appl. Therm. Eng. **122**, 579–592 (2017). https://doi.org/10.1016/j.applthermaleng.2017.04.085
9. Esapour, M., Hosseini, M.J., Ranjbar, A.A., Pahamli, Y., Bahrampoury, R.: Phase change in multi-tube heat exchangers. Renew. Energy **85**, 1017–1025 (2016). https://doi.org/10.1016/j.renene.2015.07.063

10. Kousha, N., Rahimi, M., Pakrouh, R., Bahrampoury, R.: Experimental investigation of phase change in a multitube heat exchanger. J. Energy Storage **23**(April), 292–304 (2019). https://doi.org/10.1016/j.est.2019.03.024
11. Seddegh, S., Tehrani, S.S.M., Wang, X., Cao, F., Taylor, R.A.: Comparison of heat transfer between cylindrical and conical vertical shell-and-tube latent heat thermal energy storage systems. Appl. Therm. Eng. **130**, 1349–1362 (2018). https://doi.org/10.1016/j.applthermaleng.2017.11.130
12. Sodhi, G.S., Jaiswal, A.K., Vigneshwaran, K., Muthukumar, P.: Investigation of charging and discharging characteristics of a horizontal conical shell and tube latent thermal energy storage device. Energy Convers. Manag. **188**(January), 381–397 (2019). https://doi.org/10.1016/j.enconman.2019.03.022
13. Kumar, M., Krishna, D.J.: Influence of mushy zone constant on thermohydraulics of a PCM. Energy Procedia **109**(November 2016), 314–321 (2017). https://doi.org/10.1016/j.egypro.2017.03.074

Classification of Different Floral Origin of Honey Using Hybrid Model of Particle Swarm Optimization and Artificial Neural Network

Kamalika Tiwari, Santigopal Pain, Bipan Tudu, Rajib Bandopadhyay, and Anutosh Chatterjee

Abstract The present study deals with the design of the hybrid structure of the particle swarm optimization (PSO)-based back propagation multilayer perceptron –artificial neural network (BPMLP-ANN) and technique used for the classification of monofloral honey samples. The transient response is collected from the multi-electrode electronic tongue (ET) system for the rapid floral classification of honey. Forty samples of five different floral types (eucalyptus, til, leechi, pumpkin, and mustard) are recorded based on the cyclic voltammetric technique. The obtained electronic tongue response matrix has been treated with various preprocessing techniques. Principal component analysis (PCA) is done to observe the capability of cluster formation. Further, the discrete wavelet transform (DWT) method has been used for feature selection and compression of data set. The resultant compressed data is used as the input variable for classification using back propagation multilayer perceptron-based neural network model. The weights are updated using particle swarm optimization (PSO) during the training of BPMLP-ANN. The result indicates that the proposed hybrid model is effective for classification of different floral origins of honey samples with increased in the classification rate up to 97%.

Keywords Honey · Electronic tongue · Particle swarm optimization · Back propagation multilayer perceptron · Artificial neural network

---

K. Tiwari (✉)
Department of AEIE, Dr. B. C. Roy Engineering College, Durgapur, India
e-mail: tiwari.kamalika@gmail.com

S. Pain
Department of Electrical Engineering, Haldia Institute of Technology, Haldia, India

B. Tudu · R. Bandopadhyay · A. Chatterjee
Department of Instrumentation and Electronics Engineering, Jadavpur University, Kolkata, India

P. Muthukumar et al. (eds.), *Innovations in Sustainable Energy and Technology*, Advances in Sustainability Science and Technology,
https://doi.org/10.1007/978-981-16-1119-3_13

1 Introduction

Honey from decade has been most popular sweetener consumed across the world. The taste and flavour of honey are the most influencing attribute for the consumers. The colour, taste, and flavour are an important factor in deciding the commercial value. Honey quality depends on the presence of individual chemical components and particularly nectar source having a distinct effect on taste and aroma of honey [1]. Nectar source also contributes to sugars, acids, and volatile components [2]. Essentially, all honey originates from nectar, but then again, the exact composition of the nectar has a definite impact on the taste of the honey. Moreover, monofloral honey has a distinct therapeutic value. Depending upon nectar source, various honeys are available. Thus, floral origin plays a imperative role in determining the commercial value of the honey. Till date, the assessment of floral type of honey is done by melissopalynological method [1] or volatile fraction measurement [3]. Both the methods show good precision and accuracy in determining the floral origin of honey. But these methods have issues like the need of expert and are time-consuming. The rapid growth of apiary industry and advancement of instrumental assessment in food analysis requires a fast and reliable technique for particularly determining the floral source. It is in this pursuit that electronic tongue technology is used for determination of the floral origins of honey samples.

In this study, the electronic tongue system is based on an array of noble metal electrodes that is developed in Instrumentation and Electronics Department of Jadavpur University [4], and it has been used in determining the floral origin of honey. The cyclic voltammetry technique along with multivariate data analysis is applied for floral assessment of honey. The electrochemical responses, i.e. transient responses obtained by cyclic voltammetry (CV) from the array of electrodes, are preprocessed using standard normalization techniques, e.g. baseline subtraction, auto scale, relative scale and range scale, etc. A comparative study among the preprocessing methods resulted in improvement of separability criterion. The effect of standard preprocessing methods is observed using principal component analysis (PCA). The DWT is performed for dimensional reduction of the response matrix for five groups of honey samples. Finally, the hybrid PSO-ANN classifier has been used for evaluating the classification capability.

Although BPMLP-based neural network is capable to classify different floral origin of honey samples, limitation still persists, such as the slow learning rate due to improper weights value. The PSO technique is used to choose the optimized value of weights so that ANN can perform better. The PSO, a population-based evolutionary algorithm, is a popular approach that is used to adjust the weights of an ANN in order to enhance the accuracy of the classifier model. Past studies report effective use of the hybrid PSO-ANN model for solving different complex problems [5–7].

The aim of this study is to develop a hybrid algorithm that can predict the floral origin of honey samples more accurately. The proposed technique is an application of the hybrid algorithm. The best weight selection is done by PSO algorithm and is applied to the back propagation neural network model for classification of the

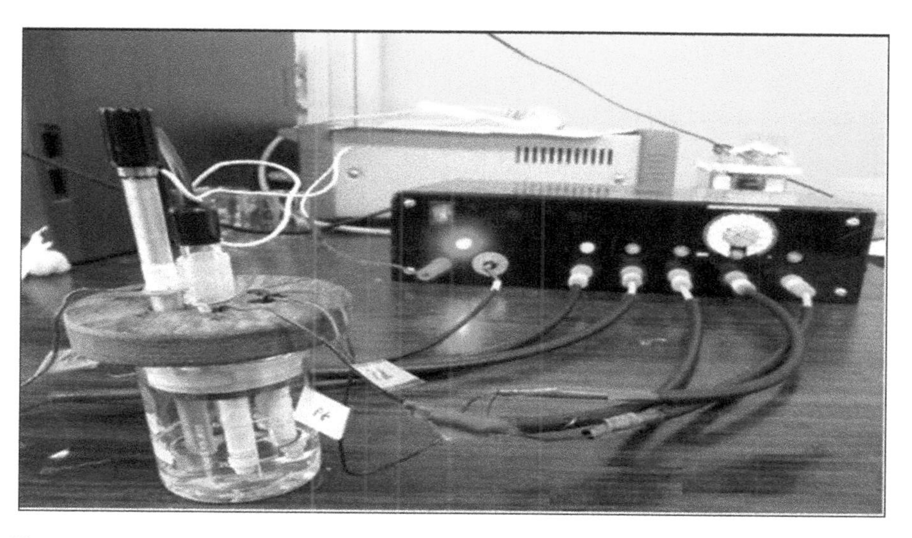

Fig. 1 Developed electronic tongue set-up

honey samples. The performance of the hybrid PSO-based ANN model shows better performance in terms of classification rate.

2 The Multielectrode Electronic Tongue Setup

In this study, all the measurement procedures are carried out with the experimental arrangement, i.e. an electronic tongue system proposed in [4]. Figure 1 shows the developed electronic tongue system. In our study, cyclic voltammetry technique is used. The electronic tongue employs a sensor array consisting of four different noble metal discs—gold, iridium, palladium, and platinum. The working electrodes are made using the wire of 1 mm diameter. They are pressed and fitted into a teflon sleeve. The two supporting electrodes are Ag/AgCl reference electrode (saturated KCL, Gamry Instruments Inc., PA) and a stainless steel counter electrode. The four working electrodes and one counter electrode are arranged in a circular fashion around the central reference electrode. The applied potential, adjustment of scan rate, data acquisition, data logging and electrode switching operation are performed through a graphical user interface using LABVIEW 8.5 from National Instruments®.

3 Material and Method

The electronic tongue responses are observed via current response against the potential applied through electrodes placed in the bob as shown in the experimental set-up shown in Fig. 1. The response collected for different monofloral honey samples is the electrochemical activity of each type of electrode and each floral type of honey against the potential applied.

3.1 Data Preprocessing

In this study, prior to pattern classification, the data obtained from different floral origins of honey samples is normalized using appropriate preprocessing method to compensate for sample-to-sample variation due to sensor drift, analyte concentration or differences in sensor scaling. So that better classification and clustering is obtained. The transient responses acquired from the electronic tongue system are normalized by auto-scaling technique prior to PCA [8].

3.2 Data Compression

The floral origin classification of honey is quite a challenging task due to the large size of transient response data set collected for each electrode type for each scan. Each sample is analysed by all the electrodes in the array of the electronic tongue, i.e. collecting one voltammogram of 1500 data points per sensor. The data matrix comprises 6000 data points (1500×4) collected from each sample. Thus, the database consists of 40 honey samples of five different floral types with 6000 data points. The classification of different floral types of honey using neural networks is not feasible on this large data set, as the number of input nodes will be very high. Hence, data compression without significant loss of information is necessary. The response matrix has been compressed with discrete wavelet transform (DWT) [9]. The mother wavelet, considered in this study, is "HAAR". The best classification results are obtained for fifth level of compression.

3.3 Data Clustering

The principal component analysis (PCA) technique has been applied to identify principal clusters in the electronic tongue responses. PCA emphasizes variation and pattern within a data set. It shows a visual impression of each floral origin of honey samples for the sensor array in the electronic tongue system. PCA gives qualitative analysis of the obtained compressed data set. The compressed data set after treated with DWT of different floral origins honey samples has been considered for the PCA.

3.4 Data Classification

Artificial neural networks (ANNs) are very powerful techniques when suitably trained to realize complicated relations over large number of input and output data. The BPMLP-ANN is a supervised method and is a generalization of gradient descent learning rule. It has the ability to approximate arbitrary functions [10]. A neural network topology comprises three computational layers (input layer, hidden layers, and the output layer). The neural network classifier is a supervised learning method due to predefined input and output attributes. The neural network can be trained to make them identify the underlying data distribution. After training process is satisfactorily completed, a trained network may classify an unknown data sample into one of the known groups. For accurate classifications, the weights of the perceptron model are adapted iteratively. The synaptic weights are adjusted in such an order to make the actual response of the network closer to desired response. The nonlinearity introduced by the activation function along with the availability of large number of adjustable weights affecting the linear response of a neuron enables a neural network to approximate very complicated and nonlinear functions.

3.5 Particle Swarm Optimization (PSO)

Particle swarm optimization (PSO) method is one of the useful algorithms to optimize complex problems [11]. This method is derivative-free and generally suited for continuous variable problems. PSO is inspired by the paradigm of birds flocking. It consists of a swarm of particles, where each particle flies through the multi-dimensional search space with a velocity, which is constantly updated by the particle's previous best performance and by the previous best performance of the particle's neighbours [12]. The algorithm depends on proper parameter settings, or else there will be a chance that the convergence process may trap in local optima.

3.6 Selection of Best Weights Using PSO

In this work, the fitness function is mean square error (MSE). The MSE is calculated using Eq. (1).

$$\text{MSE} = \frac{1}{2}\sum_{i=n}^{i=1} e^2 \tag{1}$$

where n is number of samples and e represents the error between observed and predicted class. The accuracy of the neural network model is improved by choosing the optimal weights using particle swarm optimization (PSO). The BPMLP-ANN parameters are tuned until very low MSE is achieved. The optimized selected weight values are considered as input to the back propagation artificial neural network (BPMLP-ANN). The PSO parameters are given in Appendix.

4 Design of Hybrid Model

In the experiment, we proposed PSO-based BPMLP-ANN topology for the multi-electrode e-tongue data set. The neural network classifiers are best suited for nonlinear systems i.e. multielectrode sensor systems [13] and are efficient for classification of food products [14]. The voltammetric electronic tongue data matrix is very large; thus, the input nodes become very large, and computation may require more time. The process is time-consuming and may lead to over fitting. In this paper, the data is compressed with wavelet transformation and then subjected to PCA. The optimal numbers of PCs have been selected as the input layer, and output layer depends upon number of classifying group, whereas the hidden node is optimized by trial and error method. The weights of the ANN model are optimized and updated by PSO for better classification. The initial weights which are used in BPMLP-ANN are the starting weights of hybrid PSO as shown in Fig. 2. In the training set, the number of input nodes of the PSO-based BPMLP-ANN model is 8 because first 8 PCs scores are considered as input variables. The number of nodes in the hidden layer considered is 5, and 1100 iterations have been carried out for training based on trial and error method. The mean square error (MSE) is gradually reduced to a small stable value after 1100 iterations. In the test set, a similar variation is observed after 1100 iterations as the MSE becomes unstable (over fitting). The "logsig" function and "linear" function are considered as activation function in the hidden layer and output layer, respectively. The numbers of output nodes is 5 as five different floral origins of honeys are selected. The learning rates are 0.1 in the hidden layer and output layer. The momentum factor is considered as 0.8.

5 Result and Analysis

5.1 Data Clustering with PCA

In this study, the electronic tongue data from five different floral types of honey samples was treated with principle component analysis (PCA) after compression with discrete wavelet transform. DWT was performed up to 5th level of decomposition of five floral origins of honey samples, i.e. the new data set is of 40×188 data matrix. PCA is performed to identify the underlying clusters in the electronic tongue data set. The score plot is obtained by using first two principal components (PC1 and

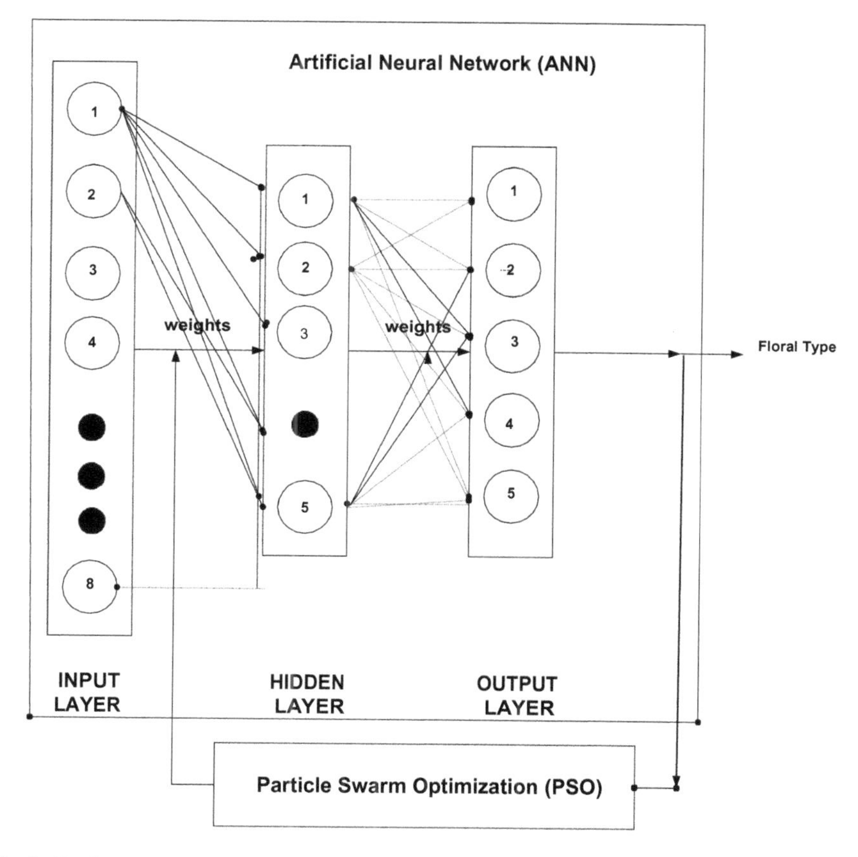

Fig. 2 Back propagation multilayered artificial neural network structure with PSO

PC2). Figure 3 shows the outcome of the principal component analysis. The cluster trend shows that PC1 and PC2 give a total variance of 95.92% for all the 40 samples used in this study. Each floral type is distinctly separated resulting into satisfactory cluster trend.

5.2 *Performance of Neural Network Classifier Using PSO*

In the experiment, BPMLP-ANN algorithm is used to estimate the classification ability of the multielectrode electronic tongue for identification of the honey samples. The data set comprises 40 samples (eight samples of each floral type), of which 25 samples (five samples from each floral origin) are considered in the training set (60% of data), while the remaining 15 samples (40% of data) are kept for the testing set. Table 1 shows the classification result using hybrid model. A higher accuracy rate

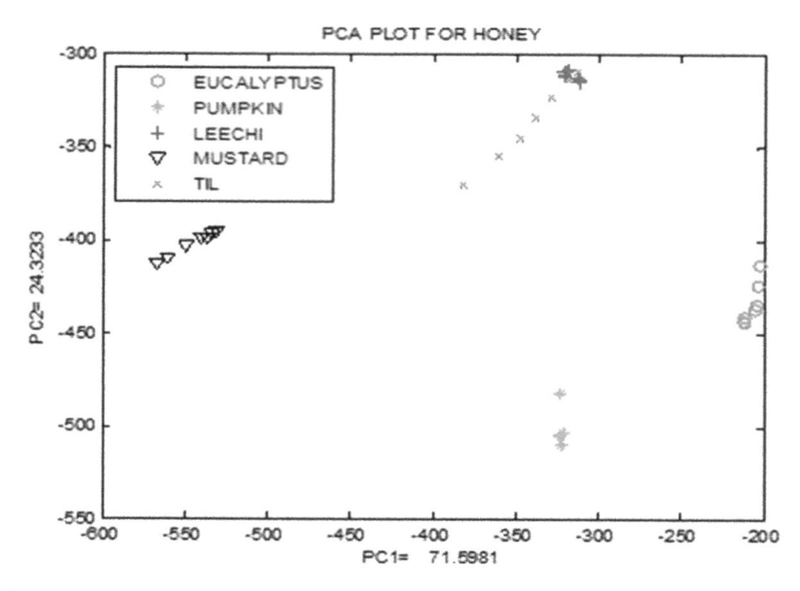

Fig. 3 PCA plot of the different floral type of honey samples analysed by the electronic tongue

Table 1 Classification result obtained using hybrid PSO-BPMLP-ANN model

Testing parameters	Nodes in hidden layer	10	7	5	5	5	Average (%)
	Iteration	800	1000	1100	1100	1100	
	Error rate	0.01	0.01	0.01	0.01	0.1	
	Population	50	50	80	100	100	
Classification result	Result 1 (%)	92	98	100	100	100	98
	Result 2 (%)	96	98	100	100	100	98
	Result 3 (%)	94	92	100	100	100	98
	Result 4 (%)	92	94	100	100	92	96
	Result 5 (%)	100	100	92	92	94	96
Average classification rate (%)		97					

is achieved using PSO by optimizing feature selection. The average honey floral origin classification rate of PSO-based BP-ANN model is 97%, i.e. all samples are identified correctly.

6 Conclusion

In this study, the complications of traditional methodology for floral origin classification of honey are solved using newly developed hybridized PSO-BPMLP-based ANN classification algorithm. PSO algorithm is used in updating the weights of the BPMLP-ANN network with less number of iterations. The performance of the algorithm is based on accuracy of the classifier. The proposed method significantly improves the classification rate of monofloral honey samples. The result indicates that multielectrode electronic tongue system along with PSO-ANN hybrid model can be efficiently used for routine analysis of different floral type of honey samples.

Appendix: PSO Parameters

Population size—50, 80,100; Maximum Iterations—800, 1000, 1100; Inertia Weight (w) = 1;

Constriction factors = 1.5.

References

1. Anklam, E.: A review of the analytical methods to determine the geographical and botanical origin of honey. Food Chem. **63**, 549–562 (1998)
2. Ball, D.W.: The chemical composition of honey. J. Chem. Educ. **84**, 1643–1646 (2007)
3. Cuevas-Glory, L.F., Pino, J.A., Santiago, L.S., Sauri-Duch, E.: A review of volatile analytical methods for determining the botanical origin of honey. Food Chem. **103**, 1032–1043 (2007)
4. Ghosh, A., Sharma, P., Tudu, B., Tamuly, P., Bhattacharyya, N., Bandyopadhyay, R.: Detection of optimum fermentation of black CTC tea using a voltammetric electronic tongue. Commun. Anal. Chim. Acta **64**, 2720–2729 (2014)
5. Tonnizam Mohamad, E., Jahed Armaghani, D., Momeni, E., Alavi Nezhad Khalil Abad, S.V.: Prediction of the unconfined compressive strength of soft rocks: a PSO-based ANN approach. Bull, Eng, Geol, Environ. (2014). https://doi.org/10.1007/s10064-014-0638-0
6. Momeni, E., Jahed Armaghani, D., Hajihassani, M., Amin, M.F.M.: Prediction of uniaxial compressive strength of rock samples using hybrid particle swarm optimization-based artificial neural networks. Measurement **60**, 50–63 (2015)
7. Isahl, O.R., Usman, A.D., Tekanyi, A.M.S.: A hybrid model of PSO algorithm and artificial neural network for automatic follicle classification. Int. J. Bio Autom. **21**, 43–48 (2017)
8. Tiwari, K., Tudu, B., Bandhopadhyay, R., Chatterjee, A.: Discrimination of monofloral honey using cyclic voltammetry. In: Proceedings of IEEE National Conference on Emerging Trends and Applications in Computer Science, **1**, 132–136 (2012)
9. Moreno-Baron, L., Cartas, R., Merkoci, A., Alegret, S., Del Valle, M., Leija, l., Hernandez, P.R., Munoz, R.: Application of the wavelet transform coupled with artificial neural networks for quantification purposes in a voltammetric electronic tongue. Sens. Actuat. B: Chem. **113**, 487–499 (2006)
10. Matreata, M.: Overview of the Artificial Neural Networks and Fuzzy Logic Applications in Operational Hydrological Forecasting Systems. National Institute of Hydrology and Water Management, Romania (2009)

11. Kennedy, J., Eberhart, R. C.: Particle swarm optimization. In: Proceedings of the 1995 IEEE International Conference on Neural Networks, vol. 4, pp. 1942–1948 (1995)
12. Eberhart, R.C., Kennedy, J.: A new optimizer using particle swarm theory. In: Proceedings of the Sixth International Symposium on Micromachine and Human Science, Nagoya, Japan, pp. 39–43 (1995)
13. Gutierrez, M., Alegret, S., Caceres R., Casadesus, J., Marf, O., Valle, Md. D.: Application of a potentiometric electronic tongue to fertigation strategy in greenhouse cultivation. Comput. Electron. Agric. **57**, 12–22 (2007)
14. Ciosek, P., Brzozka, Z., Wroblewski, W.: Classification of beverages using a reduced sensor array. Sens. Actuat. B **103**, 76–83 (2004)

Magnetoelastic Transition in Energy Efficient Magnetic Refrigerant $Ni_{50}Mn_{32}Sn_{18}$ Heusler Alloy

A. A. Prasanna

Abstract NiMnSn Heusler alloys find applications in magnetic cooling devices, magnetomechanical actuators/transducers, magnetic sensors, or spintronics. Adiabatic temperature change (ΔT_{ad}) in nanocrystallites of Heusler $Ni_{50}Mn_{32}Sn_{18}$ alloy undergoing a first-order magnetoelastic martensite (M) $\rightarrow$ austenite (A) transition has been studied in terms of the heat capacity (C_P) in warming the sample from 2 to 300 K at five different magnetic fields (B) up to 14 T. The M $\rightarrow$ A transition temperature shifts from 152 to 129 K on increasing the B-value, showing a large inverse magnetocalory in the transition, with a maximum $\Delta T_{ad} \sim 28$ K, or an isothermal magnetic entropy change ~14 J/kg-K at $B = 10$ T. The M $\leftrightarrow$ A transition in the caloric signal measured over 90–215 K results in a peak at 136 K during cooling and at 152 K during warming.

Keywords Magnetocaloric effect · Martensite transition · Heusler alloy

1 Introduction

The ferromagnetic Heusler alloys $Ni_{50}Mn_{25-x}Sn_x$ ($10 \leq x \leq 18$) undergoing a first-order magnetostructural transition (FOMT) have drawn considerable attention of researchers due to their functional properties, viz., magnetic shape memory, large magnetocalory (MC), and giant magnetoresistance [1–10]. Well-known applications include magnetic cooling devices, magnetomechanical actuators/transducers, magnetic sensors, or spintronics. The functional properties arise when a structural transition from a high temperature austenite (A) phase of a cubic crystal structure to a tetragonal/orthorhombic martensite (M) phase coincides with a magnetic transition, i.e. the FOMT. The FOMT is mediated through magnetoelasticity arising from a shear-like atomic displacement evolving a large strain energy which is minimized when the M-phase splits up into a number of crystallographic domains separated by twin boundaries [11]. The magnetoelastic coupling that arises on applying a

A. A. Prasanna (✉)
Malnad College of Engineering, Hassan 573202, Karnataka, India
e-mail: aap@mcehassan.ac.in

P. Muthukumar et al. (eds.), *Innovations in Sustainable Energy and Technology*, Advances in Sustainability Science and Technology,
https://doi.org/10.1007/978-981-16-1119-3_14

magnetic field (B) induces a twin boundary motion in shifting the M $\leftarrow$ A (or M $\rightarrow$ A) transition temperature T_M (or T_A). A few studies are available on composition and pressure dependence of the T_M and T_A in $Ni_{50}Mn_{25-x}Sn_x$ ($11 \leq x \leq 16$) alloys [12–14], but the effect of B-value is rarely studied. In this study, we explored the role of field-dependent T_A on an adiabatic temperature change (ΔT_{ad}), which is a measure of MC in terms of the heat capacity (C_P), in nanocrystallites (NCs) of a specific alloy $Ni_{50}Mn_{25-x}Sn_x$ ($x = 18$).

2 Experimental

The NCs of $Ni_{50}Mn_{32}Sn_{18}$ alloy were grown in a cylindrical disc (15 mm diameter and 8 mm width) cast from an arc melted mixture of the metals in the stoichiometry in a copper mould under argon atmosphere. The final chemical composition in this alloy was confirmed by using inductively coupled plasma optical emission spectroscopy and energy dispersive X-ray analysis performed on a scanning electron microscope. The M $\leftrightarrow$ A phase transition was studied in terms of the heat outputs while heating and cooling a specimen at 10 K/min in a differential scanning calorimeter (DSC Q100, TA Instruments). The C_P was measured using a commercial physical properties measurements system (quantum design).

3 Results and Discussion

Figure 1a shows the M $\leftrightarrow$ A transition with a well-defined peak in the caloric signal measured during heating followed by cooling the $Ni_{50}Mn_{32}Sn_{18}$ alloy of NCs, ~7 nm average size determined from the broadening of X-ray diffraction (XRD) peaks. An austenite cubic $L2_1$ crystal structure identified in terms of the XRD pattern at room-temperature describes the lattice parameter 0.6009 nm with density 8.1 g/cm^3. As marked over the thermogram, the M and A start and finish temperatures M_s, M_f, A_s, and A_f are 124, 148, 143, and 162 K, respectively. A thermal hysteresis $T_A - T_M \cong 152 - 136 = 16$ K signifies the features of the FOMT. A volume change, which can incur in the M $\leftarrow$ A transition on an increased elastic energy, causes a large enthalpy change $\Delta H^{M \leftarrow A} = 1.64$ J/g or an entropy change $\Delta S^{M \leftarrow A} = \Delta H^{M \leftarrow A}/T_M \cong 11.3$ mJ/g K. Since a paramagnetic M-state with a low thermal conductivity (κ) displays a slower enthalpy exchanger than a ferromagnetic A-state with a higher κ-value [2], a lowered $\Delta H^{M \rightarrow A} = 0.54$ J/g or $\Delta S^{M \rightarrow A} = \Delta H^{M \rightarrow A}/T_A \cong 3.6$ mJ/g-K, occurred in the reverse M $\rightarrow$ A transition on heating the NCs from a set point at 90 K.

To explore the effect of B-value on the T_A-value and concomitant change in ΔT_{ad}, we studied the field-dependent C_P-value by warming the $Ni_{50}Mn_{32}Sn_{18}$ NCs over 2–300 K temperature at five different B-values up to 14 T as shown in Fig. 1b. On increasing the B-value, a linear decrease in the T_A-value (inset) from 152 to 129 K

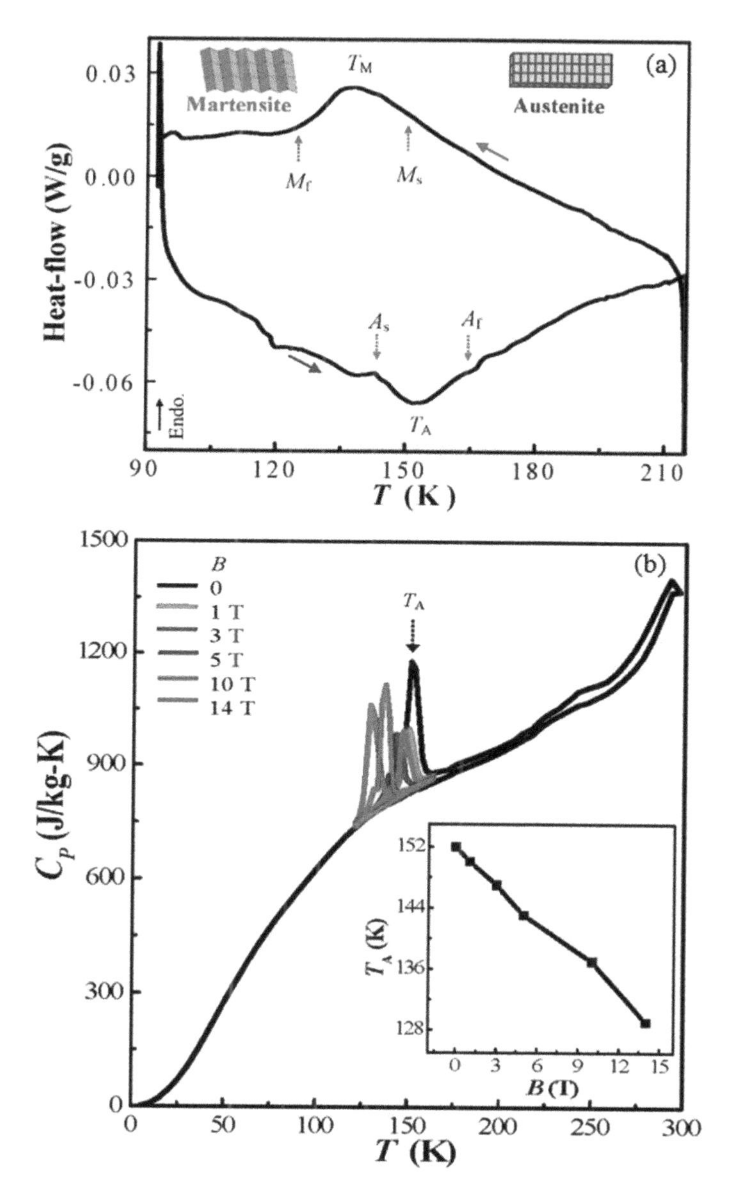

Fig. 1 **a** Heat outputs showing M ↔ A transition during heating followed by cooling $Ni_{50}Mn_{32}Sn_{18}$ NCs, with **b** the C_P-T plots showing a field dependence of $T_{A.}$, which falls down linearly as the field increases in a plot in the inset

or $\Delta T_A/\Delta B = -1.64$ K/T, incurs in stabilizing the A-phase during the M $\rightarrow$ A transition, which is possible only if the NCs have high magnetization (σ) in the A-phase relative to the M-phase. From the Clausius–Clapeyron equation, we obtain $\Delta T_A/\Delta B = -\Delta\sigma/\Delta S^{M\rightarrow A}$, where $\Delta\sigma$ is the difference between σ-values in the M- and A-phases. So, the value $\Delta T_A/\Delta B$ is large if $\Delta\sigma$-value is large and/or $\Delta S^{M\rightarrow A}$ is small in the transition. A value $\Delta\sigma$ ~ 5.9 Am^2/kg computed using the values $\Delta S^{M\rightarrow A}$ = 3.6 mJ/g-K and $\Delta T_A/\Delta B = 1.64$ K/T in the above relation is well supported by the value $\Delta\sigma$ ~ 2.3 Am^2/kg found in an σ-T plot at $B = 5$ mT. The large value of $\Delta T_A/\Delta B$ signifies not only a large $\Delta\sigma$-value, but also a concomitantly large volume change in a magnetoelastic coupling in the FOMT, useful for devising a large MC for possible applications.

To quantify the effect of field dependence of the T_A-value on the MC in the NCs, we computed magnetic entropy change ΔS_m and the ΔT_{ad}-value from the equations proposed by Pecharsky et al. [15], and the results are plotted in Fig. 2a, b. A positive ΔS_m or negative ΔT_{ad} indicates an inverse MC; a change in the signs above ~145 K signifies a conventional MC in the A-phase. The peak values of ΔS_m and ΔT_{ad} found to increase with the B-value up to 10 T, reaching 14 J/kg-K and 28 K, respectively, followed by the saturation signifying that no substantial $\Delta\sigma$-value is possible above 10 T. A similar alloy $Ni_{49.5}Mn_{25.4}Ga_{25.1}$ having a smaller $\Delta\sigma$ ~ 2 Am^2/kg with almost no volume change in the M $\leftrightarrow$ A transition exhibits reasonably lower values $\Delta T_A/\Delta B$ ~ 0.3 K/T and ΔS_m ~ 10 J/kg-K at $B = 1$ T [16].

4 Conclusions

The NCs of Heusler $Ni_{50}Mn_{32}Sn_{18}$ alloy exhibit a strong field dependence of T_A-value as a result of magnetoelastic coupling in the M $\rightarrow$ A transition. The NCs exhibit a large magnitude $\Delta T_A/\Delta B = 1.64$ K/T in terms of the shift in the peak position, or T_A, in the C_P-T plots towards lower temperatures on increasing the B-value up to 14 T. This suggests a field-controlled M $\leftrightarrow$ A transition, which is crucial for applications, particularly shape memory devices, magnetic sensors, and magnetic cooling systems.

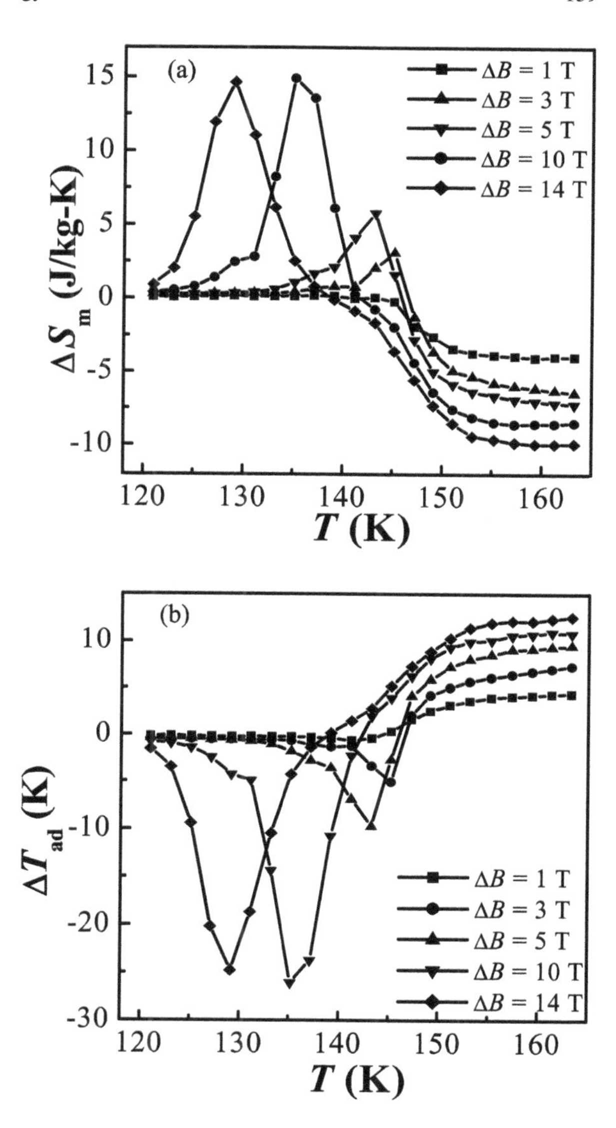

Fig. 2 Temperature variations of **a** ΔS_m and **b** ΔT_{ad} showing an inverse MC below T_A with maximum values at $\Delta B =$ 10 T after saturation. Conventional MC is seen above T_A

References

1. Krenke, T., Duman, E., Acet, M., Wassermann, E.F., Moya, X., Mañosa, L., Planes, A.: Inverse magnetocaloric effect in ferromagnetic Ni-Mn-Sn alloys. Nat. Mater. **4**, 450–454 (2005)
2. Babita, I., Ram, S., Gopalan, R., Chandrasekaran, V.: Dynamic inverse-magnetocaloric and martensite transition in $Ni_{49}Mn_{38}Sn_{13}$ nanocrystals in low magnetic fields. Phil. Mag. Lett. **89**, 399–407 (2009)
3. Prasanna, A.A., Ram, S., Effect of crystallite size on Vickers microhardness in nanostructured Heusler $Ni_{39+x}Mn_{50}Sn_{11-x}$ ($x \leq 2$) alloys. IEEE Xplore 424–427 (2011). ISBN: 978-1-4673-0072-8

4. Prasanna, A.A., Ram, S., Ganesan, V., Samantham, S.S.: Attenuating large magneto-entropy, heat-capacity and adiabatic temperature change in Heusler $Ni_{41-x}Mn_{50}Sn_{9+x}(x\leq 1.5)$ alloys. J. Emerg. Trend. Eng. Appl. Sci. **3**, 601–607 (2012)
5. Prasanna, A.A., Ram, S.: Local strains, calorimetry, and magnetoresistance in adaptive martensite transition in multiple nanostrips of $Ni_{39+x}Mn_{50}Sn_{11-x}(x\leq 2)$ alloys. Sci. Technol. Adv. Mater. **14**(13), 015004 (2013).
6. Prasanna, A.A., Ram, S., Fecht, H.-J.: Consecutive magnetic and magnetocaloric transitions in a Heusler $Mn_{50}Ni_{41}Sn_9$ alloy of herringbone nanostructure. J. Nanosci. Nanotechnol. **13**, 5351–5359 (2013)
7. Prasanna, A.A., Ram, S., Das, D.: Giant Hall resistivity at low magnetic fields in nanocrystalline $Ni_{50}Mn_{32}Sn_{18}$ Heusler alloy. Am. Inst. Phys. Conf. Proc. **1447**, 980–981 (2011)
8. Prasanna, A.A., Ram, S., Ganesan, V., Rao, S.S.: Large adiabatic temperature change in magnetoelastic transition in nanocrystallites of Heusler $Ni_{50}Mn_{32}Sn_{18}$ alloy. In: Jayakumar, S., Vaideki, K., Balaji, R. (eds.) Functional Materials. McMillan Publishers Ltd., New Delhi, pp. 195–198 (2011). ISBN: 978-935-059-046-1
9. Neibecker, P., Gruner, M.E., Xiao, Xu., Kainuma, R., Petry, W., Pentcheva, R., Leitner, M.: Ordering tendencies and electronic properties in quaternary Heusler derivatives. Phys. Rev. B **96**, 165131 (2017)
10. Tan, C., Tai, Z., Zhang, K., Tian, X., Cai, W.: Simultaneous enhancement of magnetic and mechanical properties in Ni-Mn-Sn alloy by Fe doping. Sci. Rep. **7**(43387), 1–9 (2017)
11. Krenke, T., Acet, M., Wassermann, E.F.: Ferromagnetism in the austenitic and martensitic states of Ni-Mn-In alloys. Phys. Rev. B, **73**(10), 174413 (2006)
12. Sutou, Y., Imano, Y., Koeda, N., Omori, T., Kainuma, R., Ishida, K., Oikawa, K.: Magnetic and martensitic transformations of NiMnX (X = In, Sn, Sb).Ferromagn. Shape Mem. Alloys, Appl. Phys. Lett. **85**, 4358–4360 (2004)
13. Krenke, T., Acet, M., Wassermann, E.F., Moya, X., Mañosa, L., Planes, A.: Martensitic transitions and the nature of ferromagnetism in the austenitic and martensitic states of Ni-Mn-Sn alloys. Phys. Rev. B **72**(9), 014412 (2005)
14. Yasuda, T., Kanomata, Y., Saito, T., Yosida, H., Nishihara, H., Kainuma, R., Oikawa, K., Ishida, K., Neumann, K.-U., Ziebeck, K.R.A.: Pressure effect on transformation temperatures of ferromagnetic shape memory alloy $Ni_{50}Mn_{36}Sn_{14}$. J. Magn. Magn. Mater. **310**, 2770–2772 (2007)
15. Pecharsky, V.K., Gschneidner Jr., K.A.: Magnetocaloric effect from indirect measurements: magnetization and heat capacity. J. Appl. Phys. **86**, 565–575 (1999)
16. Marcos, J., Planes, A., Manosa, L., Casanova, F., Batle, X., Labarta, A., Martinez, B.: Magnetic field induced entropy change and magnetoelasticity in Ni-Mn-Ga alloys. Phys. Rev. B., **66**(6), 224413 (2002)

Investigating the Characteristics and Choice of Electric Scooter Users: A Case Study of Tiruchirappalli City

Sandeep Singh, B. Priyadharshni, Challa Prathyusha, and S. Moses Santhakumar

Abstract India has entered an era in which its demand for energy is at an all-time high. Due to the presence of internal combustion engines which burn petroleum for operation, greenhouse gases are emitted from convection vehicles. In recent decades, the deployment of Electric Vehicle (EV) has evoked curiosity in Indian riders. The scarcity of charging stations to counter power supply demand, battery technology limitations, and high EV purchase costs are some of the obstacles facing the deployment of electric vehicles. In this context, in Tiruchirappalli, India, the study examines and analyses the different factors influencing the adoption of Electric Scooter (E-Scooter) to increase the use of electric vehicles. Quantitative data is used in order to achieve the research goals. The data analysis method includes factor analysis and principal component analysis to explore the enactment of electric scooter. The high cost associated with the E-Scooter was found to be primarily responsible for preventing the adoption of E-Scooters in Tiruchirappalli region, India. The study suggests that the promotion of incentive benefits of E-Scooters can increase the demand for E-Scooter purchase. Eventually, this will promote adoption of green technology and drive for eco-friendly practices.

Keywords Electric vehicles (EV) · Electric scooter (E-Scooter) · Users' perception · Factor analysis · Principle component analysis

1 Introduction

The adoption of Electric Vehicles (EV) is motivated by the global strategy to minimize carbon emissions set out in the framework of the Sustainable Development Goals (SDG). The electrification of the transport sector in order to decarbonize the energy

S. Singh · B. Priyadharshni · S. Moses Santhakumar
National Institute of Technology Tiruchirappalli, Tiruchirappalli, India

C. Prathyusha (✉)
REVA University, Bengaluru, India
e-mail: prathyushareddy1990@gmail.com

P. Muthukumar et al. (eds.), *Innovations in Sustainable Energy and Technology*,
Advances in Sustainability Science and Technology,
https://doi.org/10.1007/978-981-16-1119-3_15

system has encouraged India's mobility to become more competitive. Because of the increase in the number of vehicles, pollution and demand for fossil fuel need to be reduced, the deployment of EV seems to be a sustainable option. E-Mobility, however, raises significant problems, such as customer acceptability, which can be tackled in terms of the higher cost of EV finance, the higher cost of EV maintenance, and the lack of charging infrastructure facilities. Various technological and economic concerns are involved in the transport and energy policy systems. In a medium-tier city like Tiruchirappalli in India, the current research paper deals with examining these contributing factors in adopting the E-Scooter.

2 Literature Review and Background

In the different studies, the major obstacles in the adoption of electric vehicles were: high initial investment costs, ownership and maintenance costs, lack of charging infrastructure, and travel delay. Ozaki and Sevastyanova [1] suggested that consumer acceptance of electric vehicles is critical in achieving sustainable transport. Ajzen [2] explained consumer behavior through the Theory of Planning Behavior (TPB) in the adoption of electric vehicles. Diamond [3] found that there are some common obstacles to the adoption of new technology due to lack of awareness by potential adopters, high initial costs, and low risk tolerance. Curtin et al. [4] mentioned that customer responses to rising plug-in EV price premiums are generally higher, that can be justified based on purely economic rationales in terms of purchase probabilities. The cost does not only assess market acceptance, as environmental and other non-economic factors often impact the probability of potential purchases.

Roche et al. [5] explained the financial advantages of EV's over the gasoline vehicle through the principle of maximum utility where the cost of EV rises as the battery size and the vehicle range increases. Jensen et al. [6] reported the negative behavior of the driver towards EV adoption through the preference of the customer. Oliver and Rosen [7] suggested that the consumer's acceptance of Hybrid EV is very limited as there is a tradeoff between the size, performance, and price of the vehicle. Singh and Prathyusha [8] build system dynamic models to forecast fuel consumption and fuel emissions under different scenarios for an urban area in India. The authors estimated the reduction in the fuel consumption and fuel emissions by splitting the private and public vehicle share as 90% and 10% in one scenario and 80% and 20% in another scenario.

Gallagher and Muehlegger [9], explored the non-financial factors for the adoption of Hybrid EV by customers, ideally related to the environment and electricity. Liao et al. [10] analyzed the attributes of the customer preference product: charging time, driver range, density of charging stations, and operating costs that influence the purchase of the EV. Prathyusha et al. [11] proposed strategies for efficient and economic transport system strategies on the sustainability grounds. Ozaki and Sevastyanova [1] discovered that market reluctance is due to a lack of awareness of

EV advantages such as savings and lifetime operating costs aimed at stimulating the purchasing of EV by policymakers and manufacturers.

Heffner et al. [12] studied the purchasing of hybrid cars in which the criteria of vehicle symbolization such as beliefs, values, and social status were specified. Integrated psychological theories and various factors related to sustainable technology are integrated by Huijts et al. [13] in which choice studies are incorporated. The environmental concern is the primary criteria for EV adoption. Singh et al. [14] proposed models to carefully decrease the usage of fuel thereby reducing the emission levels. The study suggests augmentation of public vehicles and restriction of personalised vehicles. In this analysis, the review of studies shows that most studies have tremendous promise in addressing the implementation of EVs because they have the advantage of addressing the shortcomings of possible socio-technical obstacles to the adoption of EVs by consumers. This analysis of literature summarises that more attention should be given to the physical barriers of the EV users.

3 Study Methodology

Electric vehicles provide riders in urban areas with alternate transportation solutions. EVs in India, however, face barriers to user perception with regard to charging time, long driving range, cost of battery replacement, and low speed. In this context, the research paper aims to understand the key factors influencing the use of electric vehicles in the city of Tiruchirappalli, India, where the data acquisition process initially involves the collection of information using questionnaires. The questionnaire was developed primarily for the purpose of collecting adequate information on the factors influencing the use of the E-Scooter and the key features affecting the decision to buy the E-Scooter. A questionnaire interview was conducted with several automobile manufacturers and professional affiliated with the EV to better understand the operation of the E-Scooter, which helps to evaluate the factors influencing its widespread adoption in the city of Tiruchirappalli. Data are collected, such as vehicle information, rider preferences, variables influencing the adoption of E-Scooter, and so on, which are used for the development of databases. To analyze the initial response of the EV users, descriptive statistical analysis was performed. Factor Analysis and Principal Component Analysis (PCA) were included in the final data analysis process to analyze the factors influencing the adoption of E-Scooters.

4 Data Collection

Based on a literature review socio-economic and vehicle details denotes the attitude of passengers and the different factors influencing the adoption of electric vehicles. The data collected includes the socio-economic information of the rider: age, gender, educational qualification, and monthly income and vehicle details: electrical unit consumption/month, EV charging period, journey distance/day, driving hours/day, fuel and E-Scooter cost/month. The various factors affecting the use of E-Scooter are 1. Hovering E-Scooter cost, 2. Hovering replacement parts cost, 3. Hovering operating cost, 4. Inadequate qualified technicians, 5. Inadequate replacement parts, 6. Dearth of operational knowledge, 7. Dearth of charging infrastructure, 8. Dearth of Government support, 9. Hovering charging time, 10. Short driving range, and 11. Safety.

4.1 Socio-economic Characteristics Electric Scooter Users

In terms of socio-economic status, there is a wide variation and disparity in the characteristics of users of electric scooters (E-Scooters). Cross-tabulation or contingency tabulation is a method used to quantitatively analyze the relationship between multiple variables by changing the grouping of one variable into another.

4.2 Gender Distribution of Electric Scooter Users

The sampled data consists of a slightly higher male (68%) representation compared to females (32%). The relationship between gender and E-Scooter purchase is measured by cross-tab analysis shown in Table 1. It was found that female respondents were more likely (79%) to buy an E-Scooter than male respondents, as 79% of female respondents choose yes, while 60.8% of male respondents choose yes, therefore as opposed to males, women tend to buy E-Scooter.

Table 1 Response to purchase of E-Scooter versus respondent's gender—Cross-tabulation

	Response		Respondent's sex		Total
			Male	Female	
Purchase of an E-Scooter	Yes	Count	31	19	50
		% within respondent's sex	60.8	79.2	66.7
	No	Count	19	5	24
		% within respondent's sex	37.3	20.8	32.0
	No response	Count	1	0	1
		% within respondent's sex	2.0	0	1.3
Total		Count	51	24	75
		% within respondent's sex	100	100	100

Table 2 Chi-Square tests statistical results for gender and purchase of E-Scooter

Test	Value	df	Asymptotic significance (2-sided)
Pearson chi-square	5.346	2	0.069

From Table 2, the results of the 2 × 2 chi-square test indicate that there is no important association between gender and E-Scooter purchase, X^2 (2, $N = 75$) = 5.346, $p = 0.069$. Therefore, gender does not influence an E-purchase Scooter's decision.

4.3 *Age Distribution of E-Scooter Users*

E-Scooter riders are aged between 17–24, 25–30, 31–40 and 41–50. With the majority of respondents (64%) between the ages of 17 and 24, the overall sample is relatively young. To evaluate the pattern of association between age and the purchase of an E-Scooter, a crosstab analysis was carried out, which is shown in Table 3.

Table 3, indicates that respondents between the ages of 31 and 40 years (76.92%) were more likely to buy an E-Scooter, followed by respondents between the ages of 17–24 years, followed by respondents between the ages of 25–30 years (62.5%) and finally respondents between the ages of 50 years (50%).

A 3 × 6 chi-square test was used from Table 4 to determine whether there was a statistically significant relevant association between age and an E-Scooter purchase. $X^2(10, N = 75) = 8.067$, $p = 0.622$, the relationship between these variables was relatively insignificant.

Table 3 Purchase of E-Scooter and respondent's age—cross-tabulation

	Response		Respondent's age (years)				Total
			17–24	25–30	31–40	41–50	
Purchase of an E-Scooter	Yes	Count	32	5	10	3	50
		% within age of respondents	66.67	62.5	76.92	50.0	66.67
	No	Count	16	3	2	3	24
		% within age of respondents	33.33	37.5	15.38	50.0	32.0
	No response	Count	0	0	1	0	1
		% within age of respondents	0	0	7.69	0	1.3
Total		Count	48	8	13	6	75
		% within age of Respondents	100	100	100	100	100

Table 4 Chi-Square tests statistical results for age and purchase of E-Scooter

Test	Value	df	Asymptotic significance (2-sided)
Pearson chi-square	8.067	10	0.622

4.4 Level of Educational Qualification of E-Scooter Users

The education status of the individuals pursuing the undergraduate (UG) degree is 42.7%, followed by 40% of postgraduate (PG) degree students. To evaluate the pattern of association between education and the purchase of an E-Scooter, crosstab analysis was conducted, which is shown in Table 5.

The findings from Table 5 show that Ph.D. respondents (100%) were more likely to buy an electric vehicle, followed by respondents at the undergraduate level (68%), followed by postgraduates (66.7%), this was followed by respondents from high school (58.33%). It can also be said that if the level of education is higher, a person is more likely to buy an E-Scooter.

A 3×5 chi-square test was used from Table 6 to determine whether a statistically relevant association existed between the level of education and the purchase of an E-Scooter. It was found from the results that there was no important correlation between the level of education and the purchase of an E-Scooter, X^2 $(8, N = 75) = 5.057, p = 0.751$. Therefore, education do not affect the purchase of an E-Scooter.

Table 5 Purchase of E-Scooter and respondent's education level—cross-tabulation

	Response		Education level					Total
			Elementary	High School	UG	PG	Ph. D	
Purchase of an E-Scooter	Yes	Count	0	7	22	20	1	50
		% within education	0	58.33	68.8	66.7	100	66.7
	No	Count	0	5	9	10	0	24
		% within education	0	41.66	28.1	33.3	0	32.0
	No response	Count	0	0	1	0	0	1
		% within education	0	0	3.1	0	0	1.3
Total		Count	0	12	32	30	1	75
		% within education	100	100	100	100	100	100

Table 6 Chi-square test statistical results for education level and purchase of E-Scooter

Test	Value	df	Asymptotic significance (2-sided)
Pearson chi-square	5.057	8	0.751

Table 7 Purchase of an E-Scooter and Respondent's monthly income—cross-tabulation

	Response		Income (Rs. in thousands)				Total
			10–30	30–50	50–70	>70	
Purchase of an E-Scooter	Yes	Count	26	12	9	3	50
		% within income	61.9	63.15	81.81	100	66.7
	No	Count	16	6	2	0	24
		% within income	38.09	31.57	18.18	0	32.0
	No response	Count	0	1	0	0	1
		% within income	0	5.26	0	0	1.3
Total		Count	42	19	11	3	75
		% within income	100	100	100	100	100

Table 8 Chi-square tests statistical results for monthly income and purchase of E-Scooter

Test	Value	df	Asymptotic significance (2-sided)
Pearson chi-square	5.511	6	0.48

4.5 *Range of Monthly Income of E-Scooters Users*

The monthly income of Rs.10,000–Rs.30,000 accounts for 56% followed by Rs.30000–Rs.50,000 accounts for 25.3%. To evaluate the relationship between monthly income and the purchase of an E-Scooter, which is shown in Table 7, crosstab analysis was performed.

It was discovered from Table 7 that respondents with a monthly income of more than Rs.70,000 (100%) were more likely to buy an electric vehicle, followed by respondents with a monthly income of Rs. 50,000–70,000 (81.81%), followed by respondents with a monthly income of Rs. 30,000–50,000 (63.15%) and finally followed by respondents with a monthly income of Rs. 10,000–30,000 (61.9%).

A 3×4 chi-square test was used from Table 8 to determine whether there is any statistically relevant association between monthly sales and E-Scooter purchases. There is no important association between monthly sales and the purchase of an E-Scooter, X $(6, N = 75) = 5.511$, $p = 0.480$, as shown by the frequencies cross-tabulated in Table 7. There is therefore no significant correlation between monthly earnings and the purchase of an E-Scooter.

4.6 *Descriptive Statistical Analysis of Vehicular Characteristics*

Descriptive statistical analysis, which can be a quantitative representation of the sampled dataset, is a brief descriptive coefficient summarizing a given data set. The

Table 9 Descriptive statistical analysis of the selected variables

Sample size N=75	Consumed electric units/month	Charging time (h)	Trip distance/day (km)	Driving hours/day	E-Scooter expense/month (Rs.)
Minimum	60	3	45	2	100
Maximum	89	5	70	6	500
Mean	73.21	3.75	92.67	4.4	279.0

measures of central tendency and measures of variability are the descriptive statistics parameters. The mean, median, and mode are measures of central tendency, while the standard deviation, variance, minimum and maximum variables, and kurtosis and skewness are measures of variability. The aspect of data analysis includes evaluating each variable to determine the impact of different variables on the adoption of E-Scooters. The analyses of different factors are as follows.

Descriptive statistical analysis requires evaluating the different parameters acquired through the method of data collection and discovering their effect on EVs. The average units consumed in a month are 73 from Table 9, which takes approximately 4 h to completely charge the E-Scooter and travels 93 km in 4 h per day. While the initial cost of purchasing an EV is high, it is totally balanced by the cost of operating and maintenance.

4.7 *Measurement Model*

In the data reduction statistical methodology, factor analysis was used to classify underlying factors expressed in the variables observed. The core goal of factor analysis is to simplify many interrelated steps in an orderly manner. Using far fewer dimensions than the original variables, factor analysis defines the data. A Principal Component Analysis (PCA) was carried out on the products obtained as the key factors influencing the adoption of electric vehicles in Tiruchirappalli region. The suitability of data for factor analysis was evaluated before conducting PCA.

The Kaiser-Meyer-Olkin value in Table 10 was 0.943, above the suggested value of 0.5. Bartlett's sphericity test achieved statistically significance ($p = 0.000$), supporting the data to be appropriate for factor analysis.

Table 10 Kaiser-Meyer-Olkin and Bartlett's test

Kaiser-Meyer-Olkin measure of sampling adequacy		0.943
Bartlett's test of sphericity	Approximate chi-square	2010
	df	55
	Sig.	0.000

Table 11 Total variance explained for the 11 factors affecting E-Scooter adoption

Component	Initial eigenvalues			Extraction sums of squared loadings		
	Total	% of variance	Cumulative %	Total	% of variance	Cumulative (%)
1	10.31	93.8	93.8	10.3	93.8	93.8
2	0.27	2.4	96.2	–	–	–
3	0.10	0.96	97.2	–	–	–
4	0.09	0.83	98.0	–	–	–
5	0.06	0.59	98.6	–	–	–
6	0.03	0.34	98.9	–	–	–
7	0.03	0.33	99.3	–	–	–
8	0.02	0.20	99.5	–	–	–
9	0.02	0.18	99.7	–	–	–
10	0.01	0.15	99.8	–	–	–
11	0.01	0.12	100.0	–	–	–

4.8 *Principal Component Analysis*

For the Principal Components Analysis, a broad sample size is required to know the connection between the variables that influence the adoption of the electric vehicle.

The Principal Component Analysis reveals from Table 11 that the presence of one component, explaining 93.8% of the variance in contrast, shows that the other components from 2 to 11 display a very small percentage of the variance. Therefore only one aspect is taken for further analysis.

4.8.1 Component Matrix with the E-Scooter

Table 12. reveals the number of strong loadings which influence the E-Scooter purchase. From the PCA, it was observed that all factors had very high loadings with the component. From Fig. 1, the Scree Plot revealed a clear break, which shows only one component for further investigation greater than 1. The detailed inspection of the one component disclosed that all the factors that affect E-Scooter adoption could be divided into the following subgroup.

Table 12 Component matrix for the factors affecting E-Scooter adoption

S. No.	Factors affecting E-Scooter adoption	Component 1 (93.8% of variance)
		Loadings
1	Hovering E-Scooter cost	0.97
2	Hovering replacement parts cost	0.97
3	Hovering operating cost	0.96
4	Inadequate qualified technicians	0.95
5	Inadequate replacement parts	0.97
6	Dearth of operational Knowledge	0.97
7	Dearth of charging infrastructure	0.97
8	Dearth of government support	0.97
9	Hovering charging time	0.95
10	Short driving range	0.97
11	Safety	0.96

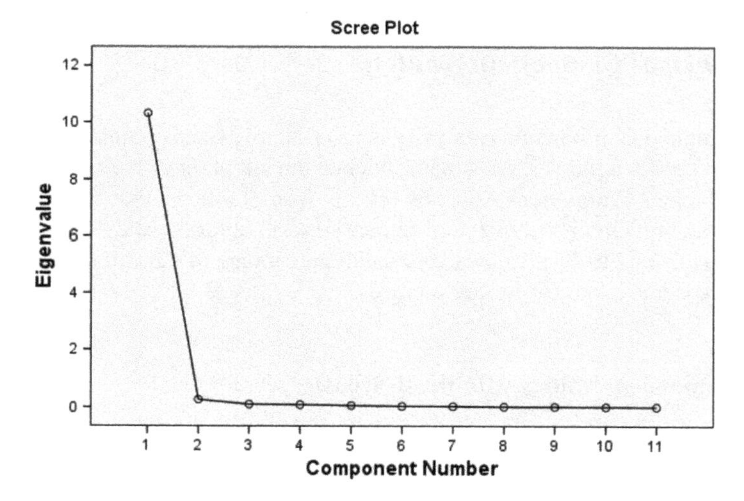

Fig. 1 Screen plot for the factors affecting E-Scooter adoption

5 Results

The buying of an E- Scooter is affected by socio-economic and vehicular variables. Based on numerous available variables, the present study examined the key features that affect the decision to purchase E-Scooter. In order to understand the characteristics of E-Scooter adoption, cross tabulation analyses have been carried out. The association between the variables is calculated by the Factor Analysis and the Principal Component Analysis is used to analyze the key factors influencing the adoption of E-Scooter.

6 Conclusions

The collected data subjects to various analyses in understanding the consumer behavior towards E-Scooter which helps the manufacturers in the design and development of electric vehicles to meet consumer demand. The launch of a low-cost E-Scooter, minimization of operating cost, trained technicians, providing charging infrastructure through charging schemes by the Government, producing short driving range considering safety and replacement parts popularizes the E-Scooter vehicles.

References

1. Ozaki, R., Sevastyanova, K.: Going hybrid: an analysis of consumer purchase motivations. Energ. Pol. **39**(5), 2217–2227 (2011). https://doi.org/10.1016/j.enpol.2010.04.024
2. Ajzen, I.: The theory of planned behavior. Org. Beh. Hum. Dec. Proc. **50**(2), 179–211 (1991). https://doi.org/10.1016/0749-5978(91)90020-T
3. Diamond, D.: The impact of government incentives for hybrid-electric vehicles: Evidence from US states. Energ. Pol. **37**(3), 972–983 (2009). https://doi.org/10.1016/j.enpol.2008.09.094
4. Curtin, R., Shrago, Y., Mikkelsen, J.: Plug-in hybrid electric vehicles. University of Michigan
5. Roche, M., Mourato, S., Fischedick, M., Pietzner, K., Viebahn, P.: Public attitudes towards demand for hydrogen fuel cell vehicles: a review of the evidence and methodological implications. Energ. Pol. **38**(10), 5301–5310 (2010). https://doi.org/10.1016/j.enpol.2009.03.029
6. Jensen, A.F., Cherchi, E., Mabit, S.L.: On the stability of preferences and attitudes before and after experiencing an electric vehicle. Transp. Res. Part D: Transp. Environ. **25**, 24–32 (2013)
7. Oliver, J.D., Rosen, D.E.: Comparative environmental impacts of electric bikes in China. J. Mar. Th. Prac. **18**(4), 377–393 (2010). https://doi.org/10.2753/MTP1069-6679180405
8. Singh, S., Prathyusha, C.: System dynamics approach for urban transportation system to reduce fuel consumption and fuel emissions. In: Proceedings of the Urban Science and Engineering, Lecture Notes in Civil Engineering, vol. 121, pp. 385–399. Springer, Singapore (2021). https://doi.org/10.1007/978-981-33-4114-2_31
9. Gallagher, K., Muehlegger, E.: Giving green to get green? Incentives and consumer adoption of hybrid vehicle technology. J. Env. Eco. Man. **61**(1), 1–15 (2011)
10. Liao, F., Molin, E., Wee, B.V.: Consumer preferences for electric vehicles: a literature review. Transp. Rev. **37**(3), 252–275 (2017). https://doi.org/10.1080/01441647.2016.1230794
11. Prathyusha, C., Singh, S., Shivananda, P.: Strategies for sustainable, efficient, and economic integration of public transportation systems. In: Proceedings of Urban Science and Engineering. Lecture Notes in Civil Engineering, vol. 121, pp. 157–169. Springer, Singapore (2021). https://doi.org/10.1007/978-981-33-4114-2_13
12. Heffner, R.R., Kurani, K.S., Turrentine, T.S.: Symbolism in California's early market for hybrid electric vehicles. Transp. Res. Part D: Transp. Env. **12**(6), 396–413 (2007). https://doi.org/10.1016/j.trd.2007.04.003
13. Huijts, N.M.A., Molin, E.J.E., Steg, L.: Psychological factors influencing sustainable energy technology acceptance: a review-based comprehensive framework. Ren. Sus. En. Rev. **16**(1), 525–531 (2012). https://doi.org/10.1016/j.rser.2011.08.018
14. Singh, S., Shukla, B.K., Sharma, P.K.: Use of system dynamics forecasting model in transportation system to reduce vehicular emissions-A mathematical approach. J. Phys.: Conf. Ser. **1531**(012118), 1–14 (2020). https://doi.org/10.1088/1742-6596/1531/1/012118

Agent-Based Path Prediction Strategy (ABPP) for Navigation Over Dynamic Environment

Samir N. Ajani and Salim Y. Amdani

Abstract Probabilistic path planning is major problem in the field of navigation and can be solved by using various navigational methods or technique. Many researchers have designed, developed and implemented the same to solve the mentioned problem. These implemented methods is categorized into classical approaches such as road map technique (RM), Cluster decomposition (CD) and artificial potential field (APF), and inspired approach such as Fuzzy logic (FL), particle swarm optimization (PSO), genetic algorithm (GA), Artificial neural network (ANN), etc. are invented. The selection of path planning algorithm is mostly depends upon the type of environment for which the path has to be generated. This environment is classified as static and dynamic. While considering the environment, all the characteristics related to the environment should be considered. Although the static environment has very less parameters to be considered while implementing the path planning algorithm, in the dynamic environment we need to consider the frames of static environment over the time series to get the best result. In this paper, we proposed an algorithm to predict the path over dynamic environment by considering the future position of the entire obstacle roaming in the environment. The proposed algorithm is based on Agent-based prediction function which will help the algorithm to generate time series obstacle position which in turn will help to generate the probabilistic path over the dynamic environment. We have also shown the probabilistic path generated with the help of mentioned algorithm along with time and path length results generated.

Keywords Prediction · Dynamic environment · Distance computation · Collision detection · Displacement · Position · Path planning · Obstacle

---

S. N. Ajani
Department of Computer Science and Engineering, Shri Ramdeobaba College of Engineering and Management, Nagpur, Maharashtra, India

S. Y. Amdani (✉)
Department of Computer Science and Engineering, Babasaheb Naik College of Engineering, Pusad, Maharashtra, India
e-mail: salimamdani22@gmail.com

P. Muthukumar et al. (eds.), *Innovations in Sustainable Energy and Technology*,
Advances in Sustainability Science and Technology,
https://doi.org/10.1007/978-981-16-1119-3_16

1 Introduction

Path planning is a navigational problem which is currently present in many regional as well as urban area. It is commonly associated in the field of military, mining, rescue operations, gaming educational along with research. Hence to solve the classic problem Path planning algorithms is required while considering the environment along with all the characteristic of the environment. The nature of environment for which the path planning [1, 2] is going to be executed will helps us for the selection of path planning algorithms.

There are so many algorithms available to solve above-stated problem, distributed as classical, hybrid, evolutionary, and nature-inspired. The classical methods like Roadmaps and cell decomposition are generally very easy to understand and implement because of the least complexity, but these algorithms are not practically effective nether in the static or in the dynamic environment. Evolutionary and nature-inspired techniques like Particle swarm optimization, genetic algorithm, Ant Colony optimization [3, 4] are more suitable for dynamic environment and becomes an excellent performer for the static environment because of the unstable/uncertain nature of environment where we need to consider the real-time state of environment. Many researchers have developed the hybrid algorithms which will work on static as well as dynamic environment. In this work, we have considered the dynamic environment for the path planning problem.

However, while implementing any path planning algorithm, we must need to incorporate future collision detection along with collision avoidance mechanism in order to get the best results. The collision detection is also one of the major challenges in its field. Many research have been carried out in order to obtaining the best collision detection mechanism [5] for the applications. In order to detect the collision of moving objects, all the moving object parameters need to consider like acceleration, speed, velocity, time, etc. Collision detection and avoidance can be implemented as model-based or a sensor-based. In the physical environment sensor-based collision avoidance mechanism will be suitable whereas in simulating environment model-based collision avoidance solved the purpose.

Finally, the path planning strategies [6] need to evaluate in terms of path predicted vs. actual path, time taken while moving the object over the predicted path, length of predicted path over the set of all candidate paths, type of environment, number of obstacles in the environment, final displacement error, average displacement error and presenting all these to understand the concept.

2 Literature Review

In [7] author Tsubasa Hirakawa, Takayoshi Yamashita, Toru Tamaki, and Hironobu Fujiyoshi describes the techniques used for path prediction based on "Bayesian Models" in which "BAYESIAN FILTERS" like particle filter, kalman filter are used to predict the future path. Here in this method internal states and observation as variables of pedestrian were used to generate a probabilistic model for predicting the path. Here in this approach iteratively a prediction step is processed by using the internal state with different observation of a pedestrian that computes the current state of a target. Here a person tracking is done by predicting its step to obtain the sequence of pedestrian.

In [8] author Huang et al. describes a path prediction approach by making use of one single picture. Here in this initially a "Patch" consisting the target is obtained for estimating the live orientation of the target. Then the total cost for covering the location of the patch is estimated by obtaining the various significant patches present around the surrounding. This estimated orientation and moving cost is added to edge and weights of the image. Then the comparison of the texture of super pixels using patches along the path that the target traced without involving any training procedure. Also a deep learning approach where Long short term memory "LSTM" and Convolution Neural Networks "CNN" is used for path prediction using deep learning framework. Here the series of coordinates of the target of last several frames were used as input to produce the target coordinates for path prediction of an target. This method applies the various feature extraction techniques using different deep learning models. Various techniques hence make use of "LSTM" for dealing with the path prediction problems, that has bi-dimensional co-ordinates, has been described by author Alahi et al. [9] need as s-pooling (social pooling) technique for avoiding collision between various pedestrians in traffic.

In [10] author discussed a path modeling technique based on Gaussian model approach. Here, in this approach, the best possible paths are shown as continuous mathematical functions with sequential steps. Path prediction process can be done in a theoretically proper probabilistic network [11]. This approach has an advantage that it has an ability to give a Gaussian probability distribution over various paths.

3 Proposed Methodology

We have presented long short-term memory-based Recurrent neural network (LSTM-RNN) to solved the probabilistic path planning problem in a dynamic environment using deep learning. This LSTM cells are supported with the agent-based prediction function. In Fig. 1, we have shown one of the cell of Long short term memory (LSTM) based Recurrent Neural Network (RNN) in the deep learning model.

Agent-based Prediction Function: It is a agent based probabilistic prediction function based on current position of robot, future predicted position of robot, and the current and future position of obstacles while traversing along the collision-free path.

$$f(n) = \text{Agent}\left(\frac{\text{Exp}\left(P_t^k\right)}{\sum_{k=1}^{N} P_t^k}\right)$$

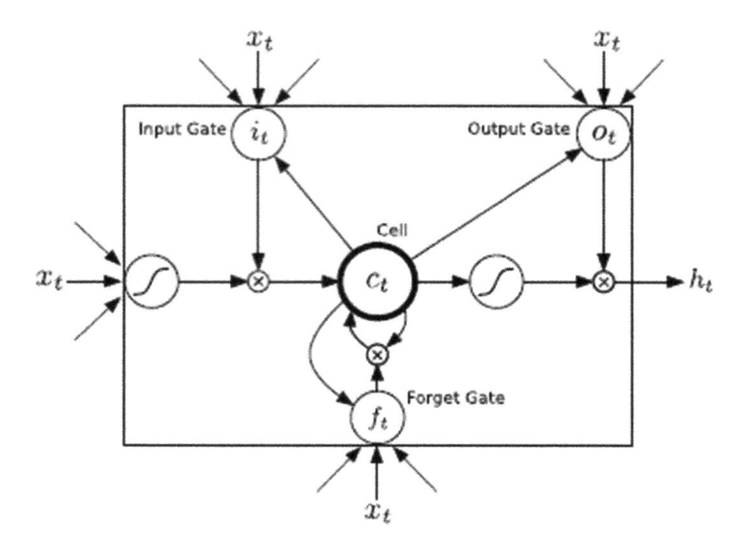

Fig. 1 LSTM cell

$f(n) \approx P_t^k$ to P_{t+i}^k Such that $O(t^2)$ along with $x_t \neq x_{t+1}$ & $y_t \neq y_{t+1}$ and $i \in \{1, 2, 3, \ldots, k\}$ (Fig. 2).

4 Proposed Algorithm

This proposed agent algorithm is based on obstacle position, object actual position, predicted position, and the error marker.

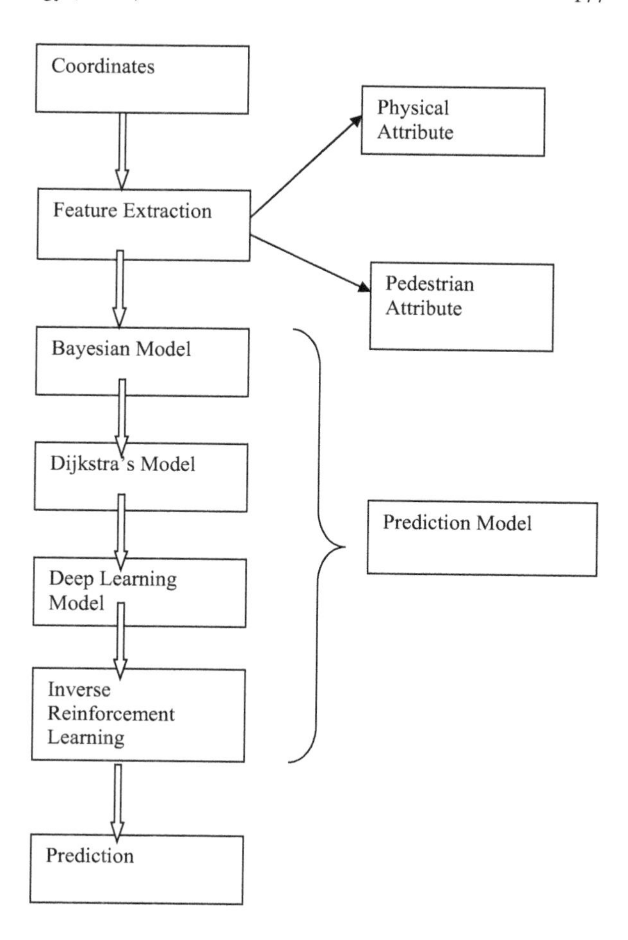

Fig. 2 Stages of Motion prediction in dynamic environment

Algorithm: Agent algorithm for Probabilistic path prediction.

Obstacle position: $x_{r\in\{1,2,\ldots,k\}}$, $y_{r\in\{1,2,\ldots,k\}}$
Actual position: x_i, y_i
Predicted Position: $x_{n\in\{1,2,\ldots,N\}}$, $y_{n\in\{1,2,\ldots,N\}}$
Error Marker: x_e, y_e

1. *Oi* $\leftarrow$ *No. of Obstacles*$_{i\in\{1,2,\ldots,k\}}$
2. $k \leftarrow \varphi$
3. $N \leftarrow \varphi$
4. *Loop for all* P_c *Exist:*
5. $P_c \leftarrow$ *Candidate path based on agent based prediction function f(n).*
6. $E(P_c) \leftarrow$ *Avg. displacement error.*
7. $P_p \leftarrow Agent\ \{MIN\ \{P_c^k\ ,\ E(P_c)\ \}\}$
8. $P_{t+i}^k \leftarrow P_p$
9. $P_{t+i}^k \leftarrow P_{t+i}^k\ U\ \{\ P_{t+i}^k, P_p\ \}$
10. $K \leftarrow k+i$,
11. $N \leftarrow N+I$, $i \in \{1,2,3,\ldots,k\}$
12. *Update* $P_p \leftarrow$ *Predicted path,* $(x_r,y_r)_o \forall\ O_i$
13. *End Loop*
14. *Compute Final Displacement Error:*
15. $x_e \leftarrow x_i - x_n$
16. $y_e \leftarrow y_i - y_n$

5 Results and Discussion

As depicted below, Figs. 3, 4, 5 and 6 shows simulation results with Agent-based Probabilistic path prediction over Dynamic environment with moving obstacles using probabilistic function based on current position of robot, succeeding probable position of robot, current along with future position of obstacles while traversing along the collision-free path (Figs. 7 and 8).

6 Conclusion

In this paper, we have presented a model-based prediction algorithm for path planning in the dynamic environment. The prediction function for obstacle position prediction and object position estimation is also presented. We have also shown the simulation results of the proposed algorithm in a systematical manner. The results shown started

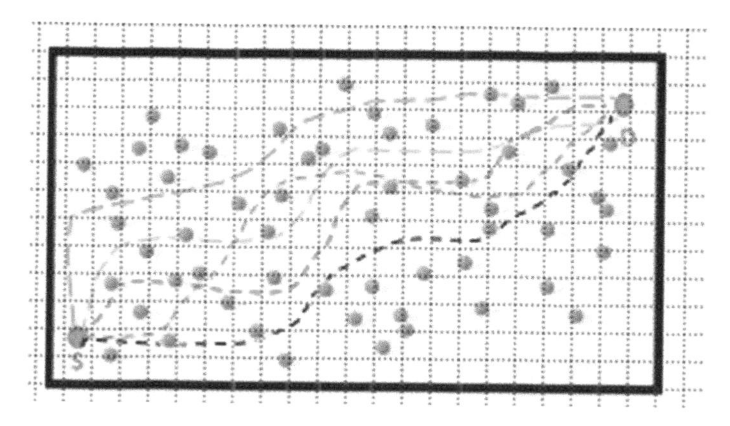

Fig. 3 Single fold algorithm with all possible path

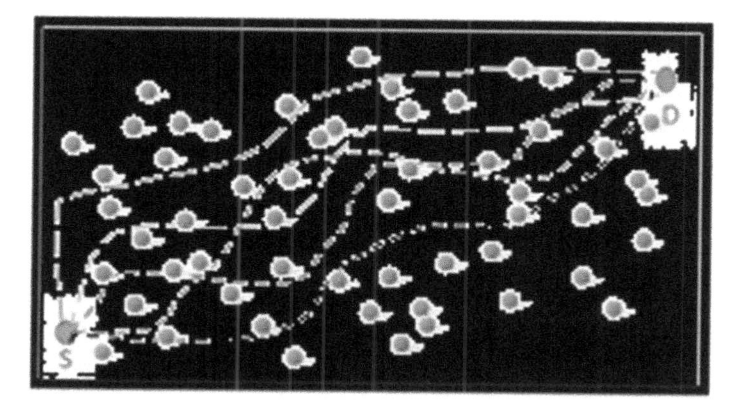

Fig. 4 Single fold algorithm showing all possible Pc (Candidate path) with $f_{\text{k}}(n)$

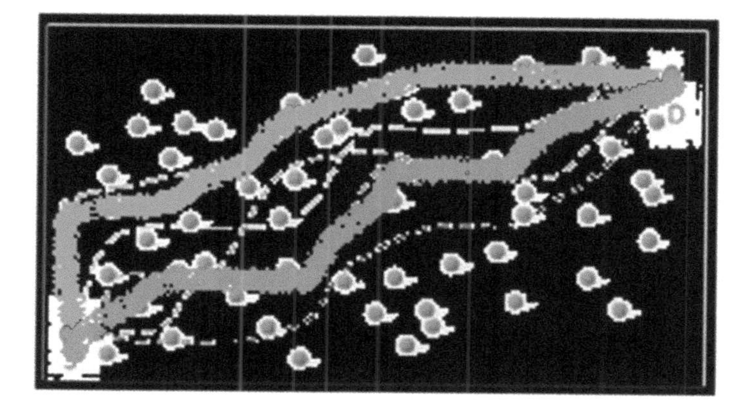

Fig. 5 Single fold algorithm showing top k possible Pc (Candidate path) with $f_{\text{k}}(n)$

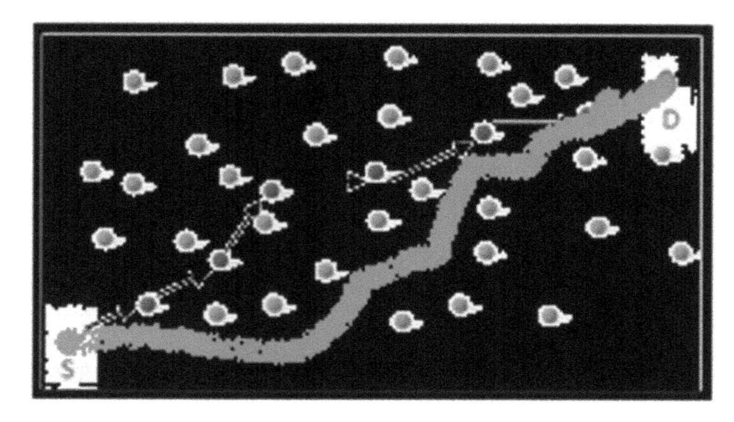

Fig. 6 Single fold algorithm showing Pp (predicted path) with $f_k(n)$

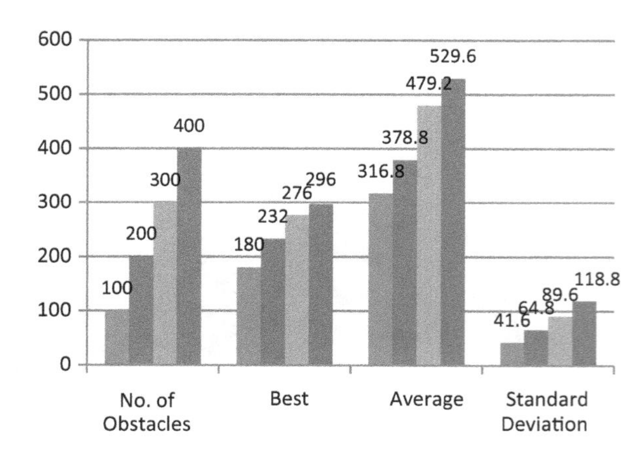

Fig. 7 Result in terms of path length (No collision)

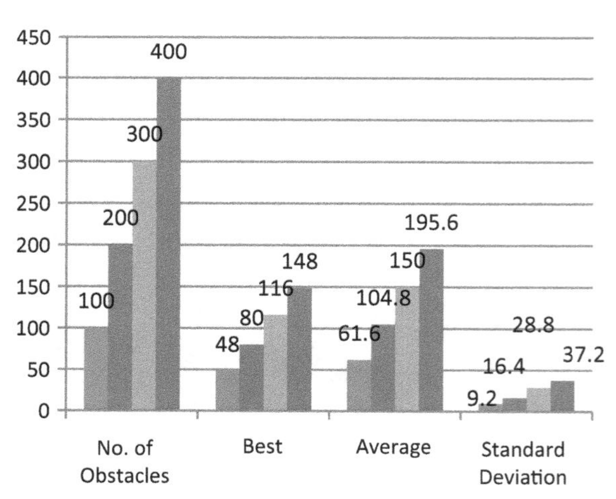

Fig. 8 Result in terms of time in ms (No collision)

with the possible paths, collision-free path, all the candidate paths without collision, top-k candidate path with less average prediction error and finally the predicated path having least prediction error is displayed. We have also shown the simulation results in the form of bar-charts representing the path length and time of traversal along the predicted path with no collision state in the dynamic environment.

References

1. Patle, B.K.: Intelligent navigational strategies for multiple wheeled mobile robots using artificial hybrid methodologies. Ph.D. thesis (2016)
2. Aoude, G.S., Luders, B.D., Joseph, J.M., Roy, N., How, J.P.: Probabilistically safe motion planning to avoid dynamic obstacles with uncertain motion patterns. Autonom. Rob. **35**(1), 51–76 (2013)
3. Alahi, A., Goel, K., Ramanathan, V., Robicquet, A., Fei-Fei, L., Savarese, S.: Social LSTM: Human trajectory prediction in crowded spaces. In: Proceedings of the IEEE Conference on Computer Vision and Pattern Recognition, pp. 961–971 (2016)
4. Arslan, O., Tsiotras, P.: Use of relaxation methods in sampling-based algorithms for optimal motion planning. In: 2013 IEEE International Conference on Robotics and Automation (ICRA). IEEE, pp. 2421–2428 (2013)
5. Chen, H., Mu, H., Zhu, Y.:Real-time generation of trapezoidal velocity profile for minimum energy consumption and zero residual vibration in servomotor systems. In: American Control Conference (ACC). IEEE, pp. 2223–2228 (2016)
6. Graves, A.:Generating sequences with recurrent neural networks. arXiv preprint arXiv:1308. 0850 (2013)
7. Otte, M., Frazzoli, W.: RRTX: real-time motion planning/replanning for environments with unpredictable obstacles. In: Algorithmic Foundations of Robotics XI. Springer, pp. 461–478 (2015)
8. Salzman, O., Halperin, D.: Asymptotically near-optimal RRT for fast, high-quality motion planning. IEEE Trans. Rob. **32**(3), 473–483 (2016)
9. Tran, H.K., Nguyen, T.N.: Flight motion controller design using genetic algorithm for a quadcopter. Meas Control **51**(3–4), 59–64 (2018)
10. Persson, S.M., Sharf, I.: Sampling-based A* algorithm for robot path-planning. Int. J. Robot. Res. **33**(13), 1683–1708 (2014)
11. Hacohen, S., Shoval, S., Shvalb, N.: Applying probability navigation function in dynamic uncertain environments. Robot Auton. Syst. **87**, 237–246 (2017)

Li-ion Battery Health Estimation Based on Battery Internal Impedance Measurement

S. Hemavathi

Abstract The diagnosis of Li-ion battery degradation is vital in electric vehicle applications for the reliable and secure operation of Li-ion batteries. This can be accomplished by measuring the two important parameters such as battery capacity and internal impedance of the battery cells over the entire lifetime of the battery. To recognize the batteries end of life (EOL), the capacity and internal impedance are principally used to estimate the battery state of health (SOH) during different conditions. In this research article, the battery SOH estimation technique is presented based on internal impedance measurement over the whole battery aging cycles. Firstly, the equivalent circuit model of Li-ion battery is developed using the charge–discharge test and electrochemical impedance spectroscopy (EIS) test. This model contains battery dynamic characteristics and parameters depending on the degradation of the battery cell. Secondly, the model parameters are identified to evaluate the internal impedance of each battery life cycle. Finally, based on internal impedance, the state of health estimator is framed and applied to measure the SOH of 18650 Li-ion battery cell for different aging cycles. The results show that the estimation of battery SOH based on internal impedance is more rapid and more reliable for E-vehicle applications.

Keywords Battery capacity · Internal impedance · Li-ion battery · State of health

1 Introduction

At present, the world faces progressively serious ecological issues, including global carbon emissions due to industrial and automotive pollutions. To determine these issues, the world move away from fuel operated vehicles and toward progressively sustainable power sources Electric Vehicles [1–4]. The advancement of appropriate

S. Hemavathi (✉)
Battery Division, Central Electrochemical Research Institute, Chennai, India
e-mail: hemavathi@cecri.res.in

Academy of Scientific and Innovative Research (AcSIR), Ghaziabad 201002, India

P. Muthukumar et al. (eds.), *Innovations in Sustainable Energy and Technology*,
Advances in Sustainability Science and Technology,
https://doi.org/10.1007/978-981-16-1119-3_17

power batteries is essential in EVs for energy storage, and Li-ion batteries are considered as the most ideal decision in EVs because of their high energy density, high cycle life, ability of fast charge and low self-discharge [5, 6]. So as to ensure the acknowledgment of these functions, a battery management system (BMS) is vital. The primary function of BMS is state estimation, cell equalization, protection control and monitor, and thermal management. One of the most important functions is to display the battery state of health (SOH) which denotes the real aging state of the battery and end of lifetime. Thus, the battery state of health can be predictable with the help of the battery capacity or battery internal impedance [7, 8].

The internal impedance of a battery cell is the main parameter that is affected by battery degradation [9]. Battery modeling is used to predict the battery parameters according to various criteria such as electrochemical model, electrical model, statistical model and analytical model [10].

In this article, the electrical equivalent circuit model is developed from electrochemical impedance spectroscopy test which is applied in Li-ion battery cell life cycle experiment. In this method, the equivalent circuit components values are measured mathematically and determine the internal impedance of the battery cell for all the battery aging cycles [11, 12]. Based on the internal impedance, the SOH of battery is estimated during operation of the cell, and results are formulated which provide the battery health state in a reliable and accurate manner [13].

The research paper is structured as follows: Sect. 2 describes the procedure to measure a Li-ion battery internal impedance. In Sect. 3, a real-time experiment is carried out which describes parameters that are related to battery degradation. Section 4 provides the estimation method for SOH measurement based on internal impedance and estimation results.

2 Internal Impedance Measurement

2.1 Theory of Internal Impedance

Real monitored internal impedance with different battery aging drives an approach to SOH estimation of Li-ion battery. As aging process happens step by step, the impedance will be changed under various frequencies. Hence, electrochemical impedance spectroscopy (EIS) is the best method, in which an AC excitation current is applied and relating AC voltage is recorded which helps to measure the real or actual internal impedance of the battery pack. EIS is a frequency response analysis of the electrochemical system which describes the electrochemical reactions and finds performance variation of battery degradation in nondestructive way. The impedance fractional order circuit can be obtained from relating EIS and battery characteristic test data.

This EIS method has different frequency of sine waves with specific amplitude to quantity resistance over a wide range of frequencies such as 1 MHz to100 kHz

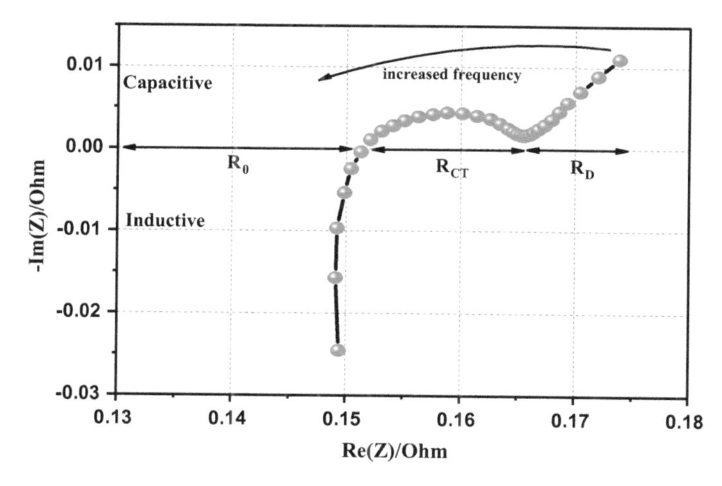

Fig. 1 Frequency response plot of Li-ion battery internal impedance

as shown in Fig. 1. Generally, the outcomes from the EIS are illustrated using a frequency response plot that is Nyquist plot with a negative *y*-axis, since most electrochemical models show gradually capacitive manners. This approach is quick and capable of distinguishing the starting point of the degradation process and aging impacts that conventional test does not perceive [14].

Clearly, the entire impedance spectroscopy comprises three primary sections such as low-frequency section, middle frequency section and high-frequency section. Low-frequency section of the Nyquist plot looks like a straight line with a positive slope and is represented by Warburg impedance, a typical phenomenon of the diffusion process. The middle frequency section looks like a semi-circle and is represented by charge transfer impedance, a typical phenomenon of the charge transfer process in electrode surfaces. The high-frequency section looks like a small curve and intersects with the abscissa axis which is represented by ohmic resistance and battery inductance caused by the porosity of electrode plated and cell geometry.

In the Nyquist plot, the high-frequency region is inductive dominated behavior which is indicated by a negative value of $-\mathrm{Im}(Z)$, and the middle frequency region is capacitive dominated behavior which is denoted as a positive value of $-\mathrm{Im}(Z)$. When $\mathrm{Im}(Z) = 0$, both capacitive and inductive observes are adjusted; this point is commonly associated with the pure ohmic resistance R_o. The contrast between the Re(Z) at min {− Im(Z)} and R_o relates to the charge transfer resistance R_{CT}. The diffusion resistance R_D is located in the low-frequency region.

The proposed structure for the time-domain equivalent circuit model of the internal impedance of the Li-ion battery is shown in Fig. 2. Where L and R_o are the battery inductance and ohmic resistance at high-frequency section, R_{CT} and CPE_1 are charge transfer resistance and constant phase element which is connected in parallel at middle frequency region, and CPE_2 is constant phase element instead of Warburg impedance at low-frequency region.

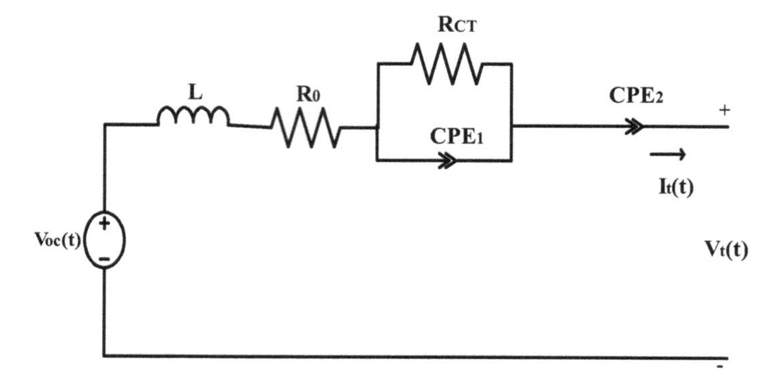

Fig. 2 Equivalent circuit model of Li-ion battery

The impedance of constant phase element can be expressed as

$$Z_{\mathrm{CPE}}(s) = \frac{1}{Qs^{\alpha}} \tag{1}$$

where Z_{CPE} is the impedance of constant phase element, Q and α are the fractional coefficient and fractional order, which belong to $0 \leq \alpha \leq 1$. It is expressed that the constant phase element (CPE) is resistance when $\alpha = 0$ and capacitance when $\alpha = 1$.

The impedance of the entire equivalent circuit is expressed as

$$Z(s) = sL + R_{\mathrm{o}} + \frac{R_{\mathrm{CT}}}{1 + R_{\mathrm{CT}} Q_1 s^{\alpha}} + \frac{1}{Q_2 s^{\beta}} \tag{2}$$

where Z is the entire equivalent circuit impedance, Q_1 and α are the fractional coefficient and fractional order of CPE_1, and Q_2 and β are the fractional coefficient and fractional order of CPE_2.

3 Aging Experimental Analysis

The proposed method is designed for state of health estimation of nickel manganese cobalt oxide (NMC) cathode chemistry Li-ion battery 18650 single cell for electric vehicle applications. Its specifications are listed in Table 1. The Li-ion battery test station is shown in Fig. 3, which contains 150 channels and measuring cell performance characteristics such as voltage, current, capacity, temperature and internal impedance. Additionally, the test station upheld the testing of various sizes and chemistries of cells at the typical and fast charging and discharging C-rates.

The battery capacity and internal impedance are important parameters to predict the battery lifetime by estimation of SOH. These two parameters are measured from

Table 1 Important parameters of NMC-based Li-ion battery cell

Parameters	Value
Nominal capacity	2600 mAh
Charging voltage	4.2 ± 0.05 V
Nominal voltage	3.7 V
Discharge cut-off voltage	2.75 V
Charging method	Constant current–constant voltage
Maximum charge current	2600 mA
Maximum discharge current	5200 mA
Operating temperature	Charge: 0–45 °C Discharge: −20 to 60 °C
Weight of cell	47 g
Dimension of cell	Diameter: 18.4 mm Height: 65 mm
Initial internal impedance	≤100 mΩ

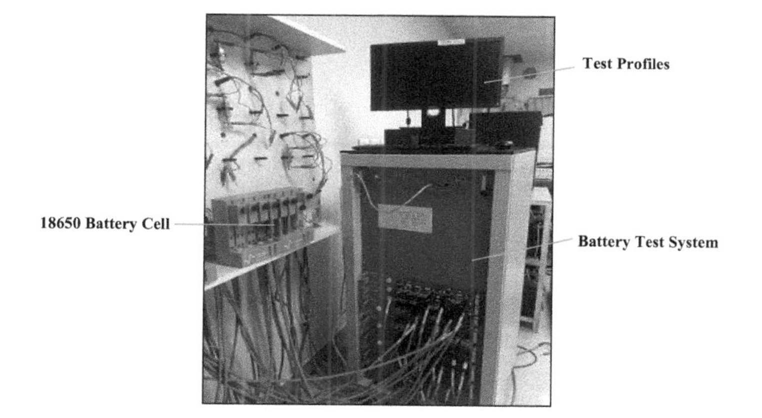

Fig. 3 Battery testing station

the aging experiment. From Fig. 3, battery characteristics are tested by performing the charge–discharge cycle test and EIS test for 100 cycles in C/2 rate at 25 °C temperature for aging process analysis. The testing outcomes and features are put away in PC with assistance of control software that is interconnected with test station [15, 16].

Charge–Discharge Test As per the datasheet, the constant current–constant voltage (CC_CV) charge and constant current (CC) discharge test are performed at C/2 rate up to 100 life cycles. The testing results obtained the battery characteristics such as voltage, current, temperature and battery capacity with respect to time. Based on these characteristics, the battery capacity is plotted with respect to the number of cycles as shown in Fig. 4. The battery capacity is degraded because of battery aging that notifies the battery health degradation.

EIS Test The electrochemical impedance spectroscopy (EIS) test is executed at each cycle to measure the internal impedance. This method is a frequency response analysis technique that measures the impedance at different frequencies as shown in Fig. 5. From the frequency analysis such as Nyquist plot, the internal impedance increases with an increase of cycle numbers such as 1, 31, 60 and 99. Therefore, the

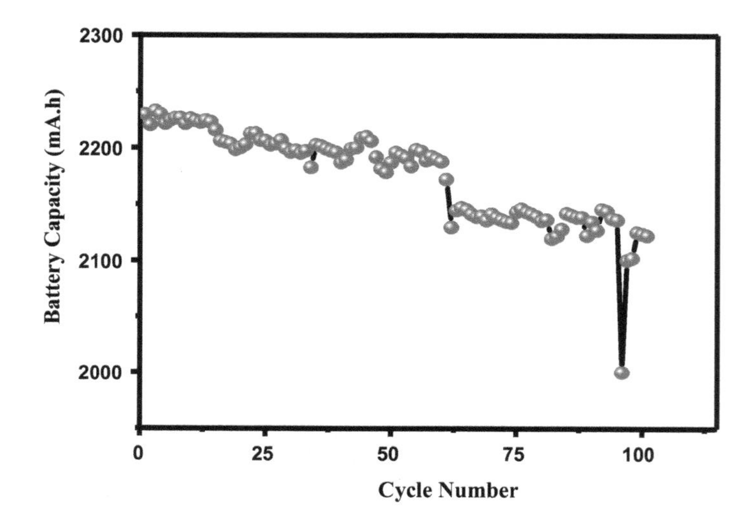

Fig. 4 Fade of battery capacity with different life cycles

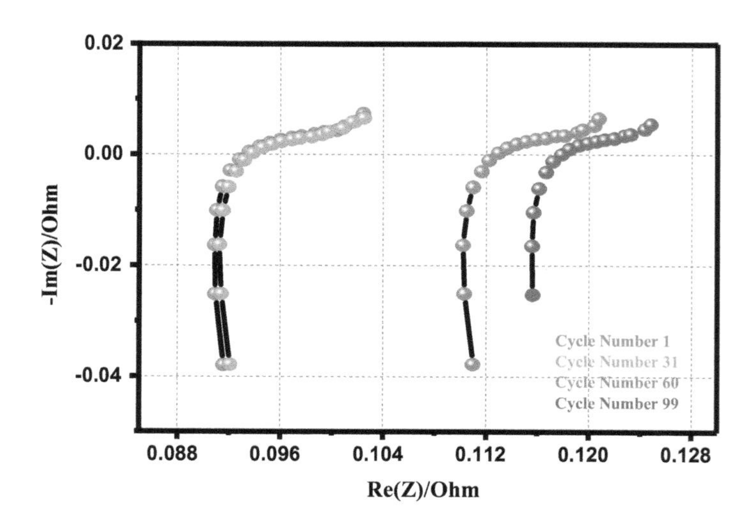

Fig. 5 Impedance frequency response plot with different life cycles

internal impedance of the battery increases because of battery aging that notifies the battery health degradation.

4 State of Health Estimation

State of health is a significant feature in Li-ion battery pack and cannot be estimated straightforwardly because of nonlinear attributes. The battery degradation causes due to the irreversible physical and chemical procedures and other electrochemical phenomena that happen inside the cell. This results in a decrease in battery capacity and an increase of battery internal impedance which changes the health of the battery. The general definition is expressed as

$$\%SOH = \frac{\text{Maximum available Battery Capacity}}{\text{Nominal Battery Capacity}} \times 100 \tag{3}$$

The battery is unfit for use when SOH drops below 80% because the battery capacity degrades exponentially after crossing 80% threshold.

SOH estimation is performed by different techniques. Most of the estimation techniques use the parameter of the internal impedance for battery degradation evaluation. Hence, the SOH should be determined by measurement of internal impedance during battery operation.

The estimation of state of health based on internal impedance is formulated as

$$\%\text{SOH} = \frac{Z_{\text{eol}} - Z_{\text{cur}}}{Z_{\text{eol}} - Z_{\text{new}}} \times 100 \tag{4}$$

where Z_{eol} is internal impedance at the end of battery life which is twice of the initial value, Z_{cur} is internal impedance at the current battery cycle, and Z_{new} is internal impedance at the start of battery life (new battery) which is determined in the datasheet. Therefore, the Z_{eol}, Z_{cur} and Z_{new} estimated precisely, and the SOH can be found with enough accuracy.

4.1 SOH Estimation Procedure Based on Internal Impedance

The battery SOH can be evaluated as per the working procedure as follows:

Step 1. At first, the charge–discharge test is performed for different battery aging cycles and identifies the battery parameters such as battery cell voltage, current and capacity with respect to the number of cycles.

Step 2. The EIS test is carried out for each cycle and determining the degradation of the battery, based on internal impedance at different frequencies.

Step 3. The equivalent circuit model of battery is developed, model parameters are identified, and the entire internal impedance of the battery cell for each cycle is determined using mathematical calculations of solving (2).

Step 4. Using the internal impedance of the current battery cycle, the SOH estimation based on internal impedance method is established using (4).

4.2 Estimation Results and Analysis

In the proposed method, the internal impedance of each aging cycle is measured from the EIS test which is observed in terms of frequency response plot as shown in Fig. 5. For time-domain analysis, the equivalent circuit model is developed as shown in Fig. 6, and model parameters are identified from the frequency response plot which helps to find the internal impedance of the current battery cycle by solving the mathematical Eq. (2). Based on this, the battery internal impedance is plotted with respect to the number of cycles as shown in Fig. 7. From this graph, the number of battery cycles increases will increase the battery internal impedance which results in battery health degradation.

When the battery internal impedance was obtained, those data were input to estimate the battery state of health using Eq. (4), and the estimated results are plotted in Fig. 8. This graph clearly shows that the state of health decreases over the aging of the Li-ion battery. When cycle is less than 60, the estimated SOH is in the range of 96–99% which indicates that the battery is in good condition as shown in Fig. 8. When the cycle number is more than or equal to 60, the State of Health of the battery drops to 80% or below, which indicates the battery is an unfit condition for that particular application and observed in Fig. 8. Therefore, this proposed method is suitable for electric vehicle applications.

The comparison results are shown in Fig. 9. The experimental outcomes recommend that degradation of battery capacity and increase of the internal impedance are

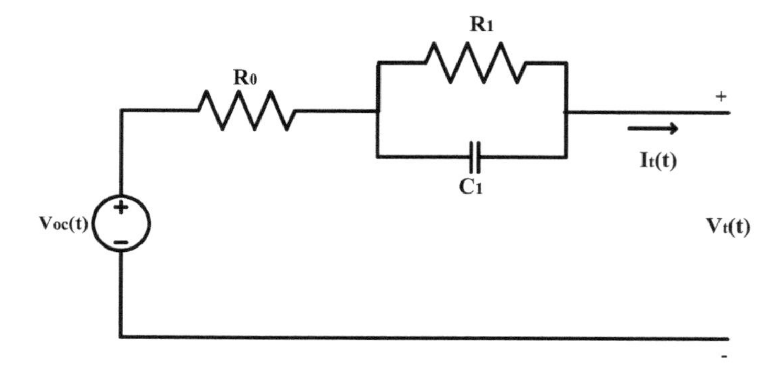

Fig. 6 Equivalent circuit of proposed Li-ion battery model

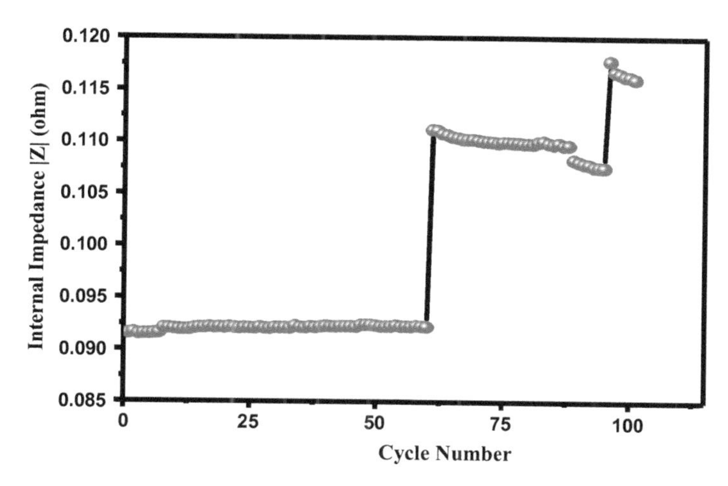

Fig. 7 Change of internal impedance with different life cycles

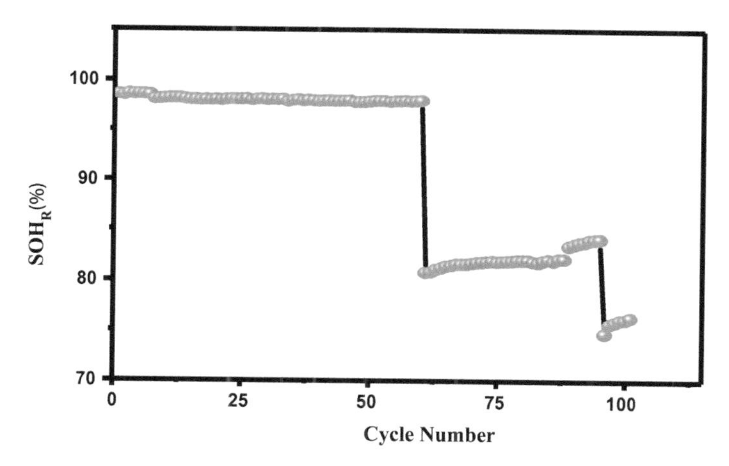

Fig. 8 SOH estimation of Li-ion battery with different aging cycles

over aging since they follow a similar development pattern. Such a linear relationship moreover suggests that they are probably accredited to a similar aging mechanism as shown in Fig. 9a. Hence, the determination of SOH using the proposed method is observed in Fig. 9b, c, which indicates the Li-ion battery health degradation caused by both the internal impedance increment and battery capacity fading.

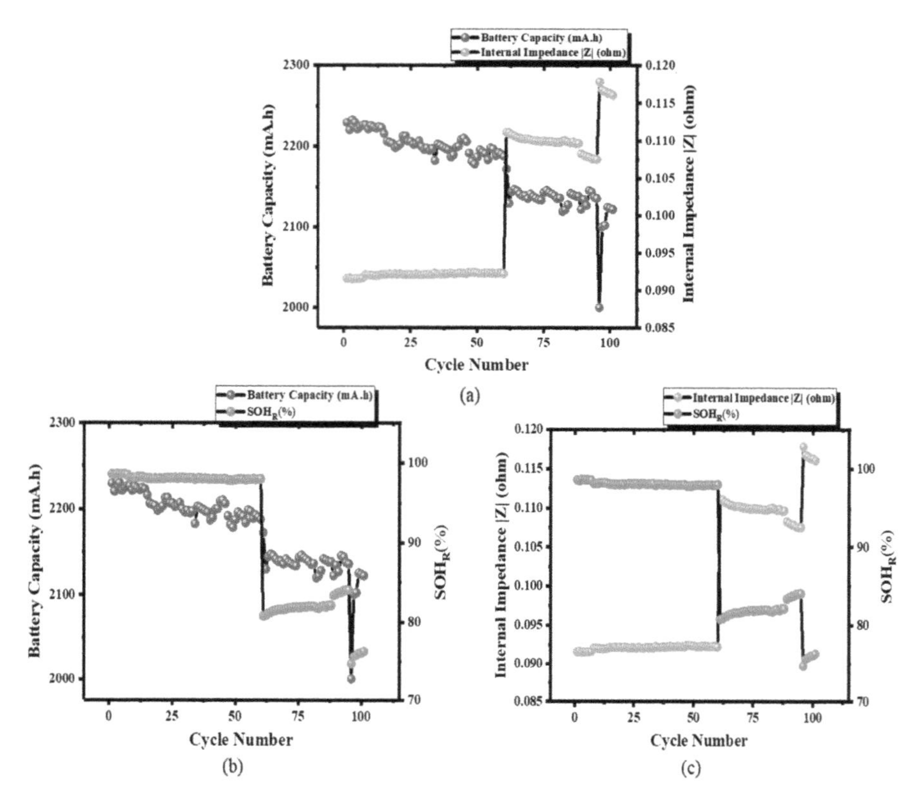

Fig. 9 Comparison results. **a** Battery capacity—internal impedance. **b** Battery capacity—battery SOH. **c** Internal impedance—battery SOH

5 Conclusion

In this article, an effective state of health estimation algorithm is developed for NMC chemistry-based 18650 Li-ion battery cell using cell internal impedance measurement. The charge–discharge test and EIS test are carried out on the battery cell over multiple aging cycles and analyzed battery dynamic characteristics and model parameters. The internal impedance is defined using equivalent circuit model of a proposed battery cell and calculated from identified model parameters. Furthermore, the method to estimate the SOH of the battery cell is formulated and calculated based on the internal impedance of the current, end of life and start of life battery cycles. The results show that the determination of SOH using the proposed method indicates the Li-ion battery health degradation caused by both the internal impedance increment and battery capacity fading. Therefore, the proposed method can provide an accurate and robust state of health estimation. CSIR-CECRI Manuscript Communication Number CECRI/PESVC/Pubs./2021-51.

Acknowledgements Author would like to thank the Research and development in battery division of Central Electrochemical Research Institute, Council of Scientific and Industrial Research in India, for the financial assistance. The work presented in this paper is a part of the Fast-Track Translation (FTT) Project MLP0304 "Design and Development of Indigenous Smart Battery Management System for Energy Storage and E-vehicle applications," funded by Council of Scientific and Industrial Research, India.

References

1. Chan, C.C.: An overview of electric vehicle technology. Proc. IEEE **81**, 1202–1213 (1993)
2. Zakaria, H., Hamid, M., Abdellatif, E.M., Imane, A.: Recent advancements and developments for electric vehicle technology. In: International Conference of Computer Science and Renewable Energies (July 2019)
3. Cheng, K.W.E.: Recent development on electric vehicles. In: 3rd International Conference on Power Electronics Systems and Applications (May 2009)
4. Zhou, X., Zou, L., Ma, Y., Gao, Z., Wu, Y., Yin, J., Xu, X.: The current research on electric vehicle. In: Chinese Control and Decision Conference (May 2016)
5. Suciu, G., Pasat, A.: Challenges and opportunities for batteries of electric vehicles. In: 10th International Symposium on Advanced Topics in Electrical Engineering, pp. 113–117 (Mar 2017)
6. Tiwari, A., Jaga, O. P.: Component selection for an electric vehicle: a review. In: International Conference on Computation of Power, Energy, Information and Communication (Mar 2017)
7. Mathew, M., Janhunen, S., Rashid, M., Long, F., Fowler, M.: Comparative analysis of lithium–ion battery resistance estimation techniques for battery management systems. Energies (2018)
8. Dai, H., Jiang, B., Wei, X.: Impedance characterization and modeling of lithium-ion batteries considering the internal temperature gradient. Energies (2018)
9. Perez, A., Benavides, M., Rozas, H., Seria, S., Orchard, M.: Guidelines for the characterization of the internal impedance of lithium-ion batteries in PHM algorithms. Intern. J. Prognostics Health Manage. (2018)
10. Barai, A., Uddin, K., Widanage, W.D., McGordon, A., Jennings, P.: A Study of the Influence of Measurement Timescale on Internal Resistance Characterization Methodologies for Lithium-Ion Cells. Springer Nature, Scientific Reports (2018)
11. Dian Wang, Yun Bao, Jianjun Shi.: Online Lithium-Ion Battery Internal Resistance Measurement Application in State-of-Charge Estimation using the Extended Kalman Filter. Energies (2017)
12. Christensen, A., Adebusuyi, A.: Using on-board electrochemical impedance spectroscopy in battery management systems. World Electr. Veh. J. **6**, 793–798 (2013)
13. Nisvo Ramadan, M., Pramana, B.A., Widayat, S.A., Amifia, L.K., Cahyadi, A., Wahyunggoro, O.: Compartive study between internal ohmic resistance and capacity for battery state of health estimation. J. Mechatron. Electr. Power Veh. Technol. **6**, 113–122 (2015)
14. Chen, L., Lu, Z., Lin, W., Li, J., Pan, H.: A new state-of-health estimation method for lithium-ion batteries through the intrinsic relationship between ohmic internal resistance and capacity. Measurement **116**, 586–595 (2018)
15. Remmlinger, J., Buchholz, M., Meiler, M., Bernreuter, P., Dietmayer, K.: State-of-health monitoring of lithium-ion batteries in electric vehicles by on-board internal resistance estimation. J. Power Sourc. **196**, 5357–5363 (2011)
16. Zhu, M., Hu, W., Kar, N.C.: The SOH estimation of $LiFePO_4$ battery based on internal resistance with grey markov chain. In: IEEE Transportation Electrification Conference and Expo (ITEC) (2016)

Modified Particle Swarm Optimization (MPSO)-Based Short-Term Hydro–Thermal–Wind Generation Scheduling Considering Uncertainty of Wind Energy

Sunil Kumar Choudhary and Santigopal Pain

Abstract Integration of renewable in hydro–thermal scheduling considering economic and environmental factors forms a multi-objective nonlinear optimization problem involving many equality and inequality constraints. Main objective of this problem is to minimize emission as well as generation cost on short-term basis maintaining all system constraints. In this research, a framework for hydro–thermal–wind generation scheduling (HTWGS) has been proposed using a modified particle swarm optimization (MPSO) algorithm. Results showed that this algorithm provides better result while various complex constraints were considered in the HTWGS problem.

Keywords Hydro–thermal–wind scheduling · Modified particle swarm optimization

Nomenclature

i, j, k	Index of thermal, hydro, wind power unit, respectively
C_T, F_T, W_T	Total cost, fuel cost, and wind cost, respectively
N_t, N_h, N_w	Total number of thermal, hydro, wind units, respectively
τ, T	Time sub-interval and scheduling period, respectively
Up	Index of upstream reservoir
$Q_{hj,\tau}$, $I_{hj,\tau}$	Discharge and inflow rate of jth hydro unit τ, respectively
$P_{ti,\tau}$, $P_{hj,\tau}$, $P_{wk,\tau}$	Thermal, hydro and wind of ith, jth and kth at τ, respectively
$V_{hj,\tau}$, $S_{hj,\tau}$	Reservoir volume and spillage of jth hydro unit τ, respectively
$P_{d,\tau}$, $P_{L,\tau}$	Total demand and transmission loss at τ
$\mathrm{OEC}_{wk,\tau}$, $\mathrm{UEC}_{wk,\tau}$	Over and under estimation cost of kth wind at τ, respectively
α_i, β_i, γ_i, δ_i, ε_i	Emission coefficient of ith thermal unit

S. K. Choudhary (✉)
Dr. B.C. Roy Engineering College, Durgapur, West Bengal, India
e-mail: sunee.world@gmail.com

S. Pain
Haldia Institute of Technology, Haldia, West Bengal, India

P. Muthukumar et al. (eds.), *Innovations in Sustainable Energy and Technology*, Advances in Sustainability Science and Technology,
https://doi.org/10.1007/978-981-16-1119-3_18

a_i, b_i, c_i, d_i	Fuel cost coefficient of *i*th thermal unit
e_i, h_i	Coefficient of the valve point effect of *i*th thermal unit
$C_{(1-6)j}$	Hydro power output coefficient of *j*th hydro unit
$P_{ti}^{\min}, P_{ti}^{\max}$	Minimum and maximum power limit of *i*th thermal unit
$P_{hj}^{\min}, P_{hj}^{\max}$	Minimum and maximum power limit of *j*th hydro unit
$Q_{hj}^{\min}, Q_{hj}^{\max}$	Minimum and maximum discharge limit of *j*th hydro reservoir
$V_{hj}^{\min}, V_{hj}^{\max}$	Minimum and maximum volume limit of *j*th hydro reservoir
$V_{hj}^{\min}, V_{hj}^{\max}$	Minimum and maximum volume limit of *j*th hydro reservoir
$V_{hj}^{\text{begin}}, V_{hj}^{\text{end}}$	Initial and final storage volume of *j*th hydro reservoir.

1 Introduction

In recent times, global warming has become a matter of great concern due to increase in power demand involving more pollution. To address this problem, an optimum operation of a thermal-renewable energy mixture is a promising option.

Inclusion of solar and wind energy into the energy sector has been proved as more cost-effective, necessitating its enclosure in the scheduling progression.

Genetic algorithm (GA) gives satisfactory results in various areas such as optimal solution of scheduling problem [1–3], hydro generator governor tuning [4], and economic dispatch [5]. However, in the literature, different classes of empirical algorithms such as genetic algorithm (GA) approach based on differential evolution (DV) [6, 7], particle swarm optimization (PSO) [8], modified dynamic neighborhood learning-based particle swarm optimization (PSO) [9], simulated annealing (SA) [10], evolutionary programming (EP) [11], modified differential evaluation (MDE) [12], and some other population-based optimization techniques have proved their effectiveness particularly in solving short-term hydro–thermal scheduling (STHTS) problems. Recently, random optimization methodologies and many other empirical algorithms based on natural phenomenon like adaptive chaotic artificial bee colony (ACABC) algorithm, artificial immune system (AIS) have given better result in solving STHTS problem. Recently, many other empirical algorithms inspired by natural phenomenon and random optimization methodology [13, 14] have been applied successfully in ST-HTWS problems.

Derived from the relationship among uncertainty budget of renewable energy, number of intermittent power supplies and upper bound of constraints-violating probability of spinning reserve capacity, the uncertain-budget decision is guided, and the blindness of decision can be reduced. The computational steps of MPSO, the contradiction between optimization depth and velocity generally existed in swarm intelligence evolutionary algorithms. The proposed technique in the test systems and its simulation results are discussed and summarized in the conclusions in this paper.

2 Mathematical Formulation of Generation Scheduling

This article demonstrates the scheduling formulation of a hydro–thermal–wind generation scheduling problem considering various economics and environmental factors. Due to its impulsive nature, renewable resources make the generation scheduling problem more challenging.

2.1 *Formulation of Multi-objective Function*

Cost involved in a hydro system is independent of its output, and hence, in the proposed HTWS scheduling, overall generation cost involves coal cost involved in thermal plant along with miscellaneous cost involved in solar and wind power.

The optimization involved in this problem is minimization of the generation cost of thermal, wind, and solar power plants along with maintaining minimum emission by considering different constraints involved in the proposed scheduling.

To achieve this, a nonlinear multi-objective function can be mathematically formulated as follows:

$$\text{Min}\ldots C_T(F_T, E_T, W_T) \tag{1}$$

$$\text{Min}\ldots C_T = \sum_{\tau=1}^{T}\left(\sum_{i=1}^{N_t}\left(P_{ti,\tau} + E_{i\tau}\right) + \sum_{k=1}^{N_w}\left(P_{wk,\tau}C_{wk} + OEC_{wk,\tau} + UEC_{wk,\tau}\right)\right) \tag{2}$$

The hydro units power output is expressed as a function of reservoir volume and head given by

$$P_{hj,\tau} = \left(c_{1j}V_{hj,\tau}^2 + c_{2j}Q_{hj,\tau}^2 + c_{3j}V_{hj,\tau}Q_{hj,\tau} + c_{4j}V_{hj,\tau} + c_{5j}Q_{hj,\tau} + c_{6j}\right) \tag{3}$$

Thus, the multi-objective function (1) can be modified as

$$\text{Minimize } C_T(F_T + h_i \times E_T + W_T) \tag{4}$$

Fuel cost of the thermal power plant can be expressed mathematically as a quadratic function of the real power output including valve point effects [2]. This can be mathematically formulated as follows:

$$F_T = \sum\nolimits_{\tau=1}^{T}\left(\sum\nolimits_{i=1}^{N_t}\left[a_iP_{ti,\tau}^2 + b_iP_{ti,\tau} + c_i + \left|e_i\sin\left(f_i\left(P_{ti}^{\min} - P_{ti,\tau}\right)\right)\right|\right]\right) \tag{5}$$

Emission from thermal power plant depends on its output by the penalty factor h_i. Overall emission of pollutant E_T can be expressed mathematically as

$$E_T = \sum_{\tau=1}^{T}\left(\sum_{i=1}^{N_t}\left[\alpha_i P_{ti.\tau}^2 + \beta_i P_{ti,\tau} + \gamma_i + \varepsilon_i \exp\left(\delta_i P_{ti,\tau}\right)\right]\right) \text{ lb/h} \tag{6}$$

Wind velocity is the deterministic factor for wind power generation. Total operating cost for a wind extraction unit consists of three components: (a) direct cost, (b) underestimation cost, and (c) overestimation cost [46]. The concerned cost function can be formulated mathematically as

$$W_T = \sum_{\tau=1}^{T}\left(\sum_{k=1}^{N_w} C_{wk} P_{wk,\tau} + \text{OEC}_{wk,\tau} + \text{UEC}_{wk,\tau}\right) \tag{7}$$

2.2 Constraints

Constraints related to the proposed HTWS problem mainly are generator capacity (operating limits), storage volume of the reservoir, discharge limit, power balance, and water balance constraints.

Dynamic water balance equation of the reservoir can be written as

$$V_{hj,\tau} = V_{hj,\tau-1} + I_{hj,\tau} - Q_{hj,\tau} - S_{hj,\tau} + \sum\nolimits_{m}^{R_{uj}} Q_{hm(\tau-t_{mj})} + S_{hm(\tau-t_{mj})} \tag{8}$$

Initial and final reservoir storage volume is expressed as

$$V_{hj,0} = V_{hj,begin} \tag{9}$$

$$V_{hj,T} = V_{hj,end} \tag{10}$$

Thermal power unit generation limit is given as

$$P_{ti}^{\min} \leq P_{ti} \leq P_{ti}^{\max}\ (i = 1, 2, 3 \ldots N_t) \tag{11}$$

Hydro power unit generation limit is given as

$$P_{hj}^{\min} \leq P_{hj} \leq P_{hj}^{\max}\left(j = 1, 2, 3 \ldots N_j\right) \tag{12}$$

Wind power unit generation limit is given as

$$0 \leq P_{wk} \leq P_{wk}^{\text{rated}}\ (k = 1, 2, \ldots N_w) \tag{13}$$

Reservoir storage volume limit is given below

$$V_{hj,\tau}^{\min} \leq V_{hj,\tau} \leq V_{hj,\tau}^{\max} \tag{14}$$

Reservoir storage discharge limit is given below

$$Q_{hj,\tau}^{\min} \leq Q_{hj,\tau} \leq Q_{hj,\tau}^{\max} \tag{15}$$

Power system power balance constraint is given as

$$\sum_{i=1}^{N_t} P_{ti,\tau} + \sum_{j=1}^{N_h} P_{hj,\tau} + \sum_{k=1}^{N_w} P_{wk,\tau} = P_{D,\tau} + P_{L,\tau} \tag{16}$$

3 Modified Particle Swarm Optimization (MPSO) Algorithm

Conventional PSO maintains a random search considering random values in velocity equation for each particle. In such a case, calculation of velocity for each particle assigns different random values.

Whereas in modified particle swarm optimization (MPSO) algorithm, a unique random value is fixed to enhance individual searching (pbest) for the population in one iteration. Similarly, each particle is assigned with different random values during global search (gbest) of velocity equation. MPSO shows improved result for individual searching, thereby providing more optimal solutions.

According to MPSO, velocity update equation can be written as

$$v_k^{(r+1)} = C_f\Big[wt\, v_i^{(r)} + c_1 rand^{(r)}\Big(pbest_k - x_k^{(r)}\Big) + c_2 rand_k^{(r)}\Big(gbest_k - x_k^{(r)}\Big)\Big] \tag{17}$$

The computational steps of the MPSO methods are as follows:

Step 1: The algorithm starts with initialization of the particles. The initial velocity is generated for all the particles.

Step 2: Compute penalty factor for all thermal power plants.

Step 3: Calculate the hydro power plant's output, and apply the respective power inequality constraints.

Step 4: Compute the fuel cost and emission of thermal power plants.

Step 5: Compute the wind power generation cost.

Step 6: Calculate the fitness of the particles, considering all costs and equality constraints. Set the present value of each particle as its best position, *pbest*.

Step 7: Check for the lowest value of particle best position. Set the value as *gbest*.

Step 8: Calculate the updated velocity of each individual by Eq. (17)

Step 9: Update each individual position.

Step 10: Calculate the new fitness value for each particle. Replace the old *pbest* value with new one, if the present value shows improvement over the previous value.

Step 11: Replace the *gbest* with the lowest value from the new *pbest*, if the present value shows improvement over the previous value.

Step 12: Repeat step 8–11 until the equality constraints fall within a specified tolerance limits or maximum number of iterations reached.

Particle giving latest gbest value provides optimum schedule of generation.

4 Simulation and Test Results

In the present study, objective function is treated along with the penalty factor. In this analysis, maximum penalty factor approach is used as it offers an acceptable solution for the problem of emission and fuel cost.

Main problem involved in applying any heuristic algorithm is the parameter setting. Range selection for such parameters is considered by considering the concerned values in the literature, and then, a fine-tuning is carried out by a trial-and-error method.

Proposed test system for the present research involves four hydro plants, three thermal plants, and two wind plants. The schematic diagram is shown in Fig. 1.

The hourly water discharge from hydro plant is shown in Fig. 2. Hourly water discharge and reservoir storage volume are tabulated in Table 1. The storage volume limitation was addressed by adjusting the water discharge from each reservoir.

Optimal demand allocation for hydro–thermal–wind system and corresponding economic and emission values from the simulation are tabulated in the following tables. Optimal hydro–thermal–wind generation scheduling for the test system is depicted in Table 2. This analysis considers various economic and emission factors obtained from MPSO method.

The simulations were carried out in MATLAB 2019a platform for 50 iterations, and results were analyzed based on the best, average, and worst case with standard deviation. It is imperative to note that MPSO provides competent and effective solution from quality and consistency point of view.

Optimal load allocation among thermal, wind, and hydro system on daily basis is shown in Fig. 3. The comparison regarding total fuel cost is shown in Table 3. It is evident that MPSO provides a better the optimal generation schedule is shown in Fig. 4.

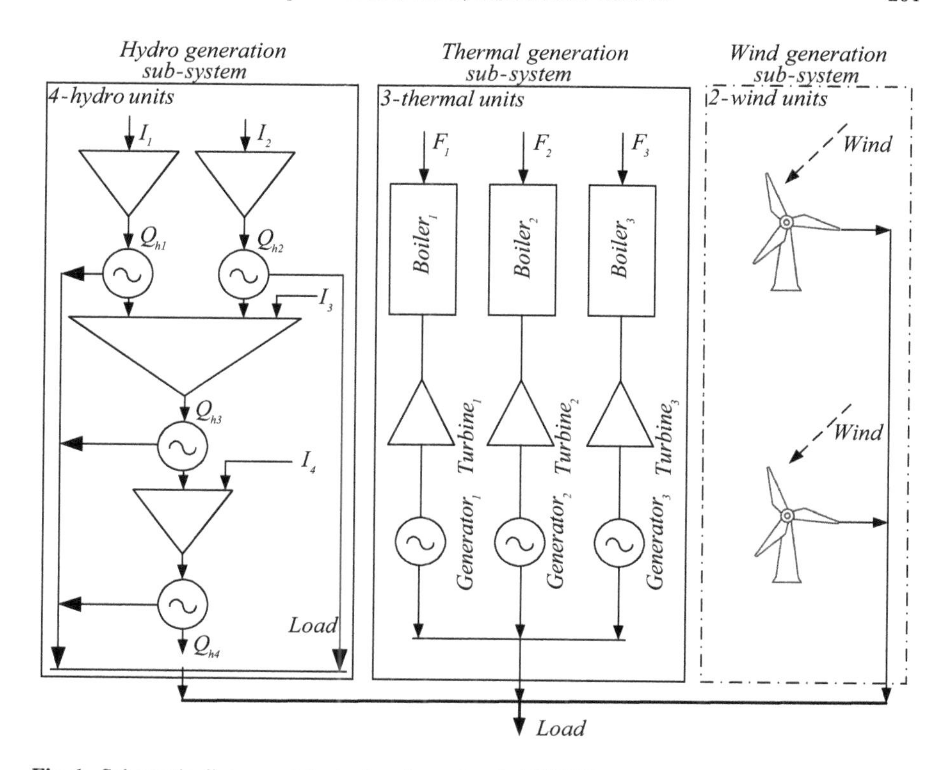

Fig. 1 Schematic diagram of the hydro–thermal–wind (HTW) test system

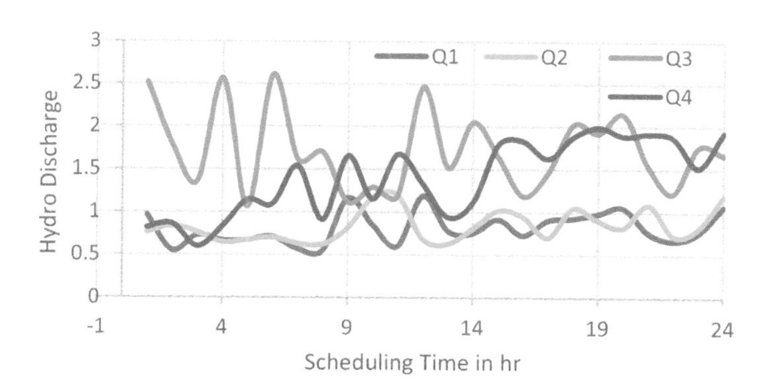

Fig. 2 Hydro plant water discharge curve of test system

5 Conclusion

Present study investigates the effectiveness of certain empirical algorithm belonging to different empirical groups for a solution of optimal generation of an HTWGS

Table 1 Hourly water discharge and reservoir storage volume obtained using the MPSO algorithm of test system

Hour	Water Discharge ($\times 10^5$ m^3/h)				Reservoir storage volume ($\times 10^6$ m^3)			
	Q_1	Q_2	Q_3	Q_4	V_1	V_2	V_3	V_4
1	0.9617	0.7633	2.5171	0.8202	1.0038	0.8037	1.5293	1.1460
2	0.5524	0.8347	1.7838	0.8641	1.0386	0.8002	1.4329	1.0836
3	0.7294	0.7643	1.3603	0.6003	1.0456	0.8138	1.4331	1.0395
4	0.6663	0.6470	2.5616	0.8300	1.0490	0.8391	1.3285	0.9565
5	0.6735	0.6756	1.0645	1.1442	1.0417	0.8515	1.4084	1.0938
6	0.7127	0.7036	2.6060	1.0924	1.0404	0.8512	1.3309	1.1630
7	0.5718	0.6349	1.6176	1.5409	1.0632	0.8477	1.3312	1.1449
8	0.5503	0.6239	1.6986	0.9114	1.0982	0.8553	1.3201	1.3099
9	1.1621	0.8146	1.1191	1.6544	1.0820	0.8538	1.3458	1.2509
10	0.8480	1.1697	1.2908	1.1552	1.1072	0.8269	1.3452	1.3960
11	0.6015	1.1994	1.1935	1.6738	1.1670	0.7969	1.4145	1.3904
12	1.1905	0.6861	2.4622	1.3302	1.1480	0.8083	1.3545	1.4272
13	0.7805	0.6280	1.5206	0.9385	1.1799	0.8255	1.4195	1.4453
14	0.7549	0.8099	2.0484	1.1162	1.2244	0.8345	1.4837	1.4628
15	0.9149	1.0200	1.6477	1.7839	1.2430	0.8225	1.4956	1.4037
16	0.7282	0.9450	1.1889	1.8279	1.2701	0.8080	1.5350	1.4672
17	0.9037	0.7044	1.4592	1.6237	1.2698	0.8076	1.5815	1.4569
18	0.9265	1.0479	2.0216	1.8794	1.2571	0.7628	1.5742	1.4738
19	0.9830	0.8998	1.9232	1.9921	1.2288	0.7428	1.5767	1.4393
20	1.0534	0.8214	2.1381	1.8888	1.1835	0.7407	1.5360	1.3693
21	0.7585	1.0850	1.5261	1.9170	1.1776	0.7222	1.6065	1.3390
22	0.6654	0.7240	1.2229	1.8673	1.1911	0.7398	1.6996	1.1337
23	0.7561	0.8097	1.7591	1.5235	1.2055	0.7388	1.6916	1.3789
24	1.0655	1.1878	1.6668	1.9274	1.2000	0.7000	1.7000	1.4000

system considering various environmental and economic factors. Modified particle swarm optimization (MPSO) method is proposed in this purpose. In the present analysis, maximum penalty factor approach was used as it converts the multi-objective economic and emission function into a single objective one. Simulations also verified that MPSO demonstrated a better performance than the other selected algorithm in terms of solution quality as well as consistency. The proposed method is very effective as it takes less time due to less computational steps involved in the analysis. Besides, the method is easy to implement which makes the algorithm suitable for addressing large-scale hydro–thermal–wind optimal scheduling problem.

Table 2 Optimal generation schedule of the hydro–thermal–wind (HTW) system obtained using the MPSO algorithm for test system

Hr.	Hydro Gen. (MW)				Thermal Gen. (MW)			Wind Gen (MW)	
	HP_1	HP_2	HP_3	HP_4	TP_1	TP_2	TP_3	WP_1	WP_2
1	84.311	60.018	8.9457	154.974	115.951	122.030	63.519	88.592	51.655
2	57.805	64.023	44.393	153.923	122.783	133.843	63.809	80.651	58.766
3	71.660	60.692	52.555	116.998	98.6265	128.984	50.935	88.748	30.799
4	67.146	54.541	0	136.567	98.6379	124.735	52.786	89.523	26.060
5	67.519	57.190	51.267	185.015	101.860	43.3162	52.042	62.125	49.663
6	70.355	59.023	0	187.139	107.191	196.366	54.469	89.761	35.691
7	59.899	54.187	45.476	225.293	102.717	217.345	124.70	78.330	42.039
8	58.628	53.839	42.981	181.227	157.493	143.593	248.55	72.739	50.946
9	95.421	66.088	50.103	245.835	167.217	201.579	131.43	89.024	43.296
10	80.923	81.495	50.351	216.458	159.862	219.653	142.90	78.652	49.699
11	63.842	80.568	52.327	262.537	102.230	209.194	209.06	79.162	41.068
12	98.491	55.386	5.3777	237.051	165.893	207.301	247.42	79.808	53.263
13	77.808	52.457	50.31	196.169	143.989	212.412	232.16	88.277	56.400
14	76.514	64.680	35.712	218.178	164.828	194.381	137.81	86.664	51.221
15	87.428	74.975	50.145	271.726	114.252	118.971	154.33	81.547	56.618
16	74.852	70.434	55.259	281.431	127.552	199.547	126.04	88.001	36.871
17	87.055	56.541	55.651	265.569	108.361	172.001	172.64	86.176	46.001
18	88.304	72.282	40.066	285.466	144.315	209.921	143.16	80.032	56.446
19	91.127	63.828	44.210	288.401	102.133	205.063	141.78	80.490	52.951
20	93.823	59.409	33.163	274.599	156.528	121.667	172.29	83.895	54.620
21	76.225	70.978	55.453	271.000	92.1141	212.701	56.250	74.990	0.2848
22	69.334	53.529	58.751	269.861	104.717	49.8435	114.99	89.471	49.491
23	76.402	58.617	52.832	249.585	98.5397	124.364	52.987	88.921	47.749
24	94.237	73.188	55.256	280.329	104.616	129.222	54.510	2.3243	6.3150

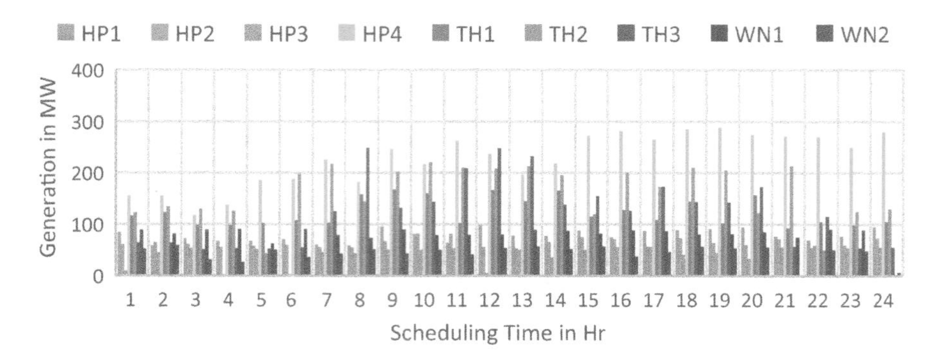

Fig. 3 Optimal power generation schedules from the MPSO algorithm over 24 h. time span of test system

Table 3 Statistical analysis of the heuristic algorithms in terms of total fuel cost

Method	Fuel cost ($/h)			
	Best	Average	Worst	Std. Dev.
MPSO	66083.66	66086.74	66089.37	1.6586
PSO	68646.80	68649.49	68652.15	1.6634
GA	71016.97	71021.02	71025.93	2.8973

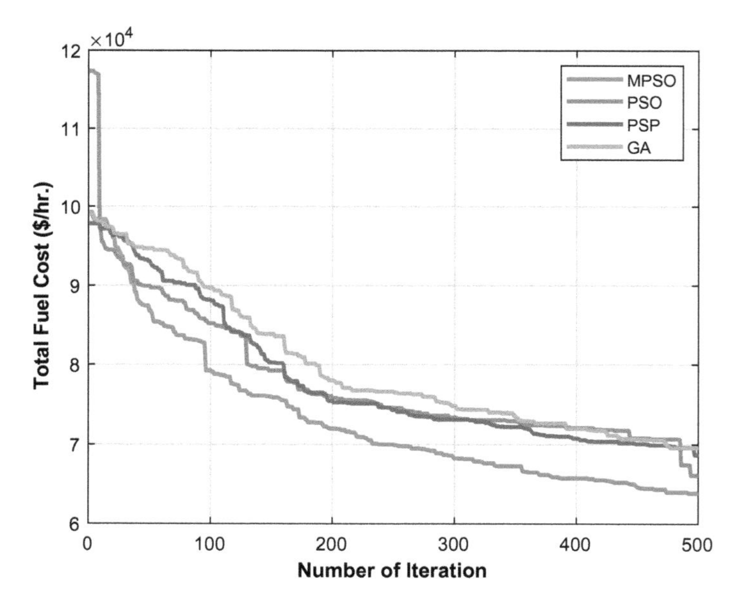

Fig. 4 Convergence characteristics of MPSO, PSO, PSP and GA algorithms in terms of total fuel cost of test system

References

1. Gil, E., Bustos, J., Rudnick, H.: Short-term hydro thermal generation scheduling model using a genetic algorithm. IEEE Trans. Power Syst. **18**(4), 1256–1264 (2003)
2. Kumar, S., Naresh, R.: Efficient real coded genetic algorithm to solve the non-convex hydrothermal scheduling problem. Int. J. Electr. Power Energy Syst. **29**(10), 738–747 (2007)
3. Lansberry, J.E., Wozniak, L.: Adaptive hydro generator governor tuning with a genetic algorithm. IEEE Trans. Energy Convers. **9**(1), 179–185 (1994)
4. Baskar, S., Subbaraj, P., Rao, M.V.C.: Hybrid real coded genetic algorithm solution to economic dispatch problem. Int. J. Comput. Electr. Eng. **29**(3), 407–419 (2003)
5. He, D., Wang, F., Mao, Z.: A hybrid genetic algorithm approach based on differential evolution for economic dispatch with valve-point effect. Int. J. Electr. Power Energy Syst. **30**(1), 31–38 (2008)
6. Zoumas, C.E., Bakirtzis, A.G., Theocharis, J.B., Petridis, V.: A genetic algorithm solution approach to the hydrothermal coordination problem. IEEE Trans. Power Syst. **19**(3), 1356–1364 (2004)
7. Zhang, J., Wang, J., Yue, C.: Small population-based particle swarm optimization for short-term hydrothermal scheduling. IEEE Trans. Power Syst. **27**(1), 142–152 (2012)
8. Rasoulzadeh-akhijahani, A., Mohammadi-ivatloo, B.: Short-term hydrothermal generation scheduling by a modified dynamic neighbourhood learning based particle swarm optimization. Int. J. Electr. Power Energy Syst. **67**(2), 350–367 (2015)
9. Wong, K.P., Wong, Y.W.: Short-term hydro thermal scheduling part. I Simulated annealing approach. IEE Proc. Gener. Transmiss. Distrib. **141**(5), 497–501 (1994)
10. Hota, P.K., Chakrabarti, R., Chattopadhyay, P.K.: Short term hydrothermal scheduling through evolutionary programming technique. Electr. Power Syst. Res. **52**(2), 189–196 (1999)
11. Lakshminarasimman, L., Subramanian, S.: Short-term scheduling of hydrothermal power system with cascaded reservoirs by using modified differential evolution. IEE Proc. Gener. Transmis. Distrib. **153**(6), 693–700 (2006)
12. Basu, M.: Artificial immune system for fixed head hydrothermal power system. Energy **36**(1), 606–612 (2011)
13. Cotia, B.P., Borges, C.L.T., Diniz, A.L.: Optimization of wind power generation to minimize operation costs in the daily scheduling of hydrothermal systems. Int. J. Electric. Power Energy Syst. **113**, 539–548 (2019)
14. Basu, M.: 'Optimal generation scheduling of fixed head hydrothermal system with demand-side management considering uncertainty and outage of renewable energy sources. In: IET Generation, Transmission & Distribution (2020)

Simulation Study on Effect of Fin Geometry on Solar Still

Begari Mary, Ajay Kumar Kaviti, and Akkala Siva Ram

Abstract Not only for humans, but also for all livelihood species on earth today, drinking water is a requirement. But the supply of drinking water is not ample; pure water is, therefore, insufficient in the universe today. Solar is also an appliance that can be used for desalination. Design modifications in solar stills can improve the productivity. One of the economical and best method integration of the fins on the still basin is used to maximize the heat transfer. But there is not much effort on the fins to increase the yield of solar still. The usage of solar fins still and how it can be used to increase solar distillate production still. Different geometries of fins like parabolic pin fin with sharp edge and blunt edge and truncated cone with the use of certain materials which improves the productivity have been studied and modeled using CATIA software. Analysis of base to check the maximum heat transfer of the base plate integrated with fins has been performed using ANSYS software and obtained that parabolic blunt edge fin and truncated cone have maximum rate of heat transfer.

Keywords Solar desalination · Parabolic fins · Truncated conic fins

1 Introduction

Potable water is a fundamental basic necessity of humanity and every living thing on earth requires a minimum of 20–50 L of natural, clean potable water per day to cook, drink, and other necessities. Drinking water ought to be unadulterated and sterile [1]. Otherwise individuals may experience the ill effects of different water-borne illnesses [2]. Desalination is one of the water decontamination processes viewed as most challenging task and the main feasible answer for determine crisp or consumable water from the accessible brackish water and saline water assets everywhere throughout the world [3]. Solar still is a device that uses the evaporation–condensation system to change over unclean saline water into the refined water by wiping out

B. Mary · A. K. Kaviti (✉) · A. S. Ram
VNRVJIET, Hyderabad 500090, India
e-mail: ajaykaviti@gmail.com

P. Muthukumar et al. (eds.), *Innovations in Sustainable Energy and Technology*, Advances in Sustainability Science and Technology,
https://doi.org/10.1007/978-981-16-1119-3_19

broke down salts just as other disintegrated debasements and suspended solids. The reason for a solar still is to catch this dissipated (or refined) water by consolidating it onto a cool surface [4].

As shown in Fig. 1, a solar still acts on two scientific concepts of evaporation and condensation. First with the dark foundation, the water that should be sanitized is set in the trough. The solar still is then allowed to remain in the light, allowing the solar radiation to be ingested by the still. As the energy is retained, as the water temperature increases and transforms into steam, it begins to heat the water and dissipates to the biased glass ceiling, leaving what is not pure in the water below in the trough [5]. It reaches the glass ceiling after the water begins to dissipate. The water on the glass gradually consolidates, causing uncontaminated beads of water. Because the glass is tilted down into the second trough, the droplets of water flow down into the trough of pure water. Since none of the minerals will evaporate with the uncontaminated water by microscopic organisms or different substances, the water beads that end up in the subsequent trough are basically refined, and now okay for drinking and cooking [6].

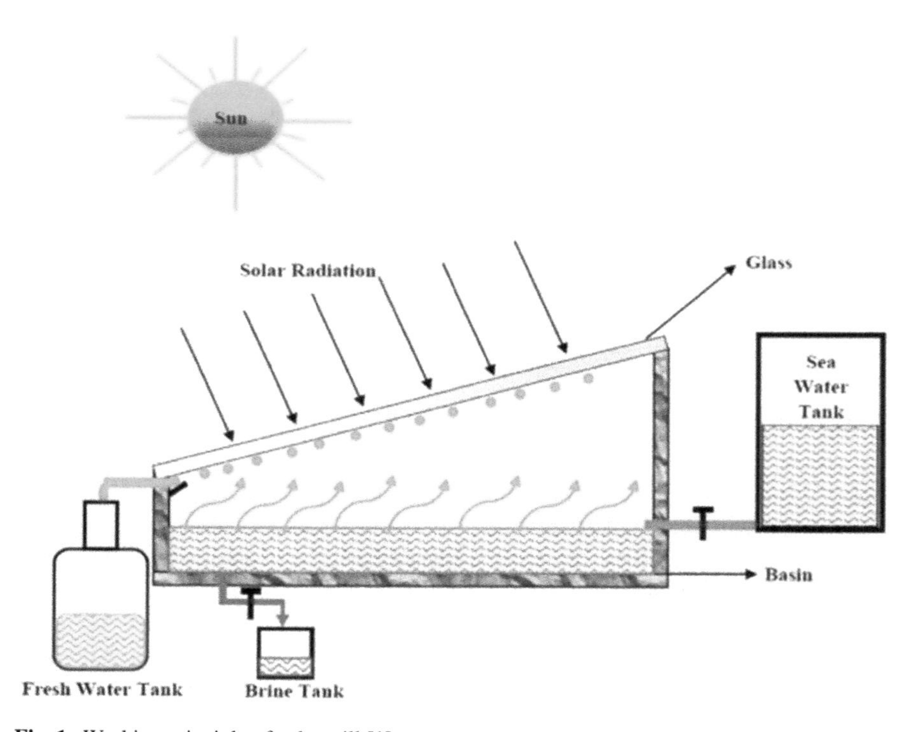

Fig. 1 Working principle of solar still [1]

2 Types of Fins Used in Solar Stills

Fins are surfaces that extended on an object in order to expand the rapid transfer of heat to or from the earth by enlarging convection. For applications such as solar thermal needs a fundamental basic parameter which also known as heat transfer [7]. The fundamental center is to examine the impact of various fin geometries like circular fins, rectangular fins, square fins, pin fins, and so on and is utilized in solar stills to improve the productivity of the water. As we know that productivity can be improved by design modifications, different geometries of fins are modeled and analyzed to get the optimum heat transfer [8].

Sebaii et al. [9] investigated the effect of finned basin solar still (FBLS) as described in Fig. 2. Despite everything utilizing FBL which are fabricated with various materials like aluminum, glass, Cu, mica, tempered steel, and saw that the balance material does not impact the efficiency. The normal efficiency for customary stills and FBLS is seen as 1467.4 and 1898.8 L/m^2. The cost of 1 L of new water rises without and with Cu fins as 0.31 and 0.28 (LE) for the stills. Productivity is improved by utilizing fins with the basin liner.

Sebaii et al. [10] fin plate solar was designed to build up the performance and heat transfer reliance rate of still efficiency and yield on fin design; the number of N_f fins, H_f height and X_f thickness were considered. The pitch rises to 0.06 and 0.12 m for 14 and 07 fins and finds that with enhancement in height of fin and decrease

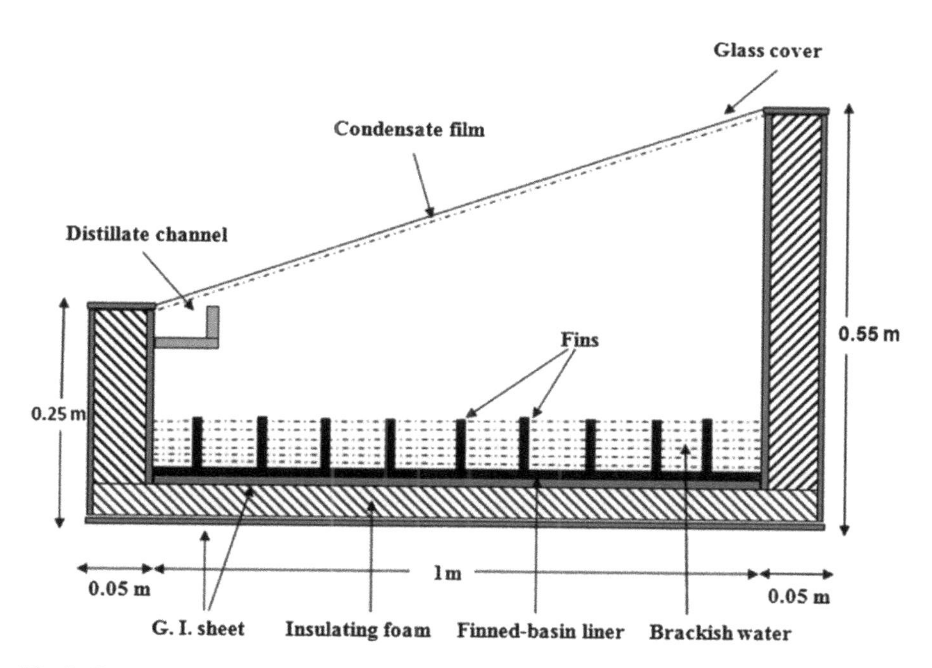

Fig. 2 Schematic representation of FBLS [9]

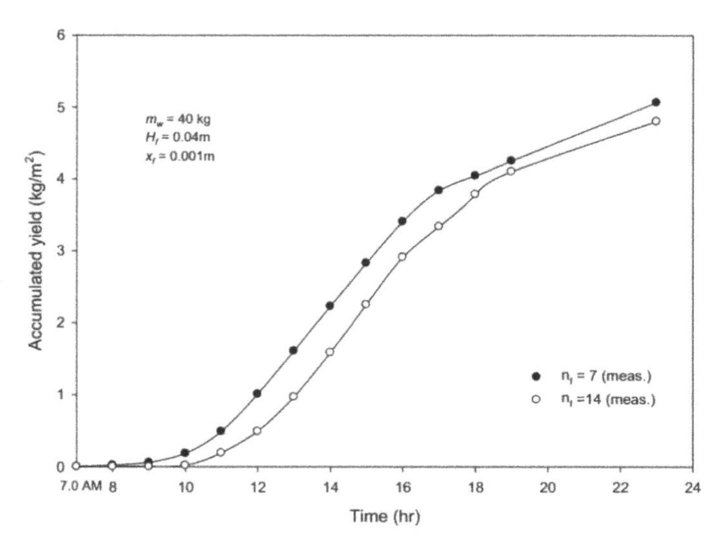

Fig. 3 Measured accumulated yield of FBLS [10]

with enhancement in thick of the fin and N_f, performance and efficiency expanded. By improving the heat transfer rate from water surface to the condensing surface leads to increasing in height of fin, efficiency can be increased, 13.7 percent greater efficiency is achieved with $N_f = 7$, $H_f = 0.04$ m and $X_f = 0.001$ m by coordinating fins. Figure 3 shows FBLS's cumulative yield.

Hardik et al. [11] investigated the hollow square fins fabricated with mild steel having side length of 25 mm, thick of 2 mm, and size of 20 mm. The fins were joined on separate basin liner which was also manufactured with mild steel material having dimensions of 300 mm × 600 mm × 2 mm apart as shown in Fig. 4. The most extreme refined water output as 0.9672 kg/m^2 day was achieved by square finned stills. For the 10 mm depth of water, the greater portion of heat was transferred to water from the basin liner, resulting in higher potable water as anticipated. Figure 5 explains the comparison of hourly variance for the various water depths.

Alaian et al. [12] analyzed the traditional and modified solar still with wick surface of pin–fin along with the evaporation rate, as shown in Fig. 6. The temperatures at the surfaces of the wick, glass, water lead to enhancing the still productivity that is demonstrated and found that as expected with the wick pin–fin still in contrast to the convectional still as depicted in Fig. 7, more than 23% of device productivity is achieved. Solar radiation fluctuation limits the increase to 11.53% in device productivity using wick surface.

Rajaseenivasan et al. [13] improved the yield output on a single solar basin with a circular and square fin inserted in the basin with various water depths such as 1, 2, 3, and 4 cm. Due to the higher exposure region, temperature of water in the fin still is consistently higher when compared to the normal still. Square fins have been found to have reached a 4.55 kg/m^3 day limit and ordinary fins have a 3.16 kg/m^3 day

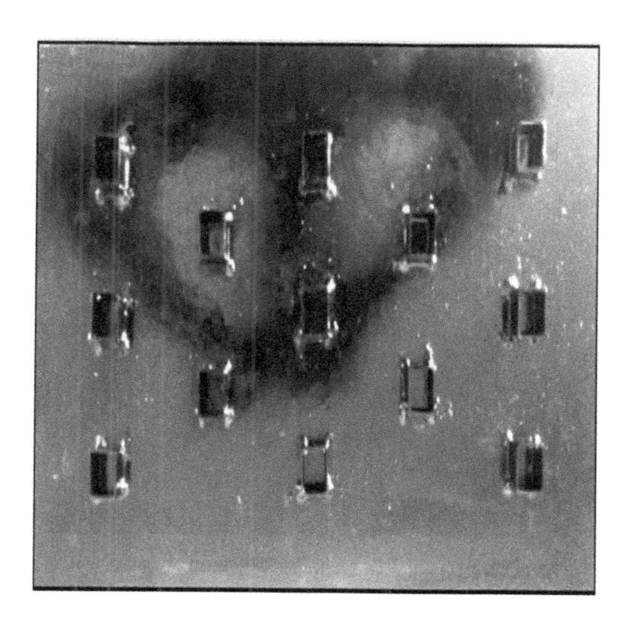

Fig. 4 Square finned absorber plate [11]

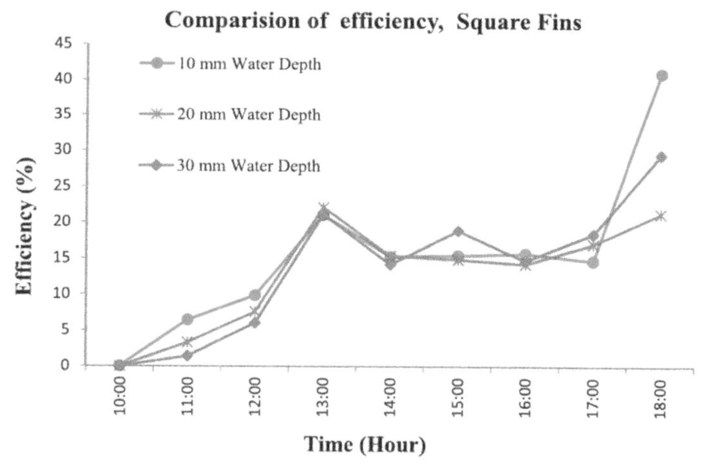

Fig. 5 Comparison of hourly variation [11]

limit. The full profitability of still enhanced is 36.7% with square fins, and it varies to 36–45.8% when fins are incorporated with the materials such as wick. Figure 8 describes about the square finned basin liner.

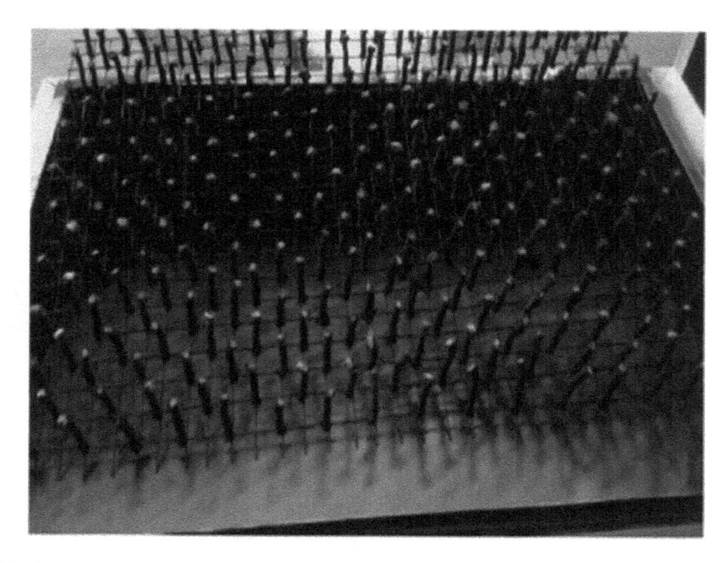

Fig. 6 Pin finned wick elements [12]

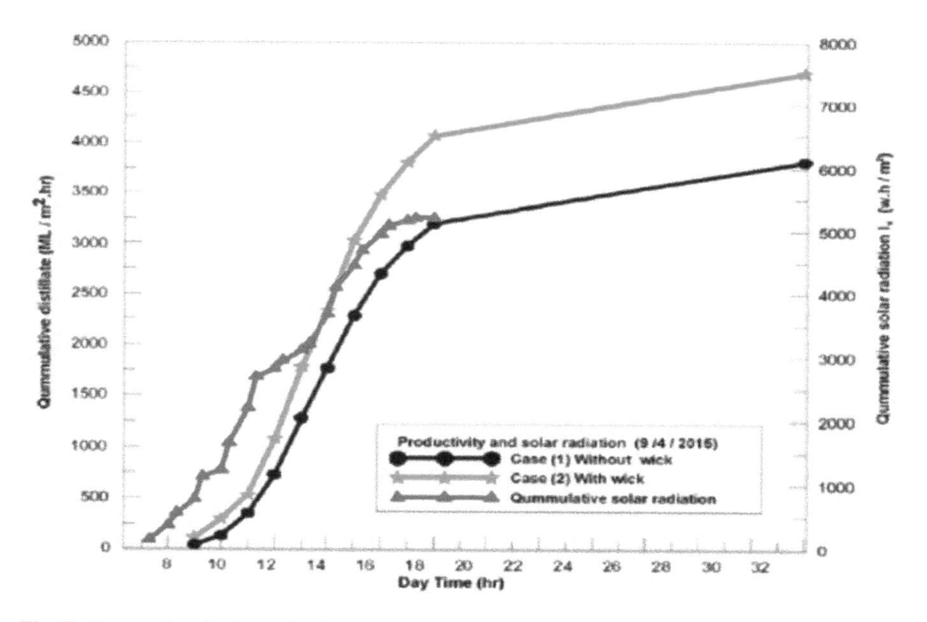

Fig. 7 Accumulated values of solar radiation [12]

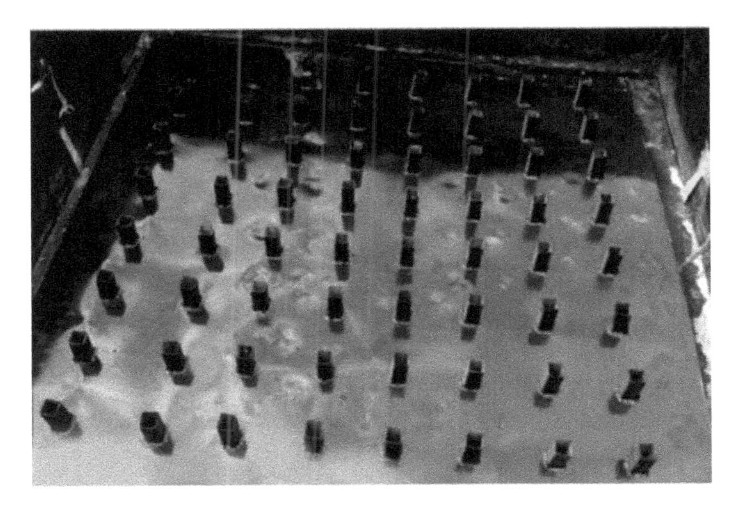

Fig. 8 Square finned basin [13]

2.1 Modeling

Modeling of the fins and solar stills were done in CATIA software. All the fin profiles were created in CATIA software. The tools used in the software are rectangle, line, spline, revolve, pad and array. The profile is created with respect to axis and revolved using revolve option. An array of fins was created and attached to the rectangular plate. After the plate is created, the double slope still is created. The glass part of the stills was made transparent. Different fin geometries with non-uniform cross section are used to improve the efficiency of the still like parabolic profile with sharp edge and blunt edge and truncated cone with change in thickness and height. Parabolic sharp edge pin fin with height and thickness of 20 × 50 mm and 30 × 50 mm as shown in Fig. 9, parabolic blunt edge pin fin with height and thickness of 20 × 50 mm and 30 × 50 mm as shown in Fig. 10, truncated cone pin fin with 20 × 50 mm and 30 × 50 mm were modeled as depicted in Fig. 11 and analyzed to check the heat transfer of the fins, i.e., array of fins as depicted in Fig. 12.

3 Analysis of Fins

Analysis of the fins was conducted to find the optimum heat transfer by using ANSYS workbench 19.2 software. The computer specifications are Intel(R) Xenon (R) CPU E3-1225 v5 @3.30 GHz as processor and 32 GB 2666 MHz DDR4 as RAM. Steady-state thermal analysis system was performed on six different plates of fins. And it is found that there is a very little temperature difference between all the plates that is 1–5 K. The heat transfer for the plate's ranges from 11.55 to 7.23 W. The maximum

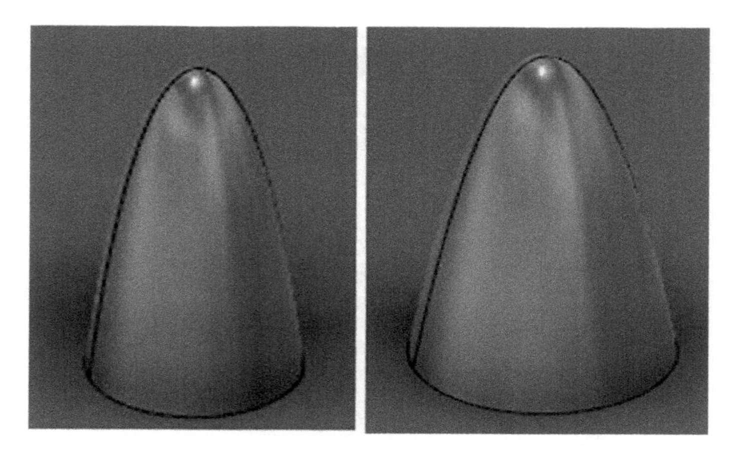

Fig. 9 Parabolic fins with sharp edge

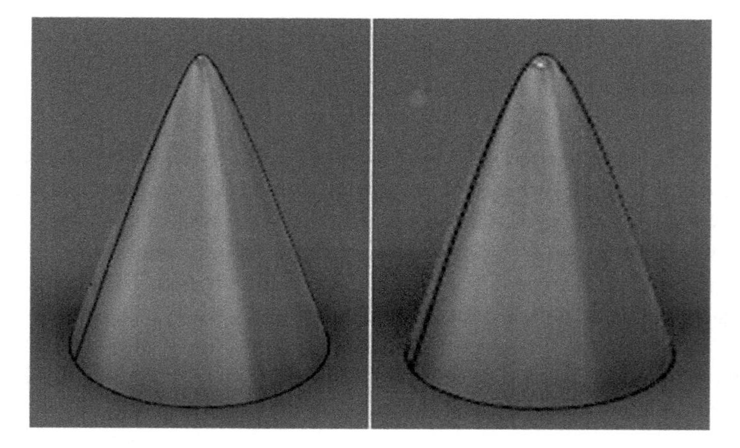

Fig. 10 Parabolic fins with blunt edge

heat transfer of 11.55 W has been obtained for parabolic blunt edge fin with 50 mm height and 30 mm thick, 10.95 W for truncated cone with 50 mm height and 20 mm thick. Figures 13, 14, and 15 describe about the heat transfer analysis of various fin profiles.

4 Conclusions

- Compared to traditional solar stills, pure water efficiency is improved by integrating fins on the basin. The rate of heat transfer from the surface of the basin liner to water was increased as fins were connected to the basin.

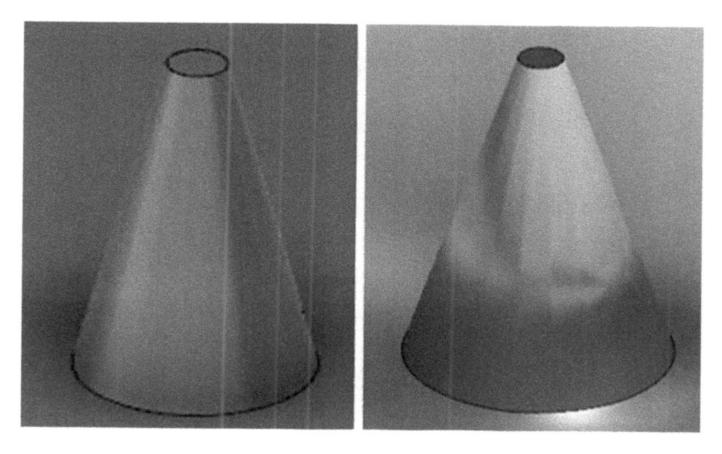

Fig. 11 Truncated conic fins

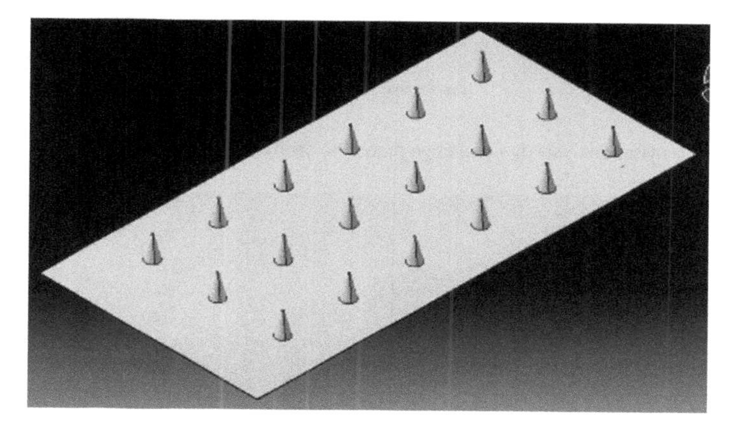

Fig. 12 Array of fins

- With increases in height of the fin and reduction in number of fins and thickness of the fin, performance and productivity have improved.
- Fins of parabolic profile with blunt edge and truncated conic profile is giving an optimum heat transfer.
- The maximum heat transfer has been obtained for parabolic blunt edge fin with 50 mm height and 30 mm thick, truncated cone with 50 mm height and 20 mm thick.

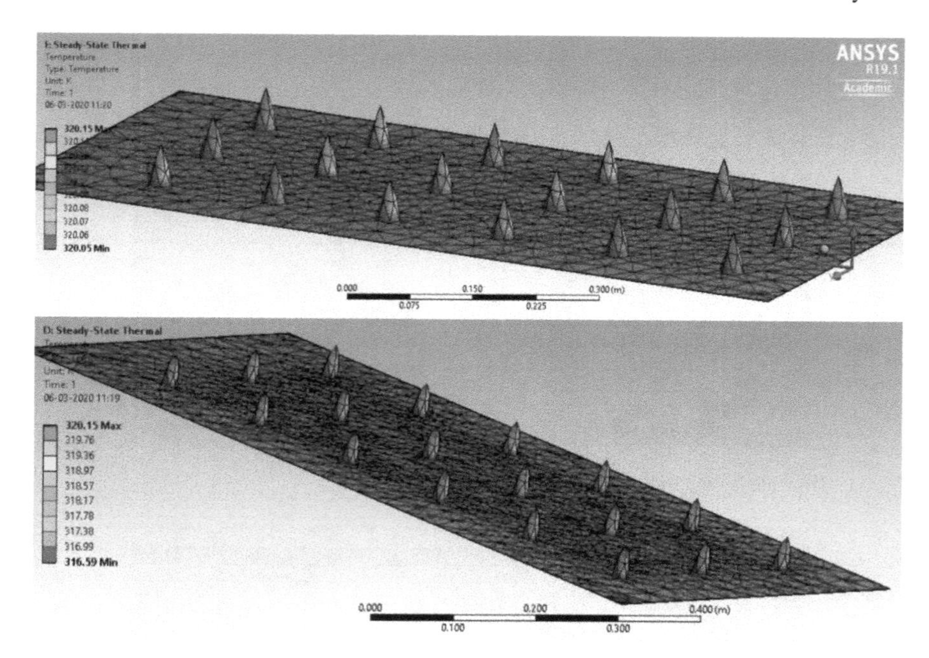

Fig. 13 Heat transfer analysis of sharp edge parabolic profile

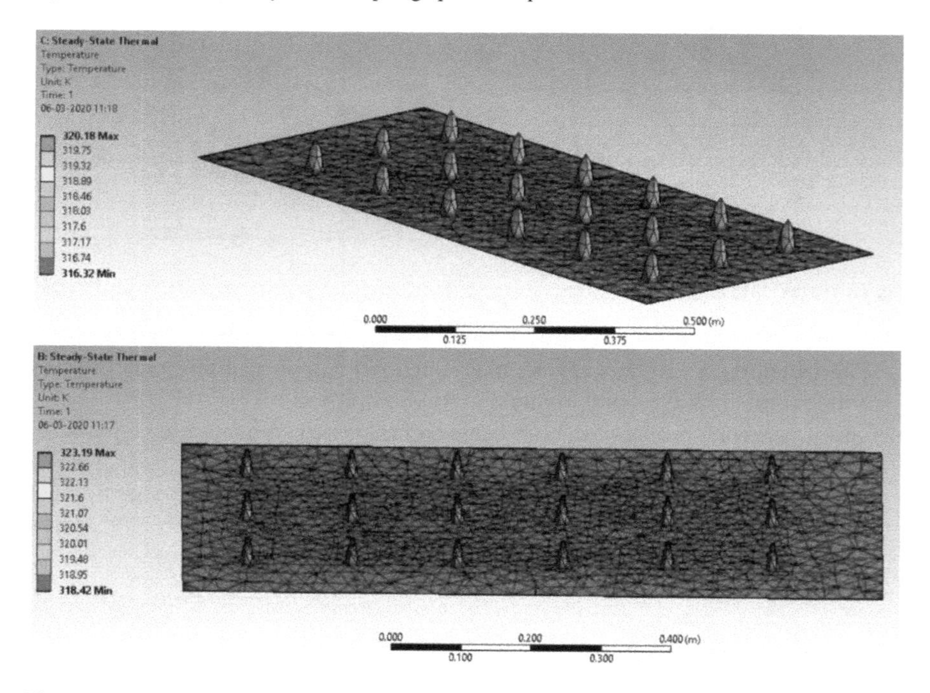

Fig. 14 Heat transfer analysis of blunt edge parabolic profile

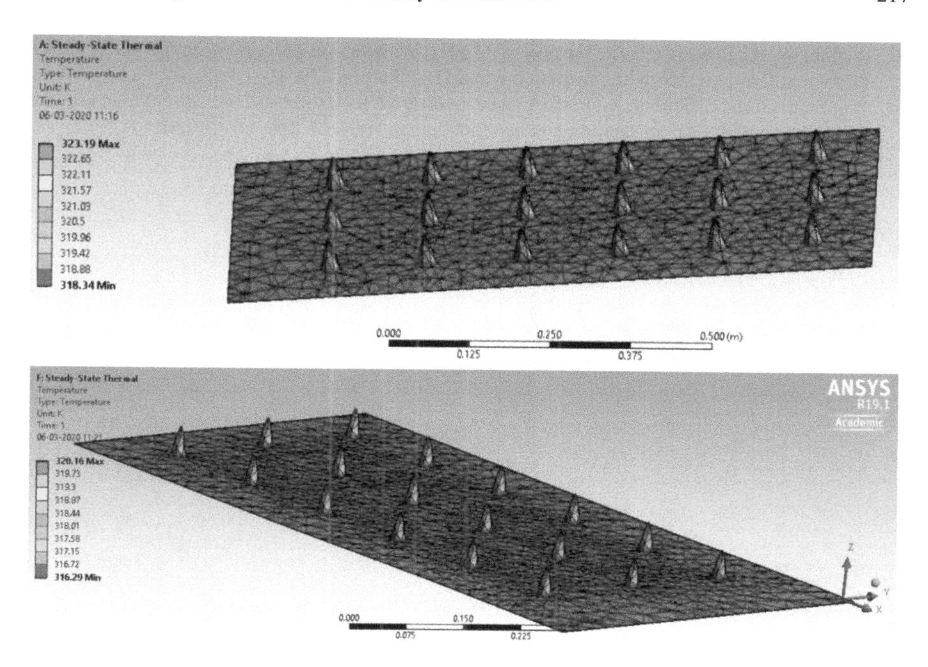

Fig. 15 Heat transfer analysis of truncated conic profile

References

1. Sharshir, S.W., Yang, N., Peng, G., Kabeel, A.E.: Factors affecting solar stills productivity and improvement techniques: a detailed review. Appl. Thermal Eng. **100**, 267–284 (2016)
2. Awasthi, A., Kumari, K., Panchal, H.: Passive solar still: recent advancements in design and related performance. Environ. Technol. Rev. **7**, 235–261 (2018)
3. Panchal, H.: Experimental analysis of diesel engine exhaust gas coupled with water desalination for improved potable water production. Int. J. Ambient Energy **38**, 567–570 (2017)
4. Kaviti, A.K., Yadav, A., Shukla, A.: Experimental investigation of solar still with opaque north triangular face. Int. J. Green Energy 1–8 (2019)
5. Durkaieswaran, P., Murugavel, K.K.: Various special designs of single basin passive solar still—a review. Renew. Sustain. Energy Rev. **49**, 1048–1060 (2015)
6. Mohan, I., Yadav, S., Panchal, H., Brahmbhatt, S.: A review on solar still: a simple desalination technology to obtain potable water. Int. J. Ambient Energy **40**, 335–342 (2019)
7. Mitra, S., Srinivasan, K., Kumar, P., Murthy, S.S., Dutta, P.: Solar driven adsorption desalination system. Energy Procedia **49**, 2261–2269 (2014)
8. Velmurugan, V., Gopalakrishnan, M., Raghu, R., Srithar, K.: Single basin solar still with fin for enhancing productivity. Energy Convers. Manag. J. **49**, 2602–2608 (2008)
9. Ei-Sebaii, A.A., El-Naggar, M.: Year round performance and cost analysis of a finned single basin solar still. Appl. Therm. Eng. **110**, 787–794 (2017)
10. EI-Sebaii AA, Ramadan MRI, Aboul-Enein S, El-Naggar, M.: Effect of fin configuration parameters on single basin solar still performance. Desalination **365**, 15–24 (2015)
11. Jani, H.K., Modi, K.V.: Experimental performance evaluation of single basin dual slope solar still with circular and square cross-sectional hollow fins. Sol Energy **179**, 186–194 (2019)

12. Alaian, W.M., Elnegiry, E.A., Hamed, A.M.: Experimental investigation on the performance of solar still augmented with pin-finned wick. Desalination **379**, 10–15 (2016)
13. Rajaseenivasan, T., Srithar, K.: Performance investigation on solar still with circular and square fins in basin with CO_2 mitigation and economic analysis. DES **380**, 66–74 (2016)

Investigations on Solar Air Collector Performance Improvement by Adoption of Response Surface Method Strategy

M. B. Gorawar, Veeresh G. Balikai, Rakesh Tapaskar, P. P. Revankar, and Vinayak H. Khatawate

Abstract In the present, world has energy identified as a vital component of our daily life for several reasons ranging from cooking energy needs to space conditioning requirements. The over-crowding of urban communities has put a huge pressure on the energy generation sector to produce more and more electric power, thereby leading to climatic degradation. The human societal development clearly marks out Industrial Revolutions occurring at periodic intervals. These events have lead to rewire the process of generation and utilization of power to meet the ever increasing energy demand. The utilization of nature-based energy sources through clean energy conversion technologies has been identified as a unique solution to mitigate climate change issues. The solar energy offers an attractive means to meet differing patterns of energy needs of human society that ranges from low grade thermal energy to high grade electric power. The promotion of clean energy has been the foremost priority of the global community of nations, under the platform of United Nations Sustainable Development goals. Hence, this work gains significance toward making a more reliable thermal energy conversion device to tap the huge potential of solar energy that has a worldwide geographical presence. The performance of solar thermal collectors depends on desired output along with climatic factors and operating parameters of working fluid like flow rate and inlet temperature. In this perspective, the overall performance regulation of solar thermal collector transforms into a multi-variable optimization problem. The technique of design of experiment (DoE) is a statistical approach of optimization that effectively handles situations encountering multi-variable selection to optimize a defined objective function. This study aims to evolve optimal operating scenarios for a solar flat absorber surface collector used for drying of agricultural produce. The experimental data collected for the solar flat absorber surface collector is utilized to improvise its operation for any general operating situation with parameters subjected to random variations. This

M. B. Gorawar · V. G. Balikai · R. Tapaskar · P. P. Revankar (✉)
Faculty, SME, KLE Technological University, Hubballi, India
e-mail: pp_revankar@kletech.ac.in

V. H. Khatawate
Faculty, Department of Mechanical Engineering, DJS College of Engineering, Mumbai, India

P. Muthukumar et al. (eds.), *Innovations in Sustainable Energy and Technology*,
Advances in Sustainability Science and Technology,
https://doi.org/10.1007/978-981-16-1119-3_20

investigation has yielded dataset on development of inputs to make an intelligent crop drying strategy based on local climatic factors.

Keywords Simulation studies · Solar thermal applications · Optimization

1 Overview on Solar Drying

The per capita energy consumption of the nation has been a strong indicator of development. The statistics of energy consumption pattern in India has shown a steady growth in this indicator … Solar energy capacity has increased from 2.6 to 34 GW during the period 2015 to 2020. The Energy Policy-2020 envisages a target of 175 GW grid-tied electric power by March 2022, through 100 GW of solar power and 60 GW of wind-based installations. The projection of National Electricity Plan (2018) indicates 275 GW of renewable power by 2027, which makes its cumulative share of 44% in total installed capacity contributing 24% in electricity generation. The government initiatives have been aligned to the changing global scenario of generating clean energy with minimum impact on environment. The nett carbon footprint has been stringently monitored in each and every activity to align it to meet the overarching limits set under the UN guidelines for sustainable development.

The solar energy aligns well with sustainability principles as it does not form any emissions during operation and adopts technology with minimum carbon footprint during manufacture and installation. The solar energy has a good compatibility for thermal applications like water heating and crop drying that can be designed with minimum number of moving parts and hence free from utilizing first grade electric power. The solar air collector is a simple device that works on the principle of 'greenhouse effect' to extract heat from sunlight, using a highly heat absorbing surface that helps capture radiant heat. The design of a solar air collector involves choice of appropriate materials of construction along with appropriate operating parameters that ensure a higher thermal efficiency of up to 75% or higher. The marginal rise in efficiency of solar thermal collector can lead to substantial reduction in material usage and also the cost of equipment. The inter-relationship between various operating characteristics also has a bearing on the efficiency of solar thermal collector and can hence influence the rate of drying or other end utility of heat collected. The appropriate choice of operating parameters is air flow rate, temperature of heated air, wind speed, solar insolation, and humidity.

The response surface method (RSM) has evolved as a power analytical tool that offers design platform to formulate, redesign, and optimize process variables. The history of system performance data under various states of operation helps to evolve inter-relationships between objective function and connected independent variables. The RSM with features of ease in estimation and application makes it more attractive for situations with limited information available on system operation. The results obtained through RSM route no doubt are approximate; however, it finds wide ranging applications in design of experiments. The historical dataset generated through observation of system performance over a wide range of independent variables can evolve a prediction strategy with respect to optimizing system to yield favorable response.

2 Literature Review

The RSM technique has wide ranging application in engineering as reported by research findings in solar air heater design and performance evaluation.

Peng et al. [1] used computations to solar collector to overcome existing limitations. The parameters of air flow and collector area recommended ranged as 50–100 and 3–6 kg/(m^2 h). The glass cover height ranged between 4 and 6 cm to attain highest collection efficiency [1].

Yadav and Prasad [2] adopted MATLAB to investigate wire roughened air heater absorber to fix collector parameters. The comparison with reported experimental data showed artificial roughness unsuitable during early transition flow (Re > 0.29 $\times 10^5$) as per conclusions drawn from the study [2].

Giovanni et al. [3] investigated high vacuum solar water heater for Messina (South-Italy) through MATLAB. The optimum elimination of 348 ton per year of CO_2 with a 4.5 year payback [3].

Earnest and Rai [4] adopted DOE to Parabolic collector for suitable useful heat and temperature. The study indicated best operation with copper reflector at 1.756 $\times 10^{-3}$ kg/s flow. The highest heat gain and exit water temperature were obtained, respectively, for receiver diameters of 26 and 30 mm [4].

Jawad et al. [5] adopted Energy Equation Solver (EES) tool for investigation of multi-cover solar water heating system through Taguchi full factorial design. The flow rate of water, number of covers absorber surfaces, and spacing between covers were optimized to be 2.5 lpm, two covers, and 2.5 cm, respectively. The experimental results when compared to EES simulation had agreements in acceptable range [5].

Kanimozhi et al. [6] reported on DOE approach for storage integrated solar water heating systems adopting honey/paraffin wax for heat storage in three varying flow rates (2–6 kg/min). The investigated charge/discharge cycles indicated heat absorption 5840 and 6408 kJ at 6 kg/min flow rate [6]. Durakovic [7] reported large use of DOE in medical and engineering apart from physics and computer science with a 13% share. The tool has emerged as a powerful tool to compare and control variables in identifying system transfer function toward optimization leading to design robustness [7].

Dharmalingam et al. [8] used Taguchi method for optimization of solar water heating on basis of nine combinations of experiments with three variables for flow characteristic number (Re), concentration of nanoparticle concentration, and incident heat flux. The optimum thermal performance was realized with Re $= 2.5 \times 10^4$, percent concentration level (3/100) and 0.8 kW/m^2 heat flux as against highest overall heat gain at 1 kW/m^2 insolation [8]

Hannane et al. [9] reported indoor solar electric system at three-level DoE adopting temperature and irradiation (input) to assess performance indicators I_{SC} and V_{OC} (output). The HIDE software results for same operating conditions indicated good agreement with experimental results [9].

Badache et al. [10] adopted full factorial method to optimize solar collector operation using inlet passage diameter, air flow magnitude, incident solar heat flux, and optical characteristics of absorber surface. The study used a three-level investigation to obtain 80% efficiency at optimum level of absorber coating and air flow [10].

3 Numerical Studies for Characterization of Collector Performance

The MATLAB-based DOE tool was adopted in optimization of solar collector on basis of identified features of absorber absorber surface temperature, flow rate of air, insolation and thermal efficiency. Table 1 details parameters for assessment of solar air heater. These climatic/operational parameters influence selection of constructional features of solar air heater. Table 2 gives detailed execution of DoE-based simulation of solar collector.

The DOE numerical studies to detect role of individual factor were performed. The multi-variable function was evolved as indicated in Eq. 1 for six functional factors in terms of operating variables. The polynomial function with defined constants

Table 1 Levels of fluid flow and heat flux used in simulation studies

Climatic/operational factors	Identified RSM level		
	Small	Moderate	Large
Mass flow (10^{-3} kg/s)	15	25	35
Solar heat flux (kW/m^2)	0.05	0.55	1.05
Wind speed (cm/s)	100	300	500
Air temperature (°C)	20	30	40

Table 2 Detailed execution of simulation studies

Stage 1: Characterization of solar device for complete operating limits of variables
Stage 2: Synchronization of system behavior with Minitab for RSM implementation Stage 3: Choice of appropriate RSM strategy in Minitab library Stage 4: Select appropriate governing factors and their levels to obtain the response equation fit Stage 5: Evolve required response equations for thermal device (use coefficients in Table 3)

Table 3 Coefficients for polynomial equation (1) to evaluate parameters related to solar collector

y	T_o	T_p	Q	η_{th}	$h1$	$h2$
Coefficients	10.3768	3.9429	−257.804	40.1433	3.3818	14.0952
a	−879.3217	−577.9849	5402.871	756.5256	0.6179	1033.691
b	0.07482	0.04739	1.1193	−0.00693	-1.6×10^{-6}	−0.00276
c	−0.09184	−0.0388	0.0387	−0.03564	-9.5×10^{-7}	−0.00164
d	1.0365	1.0423	1.2114	0.06819	-6.6×10^{-5}	−0.11162
e	7769.2344	5602.69	−105,834	−8671.09	1.4173	1754.95
f	-9.889×10^{-7}	-7.13×10^{-7}	-3.1×10^{-5}	2.21×10^{-6}	5.03×10^{-11}	8.32×10^{-8}
g	−0.0064	−0.0048	−0.1339	−0.00795	-4.6×10^{-8}	-8.5×10^{-5}
h	−0.0001	-9.44×10^{-5}	−0.0026	−0.0002	1.22×10^{-7}	0.0002
i	−0.7153	−0.5048	8.8742	0.1550	-8.9×10^{-6}	−0.0125
j	1.4595	1.9528	21.2586	0.5261	1.24×10^{-8}	-0.0043
k	0.7980	0.3810	9.1754	0.5250	-0.0011	-1.8398
l	-7.443×10^{-5}	−0.00011	−0.0027	-4.2×10^{-5}	1.59×10^{-9}	2.65×10^{-6}
n	-3.496×10^{-5}	-9.369×10^{-6}	−0.0010	-1.3×10^{-5}	4.34×10^{-9}	7.13×10^{-6}
o	0.00026	-2.107×10^{-5}	0.00012	2.24×10^{-5}	4.4×10^{-9}	8.31×10^{-6}

generated through combined numerical and experimental data generated for solar air heater. The MATLAB tool exhibited fast learnability and robustness for the efficient DoE algorithms. The simulation studies lead to minimized material of constructional owing to better utilization of available insolation. The carbon footprint of solar thermal process reduced by virtue of suitable measures evolved by detailed analysis on basis of computations and experiments.

The solar air heater operation is governed by climatic factors and constructional parameters that are subjected to multi-variable optimization indicated in Table 3 and Eq. 1.

$$y = am + bI(t) + cW + dT_a + em^2 + fI(t)^2 + gW^2 + hT_a^2 + imI(t) + jmW + kmT_a - lI(t)W + nI(t)T_a + oWT_a \quad (1)$$

The simulation indicated air flow (m) to have strong influence on heated air temperature (T_o), absorber surface temperature (T_p), heat gain(Q), thermal efficiency (η_{th}), and convective coefficient ($h1$ and $h2$). The higher air flow exhibited better efficiency on account of larger temperature gradient to ensure high heat gain capability. The low air flow facilitated increased heat transfer duration and hence delivered heated air at high temperature. The enhanced sunlight intensity and inlet air temperature facilitated more heat availability to enhance heated air temperature on account of additional heat flux of beam radiation and improved heat entrapment by collector design. In contrast to this, high inlet air temperature minimized effective fluid-surface temperature difference to lower heat gain and thermal efficiency of the solar device. The magnitude of prevailing wind flow in the vicinity of solar device had a bearing on convective heat loss coefficients defined for the top, side, and bottom surfaces of collector, thereby reduced collector thermal performance.

The Minitab-ANOVA analysis for a confidence level of 9.5/10 had every other parameter except wind flow to influence all other variables excluding thermal efficiency that had magnitude lesser than 0.05 for p-value. The wind velocity and ambient temperature were less significant with higher p-value (>0.05). The simulations showed Adj R^2 to be, respectively, 98/100 and 100/100 with respect to η_{th} and 'h'-coefficients. Similarly, data analytics of all major response variables had 'coefficient of regression' to lie within 0.98–1.00 on the scale of 0–1. This indicator has clearly established the reliability of the generated relationship between source and response parameters which can be adopted for real-time implementation of studies on solar air collector. The simulation studies are hence authenticated for adoption in building working prototypes of solar thermal conversion devices applicable to air heating.

4 Experimental Characterization of Solar Collector

The experimental studies were conducted to characterize solar air heater performance with respect to conditions of air, solar intensity, and prevailing wind pattern. These studies assessed heated air temperature, absorber temperature, device heat loss factors, heat gain, and device conversion efficiency. The customer requirement is based on end utility of collector delivered air and dictated optimum condition of operation. Figure 1 indicates experimental test rig used to conduct studies at varied climatic conditions. The absorber surface constructed in galvanized iron (GI) had high absorptivity (α) due to black chrome coat, while the device glass cover exhibited high transmittivity (τ) to incident sun beam. Therefore, the appropriate combination of these device materials enhanced the overall trasmittivity-absorptance product of thermal energy conversion device. The sixty zig-zag vertical ribs (15 cm long and 2 mm thick) placed over GI absorber assisted in heat exchange between air and absorber surface. The heat loss from the adjoining surfaces at side and bottom regions of collector was arrested using glass wool insulation material of appropriate thickness. The transparent glass covers formed the apertures to admit sun beam into the collector, on account of their high transmittivity to short-wave radiations and blocking the thermal radiations emitted by absorber surface. The toughened glass covers with these unique optical properties will ensure reduced convective losses.

Figure 2 shows variations of climatic factors of insolation, air temperature, and wind speed measured on a continuous basis. The solar insolation varied as per standard solar diurnal cycle exhibiting highest magnitude at around 12.00 local solar time, and the magnitude showed declining trend in the session between solar noon toward 15.00 h. The insolation directly influenced overall heat gain owing to higher input

Fig. 1 Solar air heater test equipment for experimental investigations

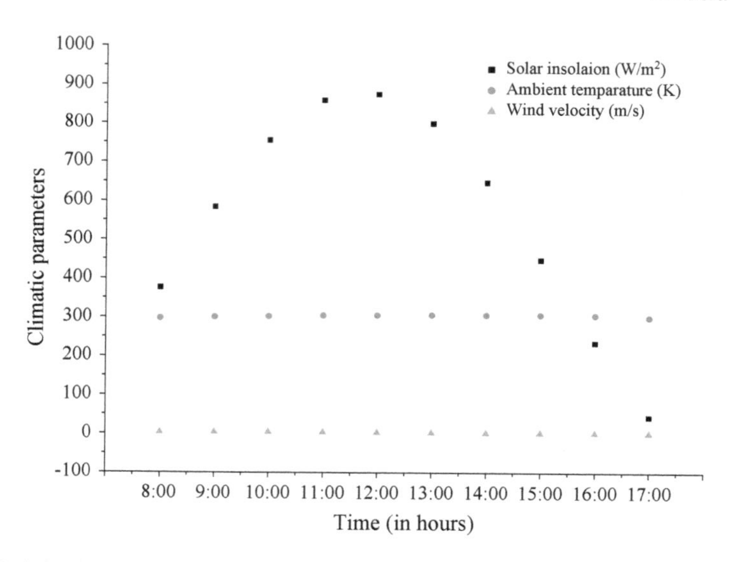

Fig. 2 Variation of climatic parameters at the test location

heat. The air temperature and wind speed had relatively meager role in magnitude of useful heat gain through heat transfer between absorber surface and air medium.

The overall heat gain by the solar thermal energy conversion devices is influenced by the input solar heat flux and conversion of radiant heat flux into heated air along with minimization of heat losses. The processes are incorporated in design of solar converter through appropriate choice of materials and sizing to facilitate greenhouse heat trapping. The mechanism of heat transfer between working fluid and device is greatly governed by thermo-physical properties of heat exchange fluid and heat source. The mechanism of heat exchange is strongly dependent on heat exchange coefficient (h) and thereby governs nett heat gain. The heat flux emerging from sun is spread over large range of wavelengths (200 to 2500 nm), predominantly in visible limits in 380 to 700 nm range. The collector glass cover functions to permit the available solar radiation and block the re-radiated long wavelength radiation back to collector. The greenhouse effect is a design attribute of solar energy collector to enhance nett heat gain. The observations as cited in Fig. 3, 4 and 5 highlights role of coefficients of heat transfer (h) on temperature of heated air and overall conversion efficiency of solar device functioning at various climatic conditions. The heat loss coefficient at side, top, and lower surfaces of collector decide the magnitude of heat lost to ambient, hence a larger 'h' means enhanced convective losses at the cost of diminishing useful heat gain. It was observed that higher value of 'h' for all air flow (low, medium, and large) heated air temperature dropped. On the other hand, heated air temperature was maximum for the low air flow on account of larger air residence time in the device that facilitated a greater temperature rise in fluid medium. The overall heat extraction process depicted through Figs. 3, 4, and 5 shows the significance of mass flow rate on overall efficiency. The variation trend clearly

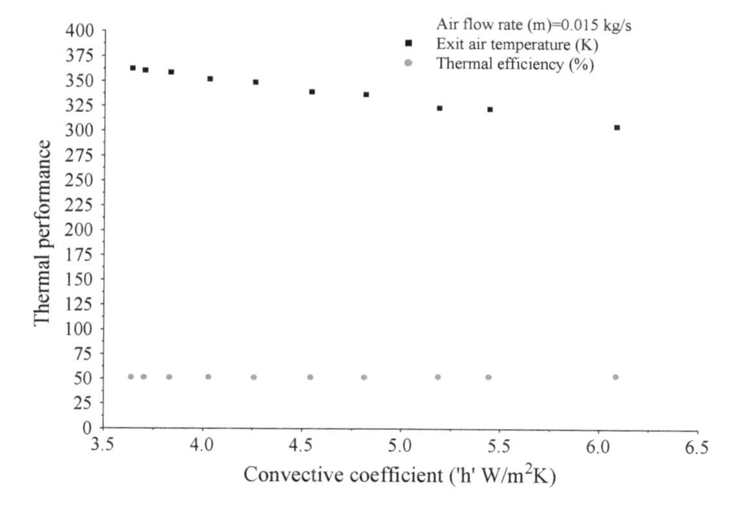

Fig. 3 Low flow rate convective heat transfer coefficients and solar collector efficiency

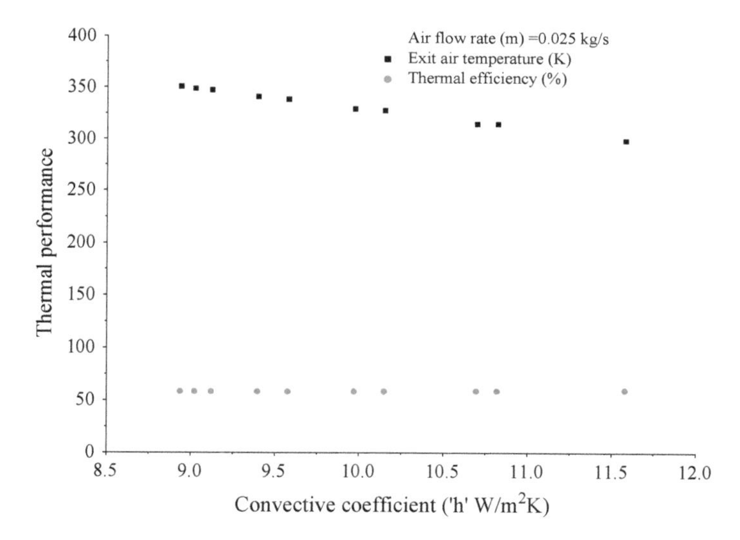

Fig. 4 Medium flow rate convective heat transfer coefficients and solar collector efficiency

exhibits high thermal efficiency for large air flow against low and medium range flow due to enhanced losses in the latter cases of flow condition. Thus, it is summarized that role of operating parameters has a distinct role to play in the overall process of heat extraction and its utilization for any intended end application. Therefore, design of solar thermal equipment involves a multi-functional formulation to optimize the performance and thereby the overall cost of end utility.

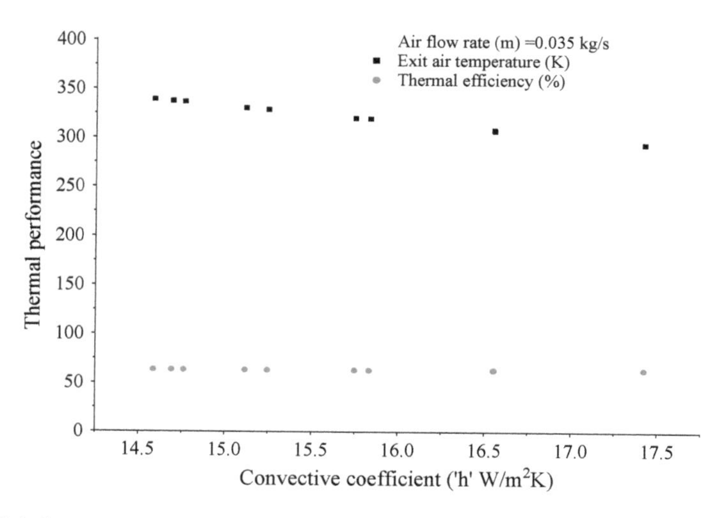

Fig. 5 High flow rate convective heat transfer coefficients and solar collector efficiency

5 Conclusions

The investigations based on RSM method and real-time experimentation on solar air heating device has lead to the following observations,

The selection of appropriate material and heat exchange mechanism for solar thermal devices has a major influence by local climatic parameters. The developed heat transfer device cannot function as a universal or generic device suitable for all regions and applications. The climatic factors play a vital role in deciding operational efficiencies and also the useful life of device that can be deteriorated in adverse climatic conditions of high humidity and severe windy conditions.

The multi-variable objective statement involving climatic and operational factors drawn for design of solar air heater necessitates the use of optimization tools like RSM to negotiate with conflicting variable behavior. The study has drawn significantly good data correlations connecting the source and response variables, between limits of 98 to 100% of Adj R^2 values to strongly support its application to real-time prototype building.

The DoE results with respect to 'p-value' within limit of 0.05 strongly supports validity of correlation along with Adj R^2 between 98 and 100% for climatic factors highlights major role of solar heat flux on effectiveness of solar thermal devices. The study also highlights the secondary role of wind speed and air temperature whose role is not as pronounced against solar radiation intensity.

The results and correlations obtained from this study are quite significant for the purpose of selection of appropriate solar thermal energy conversion device based on the data gathered with respect to climatic parameters synchronized with the end application heat requirement calculations.

Solar air heater finds viable applications in agricultural post-harvest processing. The use of solar energy for thermal applications can make a substantially curtail carbon emissions through deferred fossil fuel usage.

References

1. Peng, D.G., Cao, Z., Fu, Y.T., Li, S.L.: Yang ZX, Applicability, optimization and performance comparison of self-preheated solar collector/regenerator. Sol. Energy **198**, 113–123 (2020)
2. Yadav, K.D., Prasad, R.K.: Performance analysis of parallel flow flat absorber surface solar air heater having arc shaped wire roughened absorber absorber surface. Renew. Energy Focus **32**, 23–44 (2020)
3. Barone, G., Buonomano, A., Calise, F., Forzano, C.: Adolfo Palombo x energy recovery through natural gas turbo expander and solar collectors: modelling and thermoeconomic optimization. Energy **183**(15), 1211–1232 (2020)
4. Earnest, V.P., Rai, A.K.: Performance optimization of solar Ptc using Taguchi method. Int. J. Mech. Eng. Technol. **9**(2), 523–529
5. Jawad, Q.A., Abdulamer, D.N., Mehatlaf, A.A.: Optimization performance of solar collector based on the fractional factorial design. J. Univ. Kerbala **15**(4), 68–76 (2017)
6. Kanimozhi, B., Ramesh Bapu, B.R., Pranesh, V.: Thermal energy storage system operating with phase change materials for solar water heating applications: DOE modeling. Appl. Thermal Eng. **123**, 614–624
7. Durakovic, B.: Design of experiments application, concepts, examples: state of the art. Period. Eng. Nat. Sci. 5(3):421–439. ISSN 2303-4521 (2017)
8. Dharmalingam, K.K., Sivagnanaprabhu, C., Chinnasamy, B., Senthilkumar: optimization studies on the performance characteristics of solar flat- absorber surface collector using Taguchi method. Middle-East J. Sci. Res. **23**(5):861–868 (2015)
9. Hannane, F., Elmossaoui, H., Nguyen, T.V., Petit, P., Aillerie, M., Charles, J.P.: Forecasting the PV panel operating conditions using the design of experiments method. Energy Procedia **36**, 479–487 (2013)
10. Messaoud, B., Stephane, H., Rousse, D.R.: A full 3^4 factorial experimental design for efficiency optimization of an unglazed transpired solar collector prototype. Sol. Energy **86**, 2802–2810 (2012)

Stability and Thermal Conductivity of Ethylene Glycol and Water Nanofluid Containing Graphite Nanoparticles

Kyathanahalli Marigowda Yashawantha, Gaurav Gurjar, and A. Venu Vinod

Abstract Heat transfer enhancement in a heat exchanger using conventional fluids such as water, ethylene glycol and engine oils is limited due to lower thermal conductivity of fluids. Addition of nanoparticles to these fluids can provide better heat transfer due to improved thermophysical properties. In this study, thermal conductivity and stability of ethylene glycol (EG) and water (35:65 in terms of volume)-based graphite nanofluids were investigated experimentally. Nanofluids containing graphite nanoparticles in weight concentrations in the range of 0.5–2 wt% were prepared using ultrasonication method. Thermal conductivity of nanofluids was measured at each weight concentration in the temperature range of 30–50 °C using transient hot wire method. Subsequently, the stability of EG and water-based graphite nanofluids was evaluated by visual observation and thermal conductivity measurement over a period of 30 days at 30 °C. The results showed that thermal conductivity enhances by adding graphite nanoparticles to the ethylene glycol and water. It was found from the study that thermal conductivity of prepared nanofluid increases with increase in temperature. Addition of 2 wt% of graphite nanoparticles in EG: water showed an enhancement of 22% and 26.67% in thermal conductivity at 30 and 50 °C, respectively. Stability study overtime reveals that there is no considerable difference in thermal conductivity for 30 days indicating a stable nanofluid dispersion. A correlation was developed to predict the thermal conductivity of EG: water-based graphite nanofluids.

Keywords Graphite nanoparticles · Water · Ethylene glycol · Ultrasonication · Thermal conductivity · Stability

1 Introduction

Nanofluids contain ultra-tiny particles less than 100 nm of solid material in powder form suspended in a base fluid to form a homogenous mixture as a single base

K. M. Yashawantha · G. Gurjar · A. Venu Vinod (✉)
National Institute of Technology, Warangal, Telangana 506004, India
e-mail: avv@nitw.ac.in

P. Muthukumar et al. (eds.), *Innovations in Sustainable Energy and Technology*, Advances in Sustainability Science and Technology,
https://doi.org/10.1007/978-981-16-1119-3_21

fluid [1]. These nanofluids are capable of transferring more heat compared to the base fluids. Researchers have used different preparation methods and evaluation techniques to prepare stable nanofluids [2–4]. The thermophysical behaviour of the nanofluids is the important aspect in heat transfer application, and the same has been investigated by many researchers [5–8]. The thermophysical properties have been used to assess the performance of heat exchanger using different nanofluids [9–11]. Researchers have used nanoparticles like CuO, Al_2O_3, Fe_2O_3 and ZnO to prepare nanofluids. Similarly, some researchers have used the non-metallic nanoparticles graphite, graphene, single-walled carbon nanotube (SWCNT) and multi-walled carbon nanotubes (MWCNT) nanoparticles to prepare the nanofluids by adding the nanopowder to the base fluid. Commonly used base fluids for different cooling applications are water, engine oil, ethylene glycol (EG), propylene glycol (PEG), glycerine, EG and water mixture, and PEG and water mixture. These base fluids are selected based on their properties and application. The base fluids find application in engine cooling, aircraft engine cooling system, refrigeration and air conditioning system, etc. EG is mainly used as a base fluid for low-temperature application because of its low freezing temperature and high boiling temperature. Hence, to operate at a wide range of temperature, EG and water mixture is a better choice.

Zhu et al. [12] synthesized a stable water/graphite suspension by optimizing the pH value using Polyvinylpyrrolidone (PVP-K30) as a dispersant and reported an enhancement of 34% at 2% volume concentration. Wang et al. [13] used a mechanical ball milling method to prepare the graphite–oil nanofluids. The prepared graphite–oil nanofluids showed enhancement of 12–36% in thermal conductivity for 0.68–1.36 vol% of concentration, respectively. Kole and Dey [14] used hydrogen exfoliated graphene (f-HEG) nanosheets to prepare Graphene–EG+ Distilled water nanofluid without adding any surfactant. The nanofluids prepared by ultrasonication were stable for more than 5 months. They reported 15% enhancement in thermal conductivity for 0.395 vol. % nanofluid at 30 °C. Mehrali et al. [15] examined the effect of the different surface area of graphite nanoplatelets (GNP) at a different concentration by varying the temperature. Their results indicated that the maximum increase in thermal conductivity was around 27.64% at a larger surface area with 0.1 wt% concentration of GNP at 35 °C. Yashawantha et al. [16] evaluated the thermal conductivity of EG graphite nanofluids for different volume concentration and reported an enhancement of 24.46% at 2% of volume concentration at 30 °C. Nanofluids with different water and EG ratios have been used as base fluid to study the thermophysical properties by various researchers. Usri et al. [17] and Chaim et al. [18] dispersed the Al_2O_3 nanoparticles in the different combination of water and EG mixtures (W:EG: 60:40, 50:50 and 40:60) expressed in base ratio (BR). The results revealed that introducing the Al_2O_3 nanoparticles in the different BR affects the thermal conductivity. The results demonstrated that the percentage of EG in the base ratio (BR) has a more significant effect on thermal conductivity with respect to temperature.

From the literature, it is evident that the addition of graphite and graphene nanoparticles is capable of providing higher thermal conductivity. However, very few works have been carried out using graphite as a nanoparticle for measuring thermophysical properties to study the feasibility for applications. It is also observed that Kole

Table 1 Properties of graphite nanoparticles and EG: water (35:65)

Characteristics	Graphite	EG: water (35:65)
Size	50 nm	–
Density	2260 kg/m^3	1048 kg/m^3
Morphology	Hex-Lamellar	–
Colour	Black	–
Purity	99.5%	99.5%

and Dey [14] and Mehrali et al. [15] focused on graphene nanosheets and graphite nanoplatelets with water as a suspension to study the thermal conductivity. Wang et al. [13] have used graphite/oil suspension, whereas Ma et al. [19] and Yashawantha et al. [16] used EG/graphite suspensions to report thermal conductivity. Further, studies involving the combination of EG and water mixtures of different base ratio with Al_2O_3 nanoparticles were reported by Usri et al. [17], Chaim et al. [18] and Sundar et al. [20]. Correspondingly, Sundar et al. [21] and Hamid et al. [22] reported studies on the EG: water mixture of CuO and TiO_2 nanoparticles, respectively. These studies reported on stability and correlation for thermal conductivity. However, stability and thermal conductivity studies on the combination of EG: water and graphite nanoparticles suspension are yet to be explored. Therefore, the present study is focused on dispersing different weight concentration of graphite nanoparticles in EG: water (35% volume of EG: 65% volume of water) using ultrasonication and to determine thermal conductivity and stability.

2 Materials and Method

2.1 Materials

The graphite nanoparticles are procured from Sisco Research Laboratories (SRL) Pvt. Ltd., India and ethylene glycol from Sigma Aldrich Chemicals Ltd., USA. The characteristics data of graphite [23] and EG: water mixture (ASHRAE [24]) are presented in Table 1. To confirm the particle size supplied by SRL, a field emission scanning electron microscopy (FESEM, SIGMA GEMINI) analysis was carried out at different magnification. Figure 1 shows the FESEM images, and it can be observed that particles size is almost <50 nm.

2.2 Nanofluid Preparation

The preparation of nanofluids is a key factor to obtain good stability as well as good thermal conductivity. The EG: water mixture of 35% volume of EG and 65%

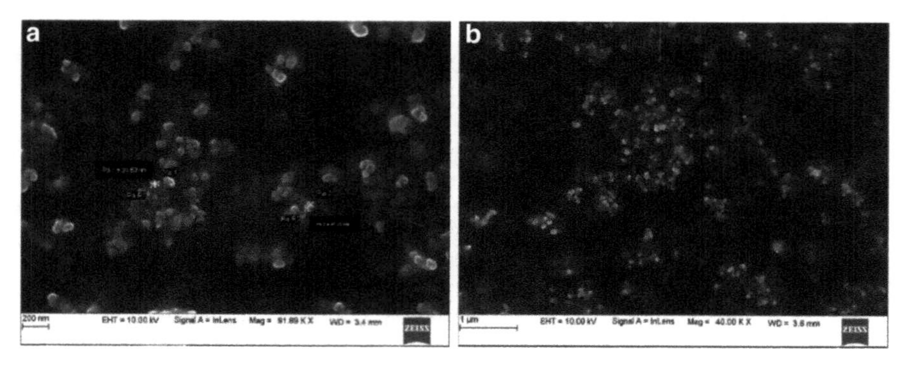

Fig. 1 **a** FESEM images of graphite nanoparticles at 91.89 KX and **b** FESEM images of graphite nanoparticles at 40 KX

volume of water is used as a base fluid. The different weight concentration of graphite nanoparticles is weighed in the precision electronic balance (accuracy of ±0.001 g). These measured graphite nanoparticles are added to EG: water mixture of 40 ml (taken in the beaker) carefully and stirred for 1 h using a magnetic stirrer at a speed of 900 rpm to ensure the initial mixing. For improving the stability, 0.05 wt% of sodium dodecylbenzene sulfonate (SDBS) surfactant was added before magnetic stirring. Further, the stirred nanofluid was ultrasonicated in Hielscher UP200H (Power: 200 W, frequency: 24 kHz) for 2 h to breakdown larger particles and to ensure the mixture stability. Similarly, nanofluids of concentration 0.5, 1, 1.5 and 2 wt% are prepared separately aforesaid method. The weight to be added for the preparation of different weight concentration is estimated using Eq. 1.

$$100\varnothing_{\mathrm{w}} = \frac{m_{\mathrm{g}}}{m_{\mathrm{g}} + m_{\mathrm{EG/w}}} \tag{1}$$

where m_{g} and ρ_{g} = mass of graphite nanoparticles and density of graphite nanoparticle, respectively. $m_{EG/w}$ and $\rho_{EG/w}$ = mass and density of EG: water mixture, respectively. $\varnothing_{\mathrm{w}}$ = the weight concentration of EG: water–graphite nanofluids.

2.3 *Thermal Conductivity and Stability*

The prepared EG: water-based graphite nanofluids at different weight concentration was investigated for thermal conductivity (k) in the temperature range of 30–50 °C. For this purpose, KD2 Pro thermal property analyser (Decagon Devices, Inc., USA) which works on the principle of transient hot wire method was used. Many researchers [8, 17, 18, 20, 25] have used this instrument to report the thermal conductivity. KD2 Pro meets ASTM D5334 and IEEE 442-1981 standard of measurement for thermal conductivity. KD2 Pro with a different sensor needle with an inbuilt microcontroller can able to measure the thermal conductivity of liquids and solids. Sensor needle is selected based on the range of measurement. KS-1 needle having a measuring range of 0.02–2.00 W/m K, which is suitable for liquids. This needle is 6 cm long and 1.3 mm in diameter. Hence, assumption can be made as an infinitely small fin for measurement. Initially, 0.2 wt% of concentration was taken in 30 ml bottle, and KS-1 sensor needle was carefully placed inside such that needle is exactly at the centre. The sample temperature was maintained at 30, 35, 40, 45 and 50 °C temperature in a constant temperature bath (TEMPO SM-1014), and thermal conductivity was measured at each temperature by taking five readings by allowing about 15 min between readings for temperatures to equilibrate. Average of these readings was used for reporting. Stability of EG: water-based graphite nanofluids was carried out by taking photographic observation overtime. Thermal conductivity measurement was performed over a time at temperature of 30 °C.

3 Results and Discussion

3.1 *Thermal Conductivity and Effective Thermal Conductivity*

The thermal conductivity (k) measurement was conducted initially for the EG: water mixture of 35:65 (v/v) at temperature 30, 35, 40, 45 and 50 °C. Figure 2 shows the KD2 Pro data and ASHRAE [24] data for EG: water at different temperatures. The data points of both ASHRAE and KD2 Pro are in the acceptable range with a deviation of 0.5% as appeared in Fig. 2.

Figure 3 shows the thermal conductivity at different weight concentration and different temperatures. Correspondingly, Fig. 4 shows the effective thermal conductivity (ratio of k of nanofluid to the k of base fluid) for different temperatures. It can be

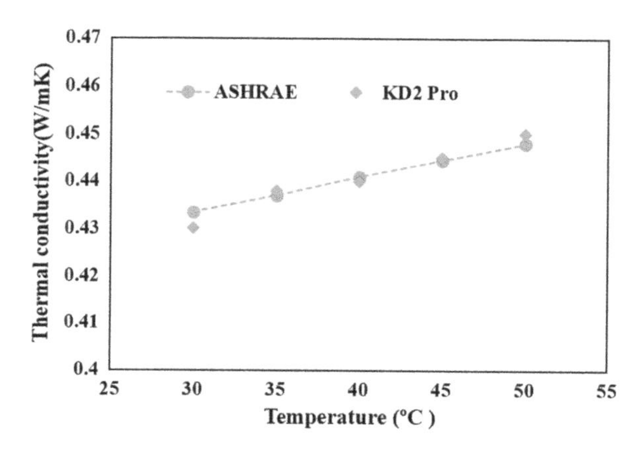

Fig. 2 Comparison of EG: water mixture and ASHRAE data for thermal conductivity

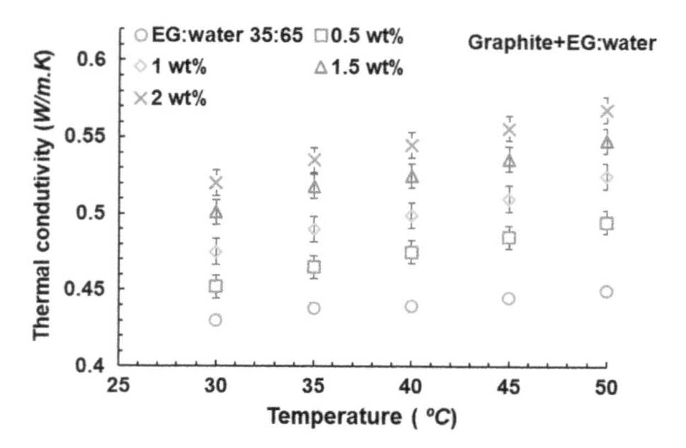

Fig. 3 Effect of temperature on thermal conductivity of EG: water–graphite nanofluids at different weight concentration

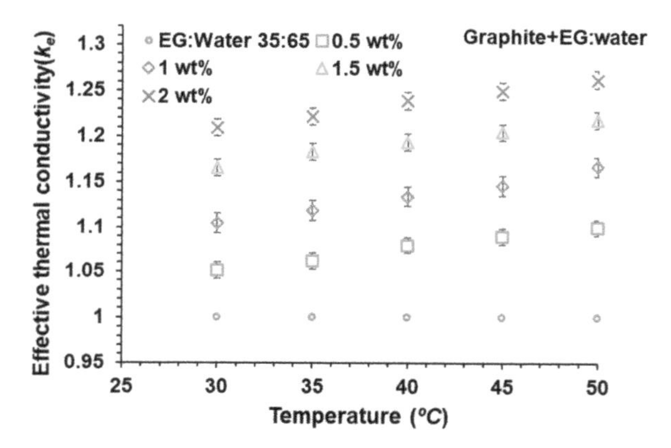

Fig. 4 Effect of temperature on effective thermal conductivity of EG: water–graphite nanofluids at a different weight concentration

observed from Figs. 3 and 4 that thermal conductivity (and effective thermal conductivity) of nanofluid increases with increase in temperature. Even for the base fluid used for the present study, a similar trend shows as can be seen in Fig. 3. However, thermal conductivity of base fluid is less than that of all nanofluid concentration considered in the study. It can be seen from Fig. 3 that at 30 °C thermal conductivity of the EG: water mixture is 0.430 W/mK. On the other hand, at the same temperature, 2 wt% weight concentration has 0.520 W/mK, which is 20.93% more than EG: water mixture. Similarly, for every concentration at different temperatures, the thermal conductivity increases. This enhancement attributes to the concentration of graphite nanoparticles in EG: water, temperature, particles thermal conductivity, shape and size of nanoparticles. The reason for the enhancement with weight concentration is due to collisions of particles inside the base fluid. When collisions take place between nanoparticles inside the base fluid, interaction between the particles increases, and as a consequence of this, Brownian motion increases, and due to this, thermal conductivity increases [18, 26].

As seen in Fig. 4, the effective thermal conductivity (k_e) for EG: water–graphite nanofluid is more compared to the EG: water at an identical temperature. The effective

Table 2 Percentage of enhancement of thermal conductivity from temperature 30–50 °C

Nanofluid	Nanofluid concentration			
Graphite + EG: water	0.5 wt%	1 wt%	1.5 wt%	2 wt%
	5.12–10	10.46–16.66	16.51–21.78	20.93–26.22

thermal conductivity at 50 °C is 1.2622 for 2 wt% of concentration, which is 26.22% greater than that of the EG: water mixture at identical temperature. Correspondingly, for all concentrations and temperatures, a similar enhancement was observed compared to the EG: water mixture at an identical temperature. This enhancement at all concentration at different temperature is due to the Brownian motion caused by a rise in temperature and decrease in viscosity with an increase in temperature. Thermal conductivity and effective thermal conductivity are strongly influenced by weight concentration due to increase in temperature. Enhancement of thermal conductivity for different weight concentration is tabulated in Table 2. The maximum enhancement of 20.93 and 26.22% was obtained at 30 and 50 °C for 2 wt% of concentration, respectively. The study carried out by the Ladjevardi et al. [27] shows that the use of water–graphite nanofluids can improve the absorbance of incident irradiation energy by 50%, whereas using water only possible was 27%. In applications such as low temperature, the EG: water-based graphite nanofluids can improve the performance of the system with lower nanoparticles concentration.

3.2 *Stability of EG: Water Based Graphite Nanofluids*

Stability of nanofluids is very crucial for any heat transfer application. Nanoparticle sedimentation in a base fluid is the main factor which affects many other properties of nanofluids such as thermal conductivity. The prepared nanofluid must ensure proper distribution of particles in the liquid. In this regard, EG: water (35:65 v/v)-based graphite nanofluids were investigated for stability by determining the thermal conductivity as a function of time. Similarly, a visual observation was also made after preparation and up to 30 days after preparation as shown in Fig. 5. Sedimentation was not observed for up to 30 days.

Nanofluids of four different weight concentration were kept at room temperature at the static condition, and thermal conductivity evaluation was conducted for more than 30 days using KD2 Pro thermal conductivity property analyser. Up to 30 days, there was no variation in thermal conductivity at room temperature at all weight concentration as appeared in Fig. 6. However, after 30 days of observation, a gradual decrease in thermal conductivity was observed. The decrease in thermal conductivity was 2.21%, 2.94% 3.19% and 3.65% for 0.5 wt%, 1 wt% and 1.5 wt% and 2 wt% nanofluids, respectively. Further decrease in stability was observed with respect to time. Hence, the prepared nanofluids were stable for at least 30 days under static condition. The study also demonstrates reduced thermal conductivity for increasing

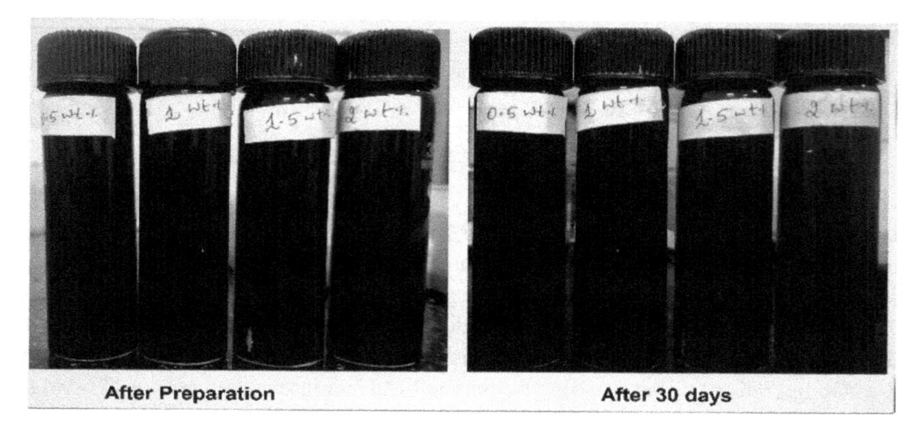

Fig. 5 Visual observation of EG: water-based graphite nanofluids after preparation and after 30 days

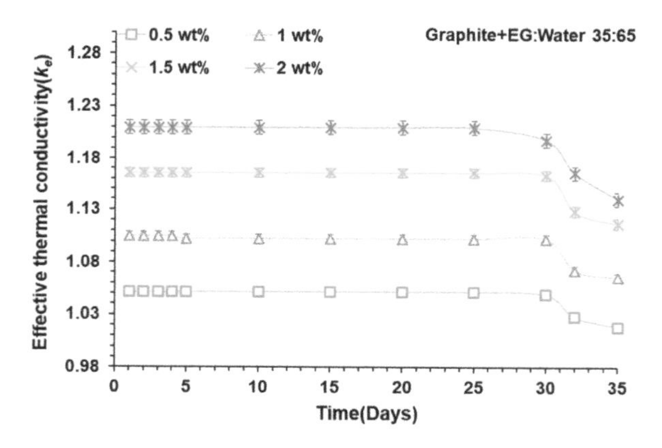

Fig. 6 Time-dependent effective thermal conductivity (k_e) of EG: water-based graphite nanofluids

weight concentration after 30 days compared to that immediately after preparation. This is due to agglomeration formation as a result of the increase in concentration. According to DLVO (Derjaguin, Verway, Landau and Overbeek) theory, stability of the dispersion of nanoparticles in liquids depends on the combined effect of Van der Waals captivating forces and electrical double-layer repulsive forces among the nanoparticles caused by Brownian motion [1]. If the particles size is large, aggregates forms as a result of clustering of particles of the same or different size as they get closer to each other (Fig. 7). On the other hand, the smaller size of particles avoids the formation of a cluster and provides longer stability of nanofluid. As the surfactant is added in this study, which develops more repulsion from absorbed surfactant effects the graphite nanoparticle to improve a more compact structure due to an increase in the particle to particle overlap area. With the addition of critical surface active agent, it was found that the graphite nanoparticles dispersed in EG: water is stabilized, and the kinetic barrier prevents the aggregation.

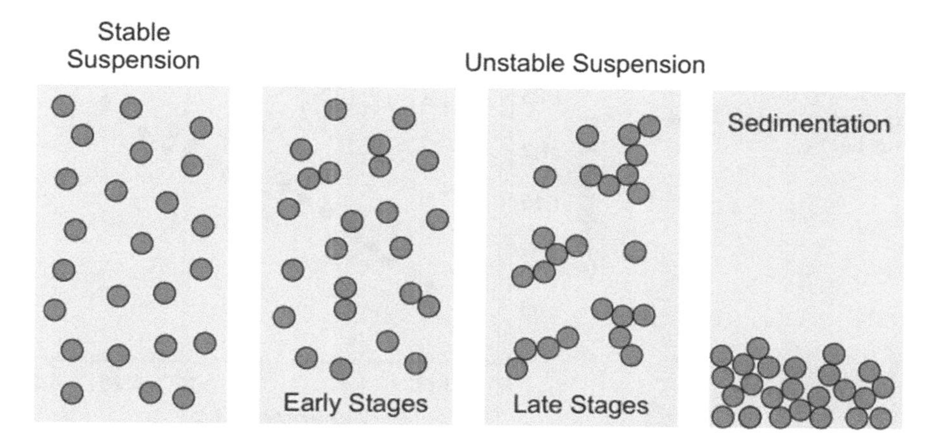

Fig. 7 Stages of graphite nanopowder sedimentation in EG: water after preparation

3.3 *Developed Correlation*

The effective thermal conductivity k_e is the ratio of the thermal conductivity of the EG: water-based graphite nanofluid to EG: water (35:65) at an identical temperature. In this study, the correlation was developed as a function of weight concentration and temperature using the regression method. The correlation is proposed in terms of weight concentration and the temperature of EG: water-based graphite nanofluid in the following form,

$$k_e = \frac{k_{gnf}}{k_{bf}} = a(\varnothing_w)^b \left(\frac{T_{nf}}{T_r}\right)^c \quad (2)$$

where k_e = effective thermal conductivity, $\varnothing_w$ = the weight concentration of graphite nanofluids, T_{nf} = temperature of nanofluid, and T_r = reference temperature in K (273 K). Constant a, b and c are to be determined from the experimental data from the regression. The effective thermal conductivity of predicted data from Eq. 2 is compared with the experimental data. Table 3 shows the regression constants a, b and c obtained from experimental data for Eq. 2, respectively. These constants are used to verify the experimental data for prediction as shown in Fig. 8. It is observed from Fig. 8 that all data points are very near to the curve fit line. The maximum and minimum deviation was observed to be 2.24% and −2.18%, respectively. Hence, the proposed Eq. 2 can predict the effective thermal conductivity with a less deviation and valid for the <2 wt% for a temperature range from 30 to 50 °C.

Table 3 a, b, c and R^2 values of developed equations

	A	*b*	*c*	R^2
Equation 2	1.0513	0.075	0.682	0.945

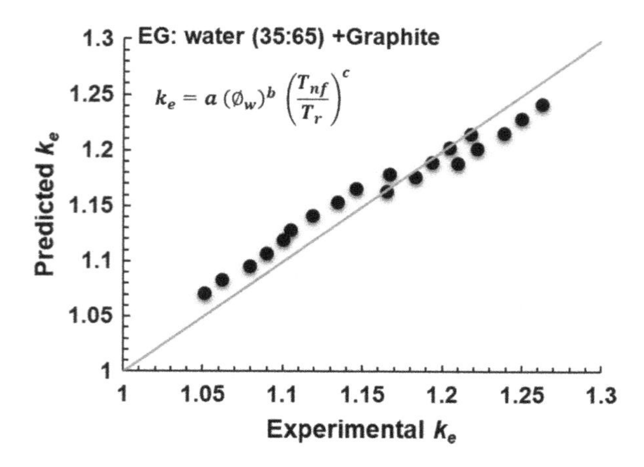

Fig. 8 Comparison of effective thermal conductivity from experimental and predicted from Eq. 2

4 Conclusion

In this study, graphite nanoparticles were dispersed in EG: water (35:65) using ultrasonication. The thermal conductivity of prepared nanofluid was investigated at a temperature range of 30–50 °C and weight concentration of 0.5–2 wt% using KD2 Pro thermal properties analyser. Stability studies of EG: water-based graphite nanofluids were investigated over a period of time for 0.5–2 wt% of weight concentration. The results indicated that the effective thermal conductivity of EG: water-based graphite nanofluid is increased with nanoparticle addition and increase in temperature. The maximum increase of thermal conductivity of 26.67% was observed at 2 wt% of concentration. Stability study showed that prepared EG: water-based graphite nanofluids were stable for 30 days without any visual sedimentation and change in thermal conductivity. Proposed correlation for effective thermal conductivity based on weight concentration (0.5–2 wt%) and temperature (30–50 °C) has shown very good prediction with experimental results.

Acknowledgements Financial support for this study was provided by a grant (ARDB/01/2031857/M/I) from the Aeronautical Research and Development Board (AR&DB)—Defence research and development organization, India.

References

1. Yu, W., Xie, H.: A review onnanofluids: preparation, stabilitymechanisms, and applications. J. Nanomater. **87**, 1–17 (2012)
2. Yang, L., Du, K.: A comprehensive review on heat transfer characteristics of TiO_2 nanofluids. Int. J. Heat Mass Transf. **108**, 11–31 (2017). https://doi.org/10.1016/j.ijheatmasstransfer.2016.11.086
3. Ukkund, S.J., Ashraf, M., Udupa, A.B., Gangadharan, M., Pattiyeri, A., Marigowda, Y.K., Patil, R., Puthiyllam, P.: Synthesis and characterization of silver nanoparticles from Fuzarium

oxysporum and investigation of their antibacterial activity. Mater. Today Proc. **9**, 506–514 (2019). https://doi.org/10.1016/j.matpr.2018.10.369
4. Ukkund, S.J., Raghavendra, M.J., Marigowda, Y.K.: Biosynthesis and characterization of silver nanoparticles from Penicillium notatum and their application to improve efficiency of antibiotics. IOP Conf. Ser.: Mater. Sci. Eng. **577**, 1–11 (2019). https://doi.org/10.1088/1757-899X/577/1/012002
5. Pastoriza-Gallego, M., Lugo, L., Legido, J., Piñeiro, M.M.: Thermal conductivity and viscosity measurements of ethylene glycol-based Al_2O_3 nanofluids. Nanoscale Res. Lett. **6**, 221 (2011). https://doi.org/10.1186/1556-276X-6-221
6. Diao, Y.H., Li, C.Z., Zhao, Y.H., Liu, Y., Wang, S.: Experimental investigation on the pool boiling characteristics and critical heat flux of Cu-R141b nanorefrigerant under atmospheric pressure. Int. J. Heat Mass Transf. **89**, 110–115 (2015). https://doi.org/10.1016/j.ijheatmasstransfer.2015.05.043
7. Pasha, J., Ramis, M.K., Yashawantha, K.M.: The effect of sonication time on alumina nanofluids with paradoxical behavior. Nano Trends J. Nanotechnol. Its Appl. **16**, 31–40 (2015)
8. Yashawantha, K.M., Afzal, A., Ramis, M.K., Shareefraza, J.U., Ramis, M.K., Ukkund, S.J.: Experimental investigation on physical and thermal properties of graphite nanofluids. In: AIP Conference Proceedings, p. 020057 (2018). https://doi.org/10.1063/1.5079016
9. Kumar, V., Tiwari, A.K., Ghosh, S.K.: Effect of variable spacing on performance of plate heat exchanger using nanofluids. Energy **114**, 1107–1119 (2016). https://doi.org/10.1016/j.energy.2016.08.091
10. Kareemullah, M., Chethan, K.M., Fouzan, M.K., Darshan, B. V, Kaladgi, A.R., Prashanth, M.B.H., Muneer, R., Yashawantha, K.M.: Heat transfer analysis of shell and tube heat exchanger cooled using nanofluids. Recent Pat. Mech. Eng. 12:350–356. http://dx.doi.org/https://doi.org/10.2174/2212797612666190924183251
11. Srinivas, T., Vinod, A.V.: Chemical engineering and processing: process intensification heat transfer intensification in a shell and helical coil heat exchanger using water-based nano fluids. Chem. Eng. Process. **102**, 1–8 (2016). https://doi.org/10.1016/j.cep.2016.01.005
12. Zhu, H., Zhang, C., Tang, Y., Wang, J., Ren, B., Yin, Y.: Preparation and thermal conductivity of suspensions of graphite nanoparticles. Carbon **45**, 226–228 (2007). https://doi.org/10.1016/j.carbon.2006.07.005
13. Wang, B., Wang, X., Lou, W., Hao, J.: Thermal conductivity and rheological properties of graphite/oil nanofluids. Colloids Surf. A **414**, 125–131 (2012). https://doi.org/10.1016/j.colsurfa.2012.08.008
14. Kole, M., Dey, T.K.: Investigation of thermal conductivity, viscosity, and electrical conductivity of graphene based nanofluids. J. Appl. Phys. **113**, 084307 (2013). https://doi.org/10.1063/1.4793581
15. Mehrali, M., Sadeghinezhad, E., Latibari, S., Kazi, S., Mehrali, M., Zubir, M.N.B.M., Metselaar, H.S.: Investigation of thermal conductivity and rheological properties of nanofluids containing graphene nanoplatelets. Nanoscale Res. Lett. **9**, 15 (2014). https://doi.org/10.1186/1556-276X-9-15
16. Yashawantha, K.M., Asif, A., Ravindra Babu, G., Ramis, M.K.: Rheological behavior and thermal conductivity of graphite–ethylene glycol nanofluid. J. Test. Eval. 49 (2021). https://doi.org/10.1520/JTE20190255
17. Usri, N.A., Azmi, W.H., Mamat, R., Hamid, K.A., Najafi, G.: Thermal conductivity enhancement of Al_2O_3 nanofluid in ethylene glycol and water mixture. Energy Procedia **79**, 397–402 (2015). https://doi.org/10.1016/j.egypro.2015.11.509
18. Chiam, H.W., Azmi, W.H., Usri, N.A., Mamat, R., Adam, N.M.: Thermal conductivity and viscosity of Al2O3 nanofluids for different based ratio of water and ethylene glycol mixture. Exp. Thermal Fluid Sci. **81**, 420–429 (2017). https://doi.org/10.1016/j.expthermflusci.2016.09.013
19. Ma, L., Wang, J., Marconnet, A.M., Barbati, A.C., McKinley, G.H., Liu, W., Chen, G.: Viscosity and thermal conductivity of stable graphite suspensions near percolation. Nano Lett. **15**, 127–133 (2015). https://doi.org/10.1021/nl503181w

20. Sundar, L.S., Ramana, E.V., Singh, M.K., Sousa, A.C.M.: Thermal conductivity and viscosity of stabilized ethylene glycol and water mixture Al_2O_3 nano fluids for heat transfer applications: an experimental study. Int. Commun. Heat Mass Transfer **56**, 86–95 (2014). https://doi.org/10.1016/j.icheatmasstransfer.2014.06.009
21. Sundar, L.S., Farooky, M.H., Sarada, S.N., Singh, M.K., Farooky, H., Sarada, S.N., Singh, M.K.: Experimental thermal conductivity of ethylene glycol and water mixture based low volume concentration of Al2O3 and CuO nanofluids. Int. Commun. Heat Mass Transfer **41**, 41–46 (2013). https://doi.org/10.1016/j.icheatmasstransfer.2012.11.004
22. Hamid, K.A., Azmi, W.H., Mamat, R., Usri, N.A.: Thermal conductivity enhancement of TiO_2 nanofluid in water and ethylene glycol (EG) mixture. Indian J. Pure Appl. Phys. **54**, 651–655 (2016)
23. https://www.srlchem.com/, https://www.srlchem.com/.
24. ASHRAE: Handbook—Fundamentals (SI Edition). American Society of Heating, Refrigerating and Air-Conditioning Engineers, Inc. (2017)
25. Yashawantha, K.M., Vinod, A.V.: ANN modelling and experimental investigation on effective thermal conductivity of ethylene glycol : water nanofluids. J. Therm. Anal. Calorim. (2020). https://doi.org/10.1007/s10973-020-09756-y
26. Hemmat Esfe, M., Karimipour, A., Yan, W.-M., Akbari, M., Safaei, M.R., Dahari, M.: Experimental study on thermal conductivity of ethylene glycol based nanofluids containing Al2O3 nanoparticles. Int. J. Heat Mass Transf. **88**, 728–734 (2015). https://doi.org/10.1016/j.ijheatmasstransfer.2015.05.010
27. Ladjevardi, S.M., Asnaghi, A., Izadkhast, P.S., Kashani, A.H.: Applicability of graphite nanofluids in direct solar energy absorption. Sol. Energy **94**, 327–334 (2013). https://doi.org/10.1016/j.solener.2013.05.012

Parametric Study of Leading-Edge Tubercle: Bio-inspired Design of Darrieus Vertical Axis Wind Turbine

Punit Prakash, Abhishek Nair, Joseph Manoj, Thomas Mathachan Thoppil, and Nishant Mishra

Abstract An aerodynamic blade with NACA0018 profile has been modified to reduce the wake by adding tubercles to the leading edge of a three-bladed Darrieus vertical-axis wind turbine (VAWT). The addition of tubercle is based on biomimicry, as seen in whales, to investigate enhancement in separation length. A 3D model of a Darrieus VAWT with tubercle blade profile turbine has been modeled and numerically investigated using a transient blade row model which of ANSYS CFX which solves continuity, momentum, turbulence eddy dissipation, and turbulence kinetic energy class of equations. A comparative analysis has been performed and Darrieus VAWT with modified blade shows promising result as compared to the baseline model.

Keywords Darrieus vertical-axis wind turbine · Biomimetics · Leading-edge tubercle · NACA0018

1 Introduction

The world energy sector is transiting from a coal-based economy toward a sustainable source of energy for power generation. As the prices for solar and wind energy generation keep on decreasing, more organizations want to add carbon neutrality and off-grid power solutions to their work culture. According to IRENA [1], the total

P. Prakash (✉) · A. Nair · J. Manoj · T. M. Thoppil · N. Mishra
Shiv Nadar University, Gautam Buddh Nagar, Uttar Pradesh, India
e-mail: pp431@snu.edu.in

A. Nair
e-mail: an233@snu.edu.in

J. Manoj
e-mail: jm886@snu.edu.in

T. M. Thoppil
e-mail: tt436@snu.edu.in

N. Mishra
e-mail: nishant.mishra@snu.edu.in

P. Muthukumar et al. (eds.), *Innovations in Sustainable Energy and Technology*,
Advances in Sustainability Science and Technology,
https://doi.org/10.1007/978-981-16-1119-3_22

installed capacity of wind energy has crossed 550 GW and the annual deployment is around 50 GW till 2018. Depending on the mechanical design, the wind turbine is classified as either horizontal-axis wind turbine (HAWT) or vertical-axis wind turbine (VAWT). The VAWTs are further classified as Savonius type and Darrieus type. The main advantage of VAWT is its single moving part (the rotor) where no yaw mechanism required and almost all of the components of VAWT requiring maintenance are located at the ground level, facilitating the maintenance work appreciably as stated by Islam et al. [2]. Brusca et al. [3] studied the performance of aspect ratio on VAWT and stated that the lower results have high power coefficients and also improves inertial moment for the turbine. Fish [4, 5] in his work discusses biomimicry of humpback whales (Megaptera Novaeangliae). The tubercles on flippers are similar to strakes of an aircraft. Large vortices generated by use of strakes and thus the stall characteristics of a wing changes. Tubercles are responsible for the formation of paired vortices in the troughs between tubercles. These vortices interact with the flow over the tubercles to keep the flow attached to the wing surface and delay stall. Paula [6] in his work states higher tubercle performance is achieved for thinner airfoil at lower Reynolds number. Sanz [7] in his experimental study stated the retardation of stall due to the tubercles leading edge and also they stated a small amplitude–wavelength ratio is associated with soft stall and gradual loss of lift. Ahmad et al. [8] stated tubercle leading-edge (TLE) blade design has increased clarity in aerodynamic study of flow separation, tonal noise, and dynamic stall. Mishra et al. [9] studied the effect of endplates and stated that addition of endplates reduces the effect of trailing vortices while compared with traditional turbines with no endplates. Seeni et al. [10] collated the work on tubercles by various researchers and came up with a few notable outcomes. Tubercles can delay stall and result in a higher angle of attack for a maximum coefficient of lift. Tubercles maintain laminar flow during surge conditions and enhance performance during turbulence when compared to straight-edged. For NACA0020 and NACA0021 profiles, they found that the lift and drag were lesser than the baseline unmodified model. An exception in this trend was observed with NACA 634-021. Zhu and Gao [11] attempted to incorporate the design changes of the leading-edge protuberance on the performance of a rudder. Two cases of Re have been taken, one at $3.2 \times e^5$ and the other at $8 \times e^5$. The drag coefficient and lift-to-drag ratio of the unchanged rudder was observed lower than that of the changed rudder. The bionic rudder results in more lift force than the unmodified rudder at angles of attack lower than 16°. Biadgo et al. [12] analyzed design changes to vertical-axis wind turbines by comparing double multiple stream tube model (DMST) and computational fluid dynamics (CFD) models and stated that DMST overestimates C_p. Bai et al. [13] investigated NACA0015 airfoil section with tubercles and concluded that increasing the values of lift-to-drag ratio and delaying the stall angles of attack are the two options for achieving an improved C_p value, so tubercles along the leading edge may energize the boundary layer and avoid the flow separation, and therefore, delay the stall.

2 Problem Statement

A comparative computational study has been performed between model A, model B, and model C as shown in Fig. 1a–c. Model A represents a straight blade VAWT with NACA0018 profile. Model B and C are the variations of chord length for a tubercle profile. The study focuses on the comparative analysis between a straight blade turbine and a tubercle blade turbine. The subparameters involved are wavelength and amplitude of the tubercle profile. The wavelength is kept constant, and the amplitude has been changed for model B and model C. The selection for amplitude is based on the angular cut of 40 and 45° of the blade profile from the ends such that the tubercle can maintain compatibility with the base model. Model B is a combination of chord length A and B as described in Table 1, Similarly, Model C is a combination of chord length A and C.

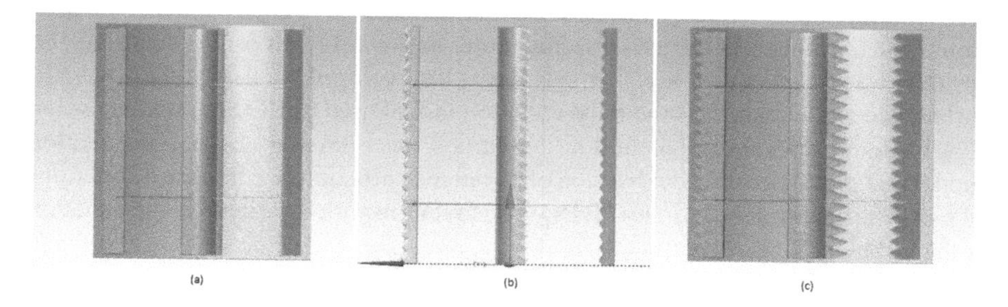

Fig. 1 Schematic overview. **a** Model A, **b** model B, **c** model C

Table 1 Specification for rotor

Parameters	Symbols	Dimension
Chord length (C)		86 mm
Chord length (B)		78 mm
Chord length (A)		60 mm
Length	L	600 mm
Diameter of turbine	D	300 mm
Diameter of central column	d_1	50 mm
Diameter of connector rod	d_2	20 mm
Connector rod location		150, 450 mm
Wavelength		30 mm
NACA profile		0018

3 Governing Physics and Solver Details

This study was performed using CFX a CFD tool in ANSYS Academic 17.2 for computational analysis on a system having Intel i5-4570 CPU @ processor with 8 GB 2666 MHz DDR4 RAM. A modular approach was used for modeling followed by meshing and finally defining the physics. The geometry was divided into two subdomains, rotor and stator. The rotor consists of the rotating domain, and the stator is the stationary portion for the fluid flow interaction. The computation concentrates only on the fluid interaction and the solid interaction with fluid has not been considered in this work. A module in CFX transient blade model has been used for combining the rotor and stator portion for setting up the boundary condition which used profile transformation and transient method of time integration with time period having a passing period 0.0698132 s and no. of time steps per period was 100 with time steps 0.0006 s and maximum number of period to be 25. The analysis uses continuity, momentum, turbulence eddy dissipation, and turbulence kinetic energy class of equations. The transient scheme uses a second order backward Euler for the solution. The residual was targeted to $1 \times e^{-4}$. $k - \epsilon$ Reynolds-averaged Navier–Stokes (RANS) turbulence model, which is the most common model used in CFD to simulate mean flow characteristics for turbulent flow conditions, has been used. It is a two-equation model that gives a general description of turbulence utilizing two transport equations of partial differential equations (PDEs). The first transported variable in the equation is the turbulent kinetic energy (k).

$$\frac{\partial(\rho k)}{\partial t} + \frac{\partial(\rho k u)}{\partial x_i} = \frac{\partial}{\partial x_i}\left[\frac{\mu}{\sigma k} \times \frac{\partial k}{\partial x_j}\right] + 2\mu_t E_{ij} - \rho \tag{1}$$

The second transported variable in equation is the rate of dissipation of turbulent kinetic energy (ϵ).

$$\frac{\partial(\rho \varepsilon)}{\partial t} + \frac{\partial(\rho \varepsilon u_i)}{\partial x_i} = \frac{\partial}{\partial x_j}\left[\frac{\mu_t}{\sigma_\varepsilon}\frac{\partial \varepsilon}{\partial x_j}\right] + C_{1\varepsilon}\frac{\varepsilon}{k}2\mu_t E_{ij} - \frac{C_{2\varepsilon}\rho\varepsilon^2}{k} \tag{2}$$

Equation (1) and (2) are used to solve RANS equations and the boundary conditions have been implemented. By implementing the natural boundary condition on the domain such as wind speed and pressure to the inlet and outlet and no slip boundary condition is implemented on the outer walls of the stator and on the wall boundary of the turbine as shown in Fig. 2a. The inlet boundary conditions are defined as $u(x, y, z)$ as 5, 0, 0 m/s and corresponding pressure as 1 Pa and the temperature is predefined in the solver as 25 °C. The corresponding outlet pressure is defined as 0 Pa for adequate pressure difference for flow. The rpm was defined as 90 rad/s to check the flow condition of the turbine blade and the variation caused by the TLE. A detailed view of various dimensions for the domain has been shown below in Fig. 2b and the description has been presented in Table 2.

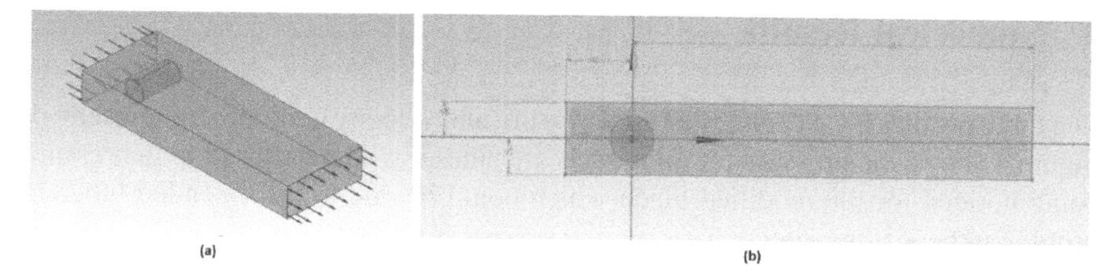

Fig. 2 **a** Boundary condition for the analysis. **b** Description of the domain

Table 2 Specifications for the domain

Symbol	Dimension (mm)
D2	320
H4	500
H5	3000
V6	250
V7	250

3.1 Meshing

The meshing is the process of discretization of boundary into smaller elements for the rotor blades. The meshing scheme has been as shown in Fig. 3 where (a)–(c) shows the meshed section of rotor for model A, B, C and (d) shows the 3D model for a stator. The mesh information has been shown in Table 3.

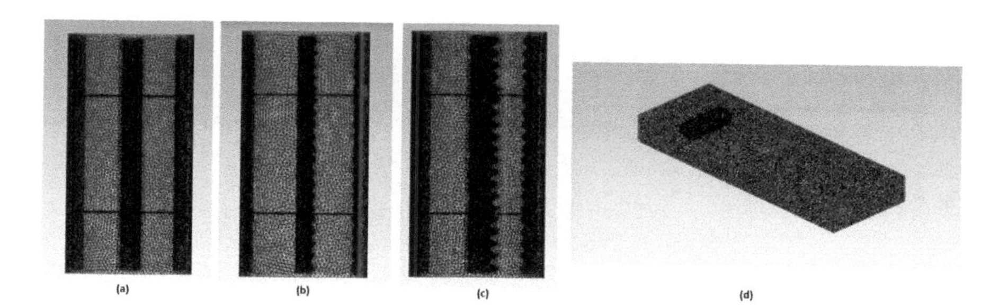

Fig. 3 Meshing for **a** Model A, **b** model B, **c** model C, **d** Stator 3D

Table 3 Mesh information

Model	Nodes	Element
Model A	1,519,463	1,021,528
Model B	2,206,408	1,475,412
Model C	2,247,116	1,510,315
Stator	108,407	71,483

4 Numerical Results

The three profiles have been tested and pressure and velocity contour have presented. Figure 4 shows the pressure for all models are similar with a minor drop in pressure being noticed for the modified blade which could be due to the extended tubercle profile on the leading edge.

Figure 5 shows velocity contour of the turbine model, a sharp rise is observed in the case of turbine B and C. This could be due to the presence of tubercle profile. The turbulence in the region increases, and this increase in turbulence causes the flow to reunite with the blade surface and thus the separation length will increase. The increase in separation length helps in lowering the drag. To study a 3D model and understand the effect of tubercle, a streamline plot of the wind flow has been shown in Fig. 6 which shows the flow tends to move toward the through of the tubercle and also it can be stated that flow shift to other planes this phenomena is difficult to

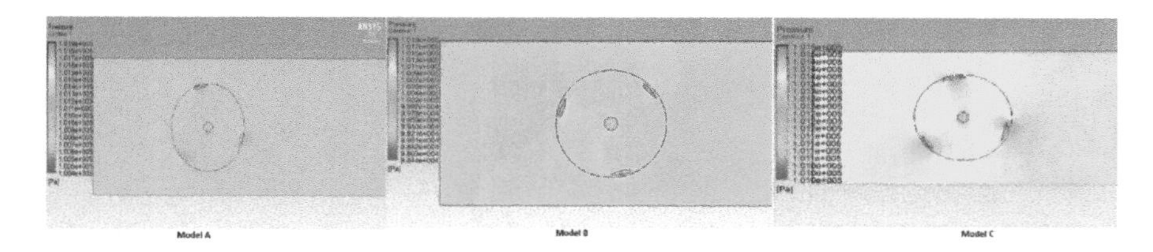

Fig. 4 Pressure contour

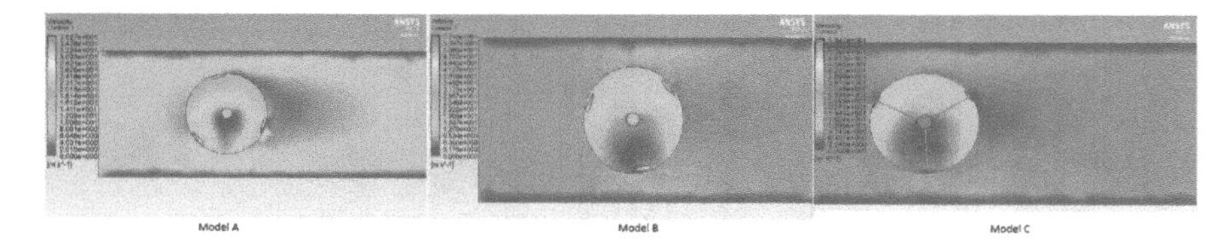

Fig. 5 Velocity contour

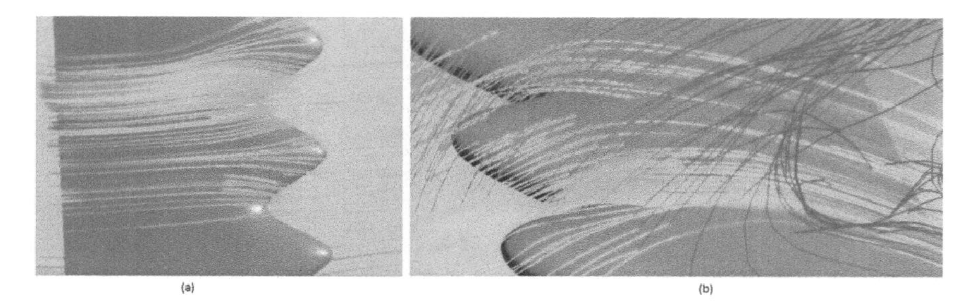

Fig. 6 **a** Streamline flow around tubercle profile. **b** Reattaching of flow

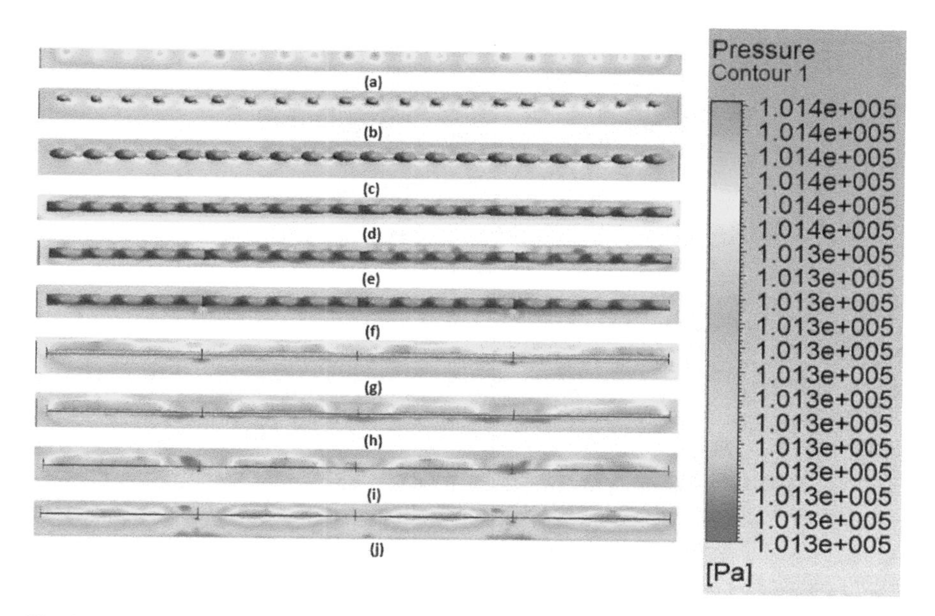

Fig. 7 Pressure contour of TLE for model B at 0° to airstream

observed in a 2D study of the flow. The phenomenon of reattaching of the flow on the tubercle blade due to the presence of tubercle profile has been shown in Fig. 6b. It can be observed that the separation length increases due to reattachment of the flow which is only possible due to an early separation caused by the tubercle profile. Non-uniformity of the flow lines is due to chaotic flow that leads to turbulence.

TLE creates a variation in pressure at the point the airstream hits the blade and the corresponding pressure profile as shown in Fig. 7, which explains the movement of streamlined flow toward the valleys of the blade profile where (a) is the location at the tip of the TLE, (b)–(i) represents the blade section at an interval of 1 cm each from the leading edge, and (j) shows the wake region after 1 cm of blade.

The plots were generated for various locations whose velocity contours have been compared with the graphs for better understanding. Planes were placed on various sections to observe the effect of TLE. The turbine comparison has been divided into three parts: near the base, middle section, and top of turbine. We observed a reduction in wake region for Model B is greater in comparison to Model A and Model C as illustrated in Fig. 8. The wake region shows considerably decreased in case of the Model B that indicates a greater separation length.

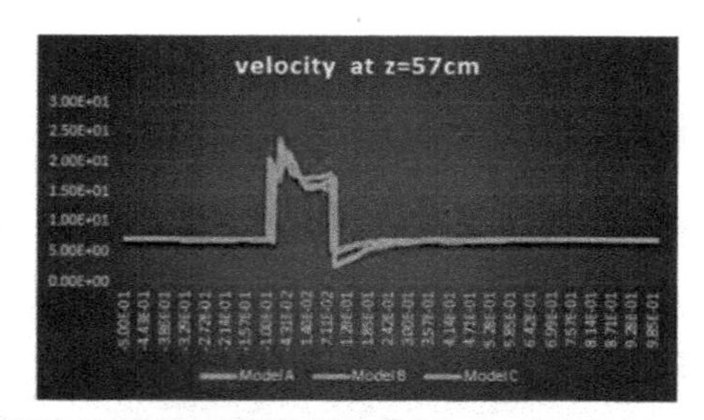

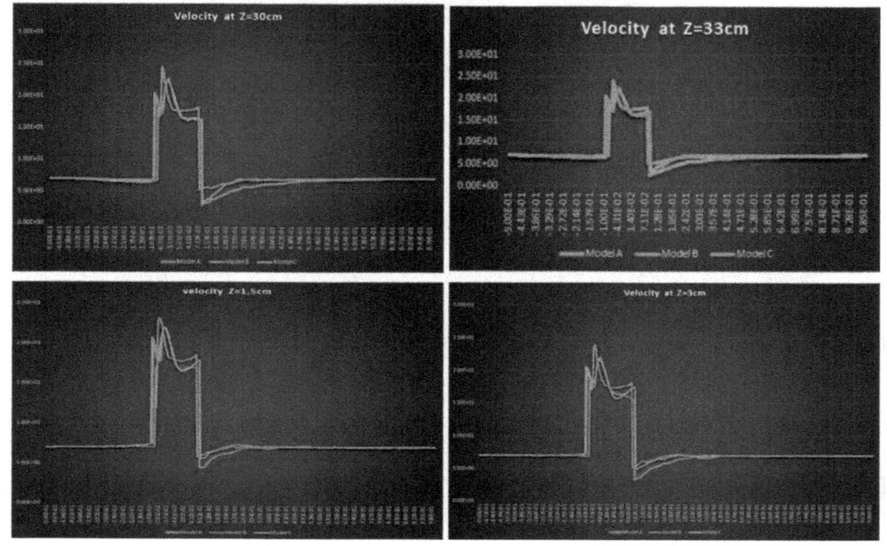

Fig. 8 Velocity on various plane of the domain

5 Conclusion

In this work, three turbine configurations marked as Model A, Model B, and Model C are being tested and the results indicate that a tubercle profile increases the separation length and decreases the wake region and subsequently the drag. The results further explain the variability of flow over the blade profile. Near the end tips of the blade, i.e., near base and top a tip loss occurred and resulted in the lower value of velocity contour which can be clearly illustrated in the result. The tubercle profile shows better results compared to the straight bladed profile. The result shows that the tubercle has an advantage over straight blade turbines but we need to check more parameters like torque to state a more decisive conclusion.

References

1. IRENA: Future of Wind: Deployment, Investment, Technology, Grid Integration and Socio-Economic Aspects. A Global Energy Transformation Paper. International Renewable Energy Agency, Abu Dhabi (2019)
2. Islam, M., David, S.K.T., Amir, F.: Aerodynamic models for Darrieus-type straight-bladed vertical axis wind turbines. Renew. Sustain. Energy Rev. **12**(4), 1087–1109 (2008)
3. Brusca, S., Lanzafame, R., Messina, M.: Design of a vertical-axis wind turbine: how the aspect ratio affects the turbine's performance. Int. J. Energy Environ. Eng. **5**(4), 333–340 (2014)
4. Fish, F.E.: Biomimetics and the Application of the Leading-Edge Tubercles of the Humpback Whale Flipper, 1st edn. Springer, Switzerland (2020)
5. Fish, F.E., Weber, P.W., Murray, M.M., Howle, L.E.: The tubercles on humpback whales' flippers: application of bio-inspired technology. Integr. Comp. Biol. **51**(1), 203–213 (2011)
6. Paula, A.A.: The Airfoil Thickness Effects on Wavy Leading Edge Phenomena at Low Reynolds Number Regime. Universidade de São Paulo (2016)
7. Sanz, M., María, C.: An Experimental Study of the Effect of Tubercle Leading Edge on the Performance of NACA Airfoils (2018)
8. Ahmad K.A., Aftab, S.M.A., Razak, N.A, Rafie, A.S.M.: Mimicking the humpback whale: an aerodynamic perspective. Prog. Aerosp. Sci. **84**:48–69 (2016)
9. Mishra, N., Gupta, A.S., Dawar, J., Kumar, A., Mitra, S.: Numerical and experimental study on performance enhancement of Darrieus vertical axis wind turbine with wingtip devices. J. Energy Resour. Technol. **140**(12), 121–201 (2018)
10. Seeni, P., Rajendran, P., Kutty, H.A.: A critical review on tubercles design for propellers. In: IOP Conference. Series: Materials Science and Engineering, vol 370 (2018)
11. Zhu, W., Gao, H.: Numerical investigation of bionic rudder with leading-edge protuberances. J. Offshore Mech. Arct. Eng **142**(1), 011–802 (2020)
12. Biadgo, A.M., Simonovic, A., Komarov, D., Stupar, S.: Numerical and analytical investigation of vertical axis wind turbine. FME Trans. **41**, 49–58 (2013)
13. Bai, C.J., Lin, Y.Y., Lin, S.Y., Wang, W.C.: Computational fluid dynamics analysis of the vertical axis wind turbine blade with tubercle leading edge. J. Renew. Sustain. Energy **7**, 033124 (2015)

Shorea robusta (Sal) Fallen Leaves Briquette—A Potential Bioenergy Fuel for Rural Community

Rajib Bhowmik and Bhaskor Jyoti Bora

Abstract Most of the rural household utilize loose biomass as fuel for cooking. Direct combustion of loose biomass deteriorates the household air quality that indirectly affects human health. Conversion of loose biomass in densified form can be used as alternate cooking fuel. In the present study, *Shorea robusta* fallen leaves are considered for briquette production. Four different pressures (4, 8, 12, and 16 MPa) were applied for densification of the crushed leaves using manual hydraulic press. Proximate analysis was performed according to ASTM standards. The results showed that the moisture content (M) was 6.21% while the fixed carbon (FC), volatile matter (VM), and ash (ASH) are 16.56%, 72.80% and 4.43%, respectively, whereas ultimate analysis depicts the elemental composition of the biomass fuel namely carbon, hydrogen, oxygen, nitrogen, sulfur is found to be 44.56%, 4.16%, 43.69%, 0.43%, and 0.59%, respectively. Higher heating value (HHV) of the *Shorea robusta* leaves was determined using bomb calorimeter apparatus and found to be 17.43 MJ/kg. As pressure increases from 4 to 16 MPa, properties like briquette density increases in the range 961.42–1592.36 kg/m^3, shatter index increases from 80.59–94.37%, durability increases from 67.26 to 92.28%, and water penetration resistance increases from 57.14 to 78.66%. Higher burning rate of 8.40 g/min can be seen for low-pressurized briquette (4 MPa) whereas high-pressurized briquette (16 MPa) has a lower burning rate of 2.67 g/min.

Keywords *Shorea robusta* leaves · Briquette · Densification · Proximate analysis · Ultimate analysis · Physical properties · Combustion properties

---

R. Bhowmik (✉)
Girijananda Chowdhury Institute of Management and Technology, Guwahati, India
e-mail: rajibaec@gmail.com

B. J. Bora
Rajiv Gandhi Institute of Petroleum Technology, Bangalore, India

P. Muthukumar et al. (eds.), *Innovations in Sustainable Energy and Technology*,
Advances in Sustainability Science and Technology,
https://doi.org/10.1007/978-981-16-1119-3_23

1 Introduction

Energy is considered as one of the most essential needs of mankind. Both energy production and utilization are the important indicators of a country's economic progress. Around 70% of the total population lives in rural areas and mainly depends on solid biomass as cooking fuel [1]. There are different types of biomass feedstock available in rural areas like agricultural and forest residue [2], coffee-pine wood residue [3], switch grass [4], cotton stalk [5], banana leaves [6], and sawdust [7] that are used for bioenergy production. Out of the different feedstock available, the end user of the rural communities mostly depends on loose biomass, as loose biomass is easily available and abundant for cooking and water heating purposes. With the use of bioenergy, CO_2 emission to the environment reduces. However, burning loose biomass for household activity produces smoke, thereby deteriorates the indoor air quality and consequently affects human health. On the other hand, loose biomass is having low bulk density and therefore it is difficult to transport, handling, and storage. Densification of loose biomass is one of the ways to reduce the carbon footprints and address the above problems.

Several studies have been conducted by researchers on the production, physical, and combustion properties of briquette from various feedstock. Chin and Siddiqui [8] conducted experiments on the characteristics of biomass briquette produced using die pressures ranging from 5 to 7 MPa. Results reveal that briquette quality was improved and burning rate decreases with increase in applied pressure. Orisaleye et al. [9] in their research work studied the effect of different input factors such as pressure, temperature, hold time, and particle size during densification of corncob. Significant factors like pressure, temperature, and particle size were identified as prime factor that directly affects the briquette density whereas holding time does not have any relation with briquette density. Li and Liu [10] performed experiment on the briquette produced by applying pressures ranging from 34 to 138 MPa using oak sawdust, cotton wood sawdust, and pine sawdust without binder and confirms that oak sawdust logs were the strongest among the briquettes. Yumak et al. [11] studied three different input parameters for densification of Soda weed (*Salsola tragus* L) without binder. Optimum condition for producing soda weed briquetting was evaluated by statistical analysis software and found that moisture ranging from 7 to 10% with an applied pressure of 31.4 MPa and temperature of 85–105 °C can be applied for producing different briquette shapes. Zhang et al. [12] utilized millet bran for briquette production and found that good-quality briquettes are produced for a pressure range of 110–130 MPa. Kpalo et al. [13] performed experiment on the production of hybrid briquettes by mixing different ratio of corncobs and bark of oil palm trunk along with 10% by weight of waste paper as binder in order to determine the performance of the produced briquette. Experimental result shows that both 50:50 and 25:75 blended briquette gave similar values of boiling time, fuel burning rate and specific fuel consumption. Among the various blended ratios, 50:50 ratio briquette showed higher thermal fuel efficiency. Navalta et al. [14] studied the mechanical and thermal properties of two particle size bagasse briquette with three different %w/w biodegradable binders and concluded that both mechanical and thermal properties of nano-lignocellulose and nano-cellulose binders showed better results as compared to lignin binder. Chungcharoen et al. [15] investigated the production rate, mechanical properties, and fuel properties of briquette produced using different operating parameters and found that high production rate was obtained from mixture 65% cashew nut shell, 25% areca nut shell, and 10% by weight of cassava flour along with compressed screw speed of 90 rpm. Lisseth et al. [3] studied the production and characterization of densified briquette produced from a mixture of coffee shrub and pinewood under different operating parameters using piston press machine. Results reveal that high-quality briquette is produced by blending of two different residues.

In the present investigation, *Shorea robusta* (Sal) fallen leaves were considered for briquette production. The physical and combustion characteristics of *Shorea robusta* leaves briquette were performed to confirm the feasibility of being used as a bioenergy source for rural household.

2 Materials and Methods

2.1 Materials

Shorea robusta (Sal) is a large deciduous tree found in India. It is one of the most important sources of hardwood timber in India. The fallen leaves from the trees are collected from the nearby forest area. Hydraulic press was used for densification process.

2.2 Shorea robusta *Leaves Characterization*

The proximate analysis is a standardized analysis for estimating fixed carbon, volatile matter, moisture content, and ash. Ultimate analysis is done to determine the elemental composition of the biomass sample. High heating value of biomass was determined using oxygen bomb calorimeter as per ASTM standard.

2.3 Briquette Production

The collected leaves contain both raw and dried leaves. The raw leaves were separated and sun-dried for several days to remove any moisture present. Dried leaves are chopped into small pieces with the help of hammer mill. A 2 mm sieve performs the screening of chopped leaves. Screened chopped leaves are allowed to stand for two

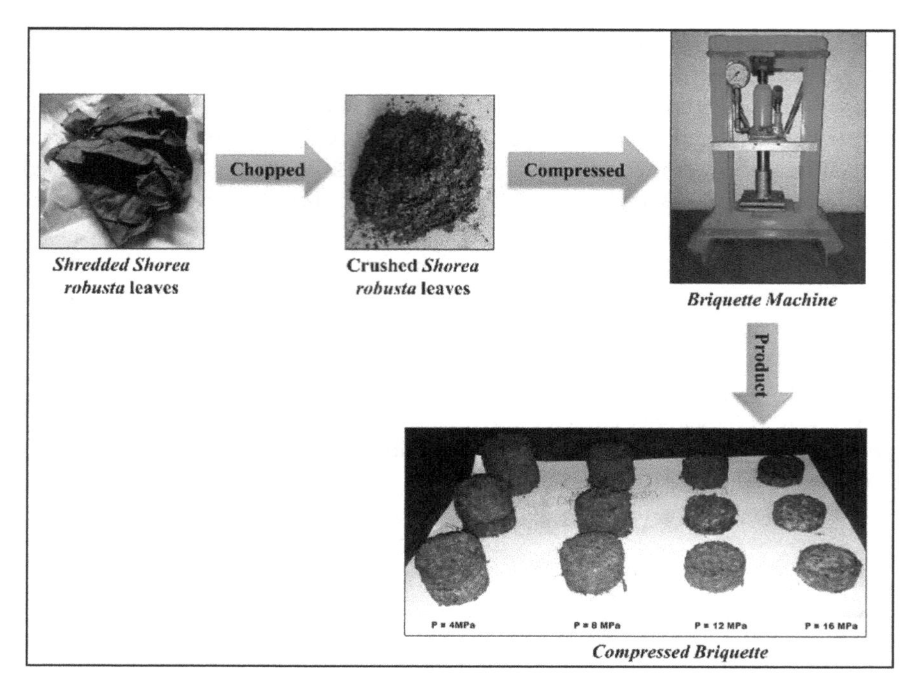

Fig. 1 Flowchart of briquette production

weeks for partially decompose. The decomposed leaves are then filled manually in a 35 mm internal diameter cylindrical hydraulic press. Different pressures ranging from 0 to 20 MPa with 5 s dwell time were applied to compress the loose biomass for briquette production. The compacted pressure was recorded from the pressure gauge. Vernier caliper determines the height and diameter of the produced briquette (Fig. 1).

2.4 *Proximate Analysis*

Moisture content

A known quantity (m_1) of briquette sample was taken in a silica crucible without lid and kept in oven maintained at 105 °C. After 1 h of oven drying, the crucible was taken out from the oven and kept in a desiccator for cooling. The oven-dried sample was weighted again (m_2). The moisture content (MC) was determined using the relation

$$\text{Moisture content}(\%) = \frac{m_1 - m_2}{m_1} \times 100 \tag{1}$$

Volatile Matter

The same crucible containing oven-dried sample (m_2) was covered with a vented lid and placed in a muffle furnace at 950 ± 20 °C for 7 min. The crucible was taken out from the muffle furnace and cooled to room temperature and weighted again (m_3). Volatile matter (VM) was determined using the relation

$$\text{Volatile Matter}(\%) = \frac{m_2 - m_3}{m_1} \times 100 \tag{2}$$

Ash Content

The crucible containing residual sample (m_3) from the volatile matter was taken and then heated without lid in a muffle furnace at 700 ± 50 °C for 30 min. The crucible was taken out from the muffle furnace, cooled in the desiccator, and weighted (m_4). Ash content (ASH) was determined using the relation

$$\text{Ash Content}(\%) = \frac{m_4}{m_1} \times 100 \tag{3}$$

Fixed carbon Content

Fixed carbon of the briquette sample was calculated using the relation.

$$\text{Fixed Carbon }(\%) = 100 - [\text{M}(\%) + \text{VM}(\%) + \text{ASH}(\%)] \tag{4}$$

2.5 *Ultimate Analysis*

CHNS analyzer was used to test the elemental composition. Oxygen content was calculated by subtracting the sum total of carbon, hydrogen, nitrogen, sulfur, and ash from 100%.

2.6 *Physical Properties*

Briquette density

Density of briquette was determined by knowing the volume of the compressed briquette. A vernier caliper measures the length and diameter of the briquette and therefore the volume was calculated. Digital weighing balance determines the mass of briquette. Density was calculated using the following relation

$$\text{Briquette density} = \frac{\text{mass of the briquette}}{\text{volume of the compressed briquette}} \tag{5}$$

Shattering Index

Transportation and handling of briquette from one place to another may result in sudden breaking of the briquette, thereby reduces the strength of the briquette. In this experiment, briquette is allowed to fall from a fixed height of 1 m as suggested by Rajaseenivasan [16]. The experiment was carried out two times and the average weight of the briquette before and after the experiment was noted. The percentage weight loss of the briquettes was determined by

$$\text{Weight loss}(\%) = \frac{\text{Initial briquette weight} - \text{Final briquette weight}}{\text{Initial briquette weight}} \times 100 \tag{6}$$

$$\text{Shatter resistance}(\%) = 100 - \text{weight loss}(\%) \tag{7}$$

Tumbling Test

Durability is one of the desirable qualities of any briquette and can be determined from tumbling test. In this experiment, initial weight of briquette (W_i) before tumbling was recorded and the briquette was placed inside a 300 mm × 300 mm × 450 mm cubical box. The cubical box was mounted diagonally on a hollow shaft. The tumbling action was carried out by rotating the cuboid for 15 min. The final weight of briquette (W_f) after tumbling action was noted. The percentage weight loss due to tumbling action is calculated by [17]

$$\text{Weight loss } (\%) = \frac{W_\mathrm{i} - W_\mathrm{f}}{W_\mathrm{i}} \times 100 \tag{8}$$

$$\text{Durability index } (\%) = 100 - \text{weight loss } (\%) \tag{9}$$

Water resistance test

It is the resistance offered by the briquette from absorbing water in the form of moisture during transportation and handling. Water absorption capacity of briquette was determined by dispersing the briquette in a container having 150 mm of water at room temperature for 30 s. The higher the water resistance of briquette, the more is the weathering resistance. The percentage of water absorbed by the briquette can be calculated by the following relation [18]

$$\text{Water gained by the briquette } (\%) = \frac{W_\mathrm{f} - W_\mathrm{i}}{W_\mathrm{i}} \times 100 \tag{10}$$

$$\text{Water resistance}(\%) = 100 - \text{water gained by the briquette}(\%) \tag{11}$$

where W_i is the initial weight of the briquette before the test (g) and W_f is the final weight of the briquette after the test (g).

2.7 *Combustion Properties*

Burning Rate

Burning rate is the rate at which specific mass of fuel is burned completely into ashes. The experiment was carried out with an insulated wire gauge placed on the weighing digital balance. Briquette was placed on the wire gauge, and it is ignited by means of a Bunsen burner placed underneath the wire gauge. The time taken for complete burning the briquette into ashes was noted using a stopwatch. After every 10 s, the weight of the briquette was recorded until a constant weight was noted. The burning rate was measured using the following relation

$$\text{Burning rate} = \frac{\text{weight of the briquette burnt into ashes}}{\text{total time taken}} \tag{12}$$

Water Boiling Test

Water boiling test was carried out as per the standard procedure [16]. In the present study, cold start test was carried out to determine the amount of briquette required to boil a definite volume of water. Initially, an aluminum pot containing 1 l of water was taken and placed on the stove. Each briquette sample of known weight was put at regular interval below the stove and ignited till the water reaches its boiling temperature and was recorded using digital thermometer. Total amount of briquette burnt to reach the boiling point was noted. The time taken to arrive at the boiling point of water was also recorded using digital stopwatch. During the experiment, specific fuel consumed was calculated with the following relation

$$\text{Specific fuel consumed} = \frac{\text{mass of fuel consumed (kg)}}{\text{total mass of boiling water (l)}} \tag{13}$$

3 Results and Discussion

3.1 *Proximate and Ultimate Analysis*

From Table 1 it can be seen that sulfur and nitrogen content in fuel was below 1%. Previous study reveals the fact that the air pollution is minimized due to combustion of briquette containing sulfur and nitrogen less than 1% [19]

Table 1 Proximate and ultimate analysis of *Shorea robusta* leaves

Raw sample	Ultimate analysis (wt%, dry basis)					Proximate analysis (wt%, dry basis)				HHV (MJ/kg)
	C	H	O	S	N	M	FC	VM	ASH	
Shorea robusta leaves	44.56	4.16	43.69	0.43	0.59	6.21	16.56	72.8	4.43	17.43

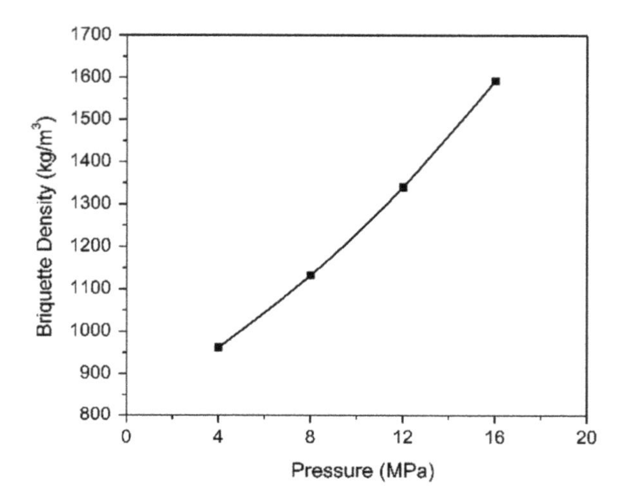

Fig. 2 Briquette density with pressure variation

3.2 *Physical Properties*

Briquette Density

Variation of briquette density with pressure ranging from 4 to 16 MPa is shown in Fig. 2. It can be observed that briquette density increases with increase in applied pressure. The highest density of 1592.36 kg/m^3 was measured at 16 MPa pressure and lowest value of 961.42 kg/m^3 can be observed for a pressure of 4 MPa. As pressure is increased, the bonding between leaves particle increases and porous structure reduces and thereby accommodates more quantity of leaves particle per unit volume. Lower-density briquettes are subjected to highly porous structure and hence may develop crack during transportation, storage, and handling.

Shattering Index

Figure 3 represents the variation of briquette shattering index with pressure. The lowest value of 80.59% shatter index can be observed for a 4 MPa pressurized briquette and a maximum shatter index of 94.37% can be observed for 16 MPa pressure. Briquette weight loss is severe in case of 4 MPa pressure briquette because at this pressure the chopped leaves are loosely bonded during compaction process. As reported by Moses and Augustina [20], shatter index should have more than

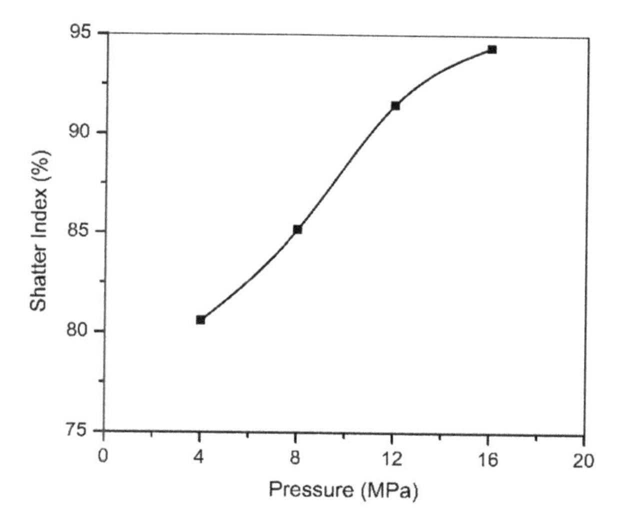

Fig. 3 Effect of applied pressure on shatter index of briquette

or equal to 95% for easy transportation and handling. From the present study, it can be concluded that briquette produced above 16 MPa can be a favorable option considering shatter index.

Tumbling test

Figure 4 shows that durability of briquette was increased from 67.26 to 92.28% for a pressure increase from 4 to 16 MPa, respectively. Mahadeo et al. [18] in their study reported that briquette produced with durability index less than 70% can cause dust emissions. Also, briquette with durability index above 80% is considered as quality

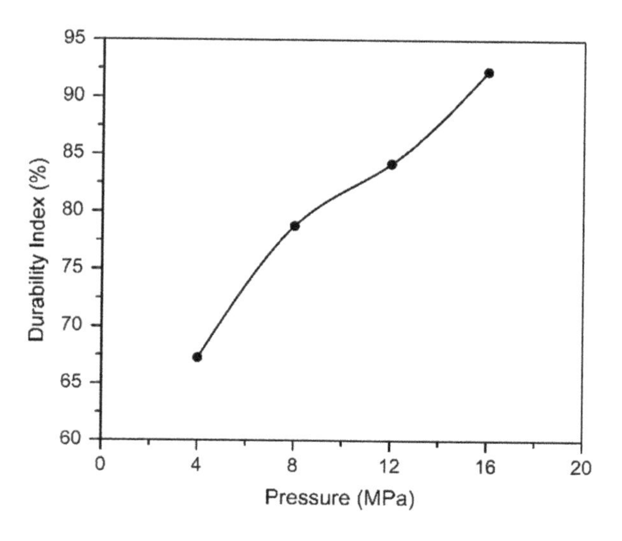

Fig. 4 Durability index versus pressure

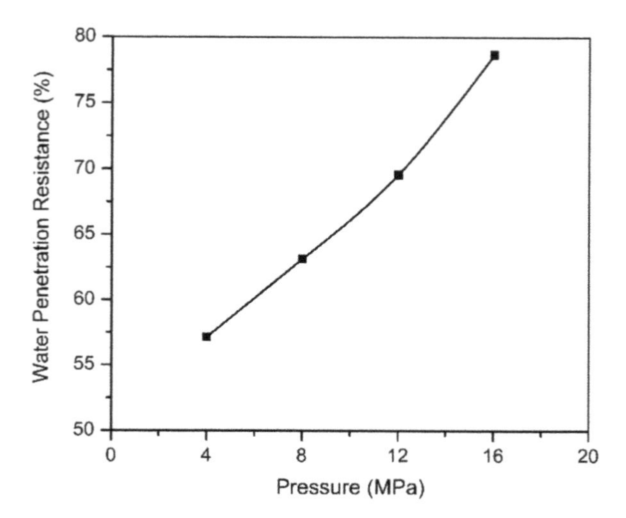

Fig. 5 Effect of applied pressure on water penetration resistance

briquette for transportation, handling, and storage [21]. Therefore, the study reveals that briquettes produced above 12 MPa pressure are good-quality briquettes.

Water Penetration Resistance Test

Good-quality briquette should have low affinity to water absorption during transportation and storage for long time. It was found from Fig. 5 that the maximum value of 78.66% resistance to water penetration can be seen for an applied pressure of 16 MPa. The reason for this increase is due to the decrease in voids between the chopped leaves particle of the compressed briquette and thereby reduces water absorption. However, results reported by Rajaseenivasan et al. [16] show similar trend of resistance to water absorption with pressure variation.

Burning Rate

Figure 6 depicts that the burning rate decreases significantly with pressure rise. Increasing the briquette density decreases voids in the briquette sample and thereby reduces the rate of mass loss during combustion. Higher burning rate of 8.4 g/min can be seen in case of low-pressurized briquette at 4 MPa. Higher-pressurized briquette of 16 MPa has a lower burning rate of 2.67 g/min because cylindrical briquette has higher packing factor that restricts airflow. As pressure increases, density of briquette increases and hence ignition time increases. As a result, specific fuel consumption decreases for densified briquette. It can be observed from Fig. 6 that high-pressurized briquette (16 MPa) has low specific fuel consumption of 0.072 kg/l while low-pressurized briquette (4 MPa) is having a higher specific fuel consumption of 0.126 kg/l.

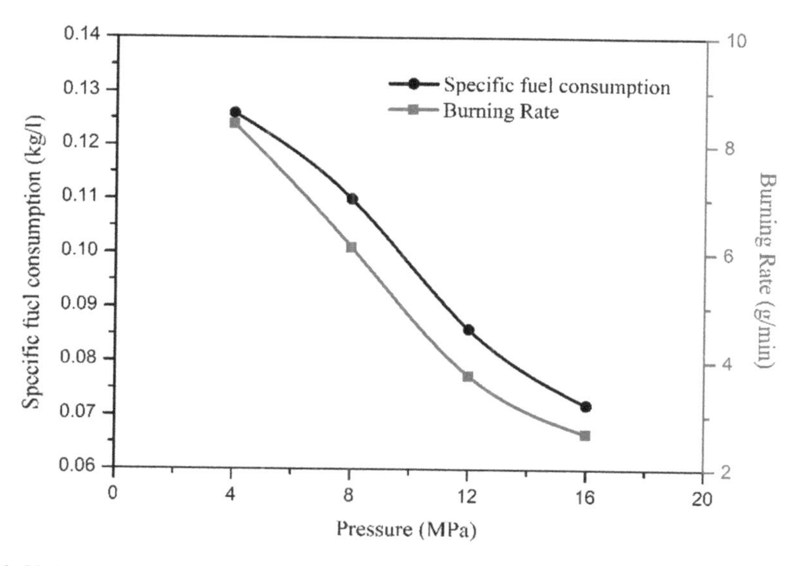

Fig. 6 Variation of burning rate and specific fuel combustion with applied pressure

4 Conclusion

From the study, it can be concluded that *Shorea robusta* leaves can be used for briquette production in rural areas as cooking fuel. The physical properties of briquette showed promising result above 16 MPa pressure and can be considered as a promising biomass fuel for the rural community. This implied that higher-densified briquettes are less inclined to damage during storage, handling, and transportation. It can also be concluded from the combustion test that the specific fuel consumption and burning rate were found to be lower for high-pressurized briquette. In addition, greenhouse gas emitted during briquette combustion contributes less as compared to open firing of unorganized biomass residue. This is because densified briquette has higher calorific value as well as higher density. Apart from this, deforestation can be prevented by proper utilization of loose biomass for briquette production that will help in developing sustainable carbon sink.

References

1. Ramachandra, T.V., Hegde, G., Setturu, B., Krishnadas, G.: Bioenergy: a sustainable energy option for rural India. Adv. For. Lett. **3**(1) (2014)
2. Okello, C., Pindozzi, S., Faugno, S., Boccia, L.: Bioenergy potential of agricultural and forest residues in Uganda. Biomass Bioenerg. **56**, 515–525 (2013)
3. Martinez, C.L.M., Sermyagina, E., Carneiro, A.D.C.O., Vakkilainen, E., Cardoso, M.: Production and characterization of coffee-pine wood residue briquettes as an alternative fuel for local firing systems in Brazil. Biomass Bioenerg. **123**, 70–77 (2019)

4. Kaliyan, N., Morey, R.V.: Constitutive model for densification of corn stover and switchgrass. Biosys. Eng. **104**, 47–63 (2009)
5. Onaji, P.B., Siemons, R.V.: Production of charcoal briquettes from cotton stalk in Malawi: methodology for feasibility studies using experiences in Sudan. Biomass Bioenerg. **4**, 199–211 (1993)
6. Maina, B.G.O., Souza, O., Marangoni, C., Rotza, D., Oliveira, A.P.N., Sellin, N.: Production and characterization of fuel briquettes from banana leaves waste. Chem. Eng. Trans. **37**, 439–444 (2014)
7. Sotannde, O.A., Oluyege, A.O., Abah, G.B.: Physical and combustion properties of briquettes from sawdust of *Azadirachta indica*. J. For. Res. **21**, 63–67 (2010)
8. Chin, O.C., Siddiqui, K.M.: Characteristics of some biomass briquettes prepared under modest die pressures. Biomass Bioenerg. **18**, 1–6 (2000)
9. Orisaleye, J.I., Jekayinfa, S.O., Adebayo, A.O., Ahmed, N.A., Pecenka, R.: Environmental effects effect of densification variables on density of corn cob briquettes produced using a uniaxial compaction biomass briquetting press. Energy Sour. Part A Recov. Util. Environ. Effect **40**, 3019–3028 (2018)
10. Li, Y., Liu, H.: High-pressure densification of wood residues to form an upgraded fuel. Biomass Bioenerg. **19**, 177–186 (2000)
11. Yumak, H., Ucar, T., Seyidbekiroglu, N.: Briquetting soda weed (*Salsola tragus*) to be used as a rural fuel source. Biomass Bioenerg. **34**, 630–636 (2010)
12. Zhang, J., Zheng, D., Wu, K., Zhang, X.: The optimum conditions for preparing briquette made from millet bran using generalized distance function. Renew. Energy **140**, 692–703 (2019)
13. Kpalo, S.Y., Zainuddin, M.F., Manaf, L.A., Roslan A.M.: Evaluation of hybrid briquettes from corncob and oil palm trunk bark in a domestic cooking application for rural communities in Nigeria. J. Clean. Prod. 124745 (2020)
14. Navalta, C.J.L.G., Banaag, K.G.C., Raboy, V.A.O., Go, A.W., Cabatingan, L.K., Ju, Y.H.: Solid fuel from co-briquetting of sugarcane bagasse and rice bran. Renew. Energy **147**, 1941–1958 (2019)
15. Chungcharoen, T., Srisang, N.: Preparation and characterization of fuel briquettes made from dual agricultural waste: cashew nut shells and areca nuts. J. Clean. Prod. **256**, 120434 (2020)
16. Rajaseenivasan, T., Srinivasan, V., Qadir, G.S.M., Srithar, K.: An investigation on the performance of sawdust briquette blending with neem powder. Alexandria Eng. J. **55**, 2833–2838 (2016)
17. Sengar, S.H., Mohod, A.G., Khandetod, Y.P., Patil, S.S., Chendake, A.D.: Performance of Briquetting machine for briquette fuel. Int. J. Energy Eng. **2**, 28–34 (2012)
18. Mahadeo, K., Dubey, A.K., Mahalle, D., Kumar, S.: Study on physical and chemical properties of crop residues briquettes for gasification. Am. J. Energy Eng. **2**, 51–58 (2014)
19. Akowuah, J.O., Kemausuor, F., Mitchual, S.J.: Physico-chemical characteristics and market potential of sawdust charcoal briquette. Int. J. Energy Environ. Eng. **3**, 1–6 (2012)
20. Moses, D.R., Augustina, D.O.: Some physical and mechanical properties of water lettuce (*Pistia stratiotes*) briquettes. Am. J. Sci. Technol. **1**(5), 238–244 (2015)
21. Fasina, O.O.: Physical properties of peanut hull pellets. Biores. Technol. **99**, 1259–1266 (2008)

Design Optimization of Vertical-Axis Wind Turbine Based on High Power Extraction by Using Computational Fluid Dynamics (CFD)

M. Ramesh, M. Senthil Kumar, R. Vijayanandh, K. Sundararaj, S. Balaji, and S. Bhagavathiyappan

Abstract Vertical-axis wind turbine (VAWT) is evergreen and needful methodology in the energy conversion through the wind. Comparatively, VAWT is more fit in all environments because of its design methodology, which turned VAWT is more implementable one in real-time conditions. Because of this peak, unique characteristics of VAWT, this work desired to deal with the conceptual design and its optimization based on high power extraction. Various power-enhancing methodologies are submitted, in which the leading-edge modifications-based enhancement technique is implemented in the various formation. Two different power optimization techniques are imposed in the base model of VAWT, which are curved cut in the leading-edge and sharp-edge cut in the leading edge. The engineering approach used in this work for the entire comparative analysis is CFD. In this regard, the conceptual designs of all the VAWT models are designed in CATIA. A CFD tool, ANSYS Fluent is used as a solver for all the cases. Finally, the optimized model is selected based on the high power extraction rate from VAWT.

Keywords CFD · Curvy cuts · Modified leading edge (LE) · Renewable energy · Sharp cuts · VAWT

1 Introduction

The production of electricity by renewable resources plays a dominant role, pivotally by solar and wind across the globe. In the innovation in the wind turbine based on its axis, horizontal-axis wind turbine (HAWT) and VAWT convert the kinetic energy from the wind to electricity by the rotor. In the consideration of maximum energy, the HAWT provides the maximum but the difficulty attained with implication, wind replenishment, besides with environment. The VAWT does not need the problems of implication, wind direction, lower altitude, and relative environment, but the lacking

M. Ramesh · M. Senthil Kumar · R. Vijayanandh (✉) · K. Sundararaj · S. Balaji · S. Bhagavathiyappan
Aeronautical Engineering, Kumaraguru College of Technology, Coimbatore, Tamil Nadu, India
e-mail: vijayanandh.raja@gmail.com

P. Muthukumar et al. (eds.), *Innovations in Sustainable Energy and Technology*, Advances in Sustainability Science and Technology,
https://doi.org/10.1007/978-981-16-1119-3_24

criteria are high efficiency. With the proper aiding for environmental needs and ease of implementation, VAWT will provide the appropriate choice for the rural and urban areas across the world. From the study, VAWT on a conventional basis produces higher torque, which reduces the overall efficiency. And also, with the effect with a higher altitude, it exhibits lesser functionality, drag, and lesser reliability while compared with HAWT. But for the low-scale setup and medium energy production sector might attain with VAWT due to its characteristics of higher stability, proximity to the ground, stress uninducement, appropriate energy production. In this work, an aerodynamical perspective to provide the better design of blades with the leading-edge modifications improves the harness of energy in terms of efficiency for the large range solicitation in implementation with the help of numerical simulation [1–5].

2 Literature Survey

The article [6] dealt about the computational fluid behavior on the VAWT blades and its comprehensive analysis through CFD with the help of Fluent. Lift coefficients, rotational force, drag coefficients, and power were played a major role in the selection factors. Both steady and transient computational analyses were executed under the various rotational velocities from 30 to 150 RPMs. The unstructured mesh-based construction was used and thereby four mesh cases-based gird convergence test was organized. The boundary conditions such as operating pressure, slip representations, and fluid velocities were perfectly provided. The initial conditions such as velocity–pressure coupling, accurate order, and turbulence model were listed. Finally, the straight and helical kinds of VAWT blade setups were analyzed and executed the optimization. The important observations are a tool used, boundary conditions, turbulence model, mesh setup, dimensions of the control volume, and selection factors. The article [7] was investigated the fluid dynamics analysis over the VAWT through ANSYS Fluent, in which NACA 0015 was primarily used in the profile of VAWT's blades. The SIMPLE scheme-based coupler was implemented in between the pressure and velocity over the VAWT and k-omega-based shear stress turbulence model was used for the eddies' capture in and around the VAWT, which was located inside the control volume. The lift and drag coefficients were compared with experimental results for validation purpose. After the validation, the comprehensive analysis was computed between VAWT with modified leading edge and VAWT without leading-edge modification. Besides the validation, the grid convergence test was executed and thereby the optimizations were executed. At last, the VAWT with straight LE was performed well than other models. The important observations are, the type and name of the airfoil used, the implemented computational tool and its computational procedures, the type pressure correction used, the turbulence model used and types of LE modifications. The article [8] was simulated the cavity type VAWT through FLUENT 14.5 CFD solver tool. The preprocessing was completed with the support of GAMBIT, and 3D unstructured meshes were implemented. In the later year of 2010, it was found that VAWT is been more comfortable cum efficient than other wind turbines. The fluid velocity was provided as 20 m/s and thereby the computational simulations were carried out with the inclusion of k-omega SST-based turbulence model. The major selection factors involved in this investigation were drag and its coefficients. The proposed cavity-based VAWT was performed well than other models and the computational results were compared with previous experimental tests. The notable observations are turbulence model, the advantage of VAWT and its lack of research activities, angle of attack, and inlet fluid velocity. The article [9] has executed the comparative numerical analysis between Darrieus-based VAWT and Savonius-based VAWT, in which the symmetrical airfoil of NACA 0012 was used as a fundamental design parameter. The 2D-based computational simulation was executed with the support of ANSYS Fluent 15.0. The 2D fine structural mesh was created through the help of ANSYS Mesh Tool 15.0. The important design dimensions such as the design of VAWT, control volume and its dimensions were provided. In the solution part, k-omega SST-based turbulence model was used, the coupled-based velocity—pressure coupling was used, and second-order-based accuracy schemes were implemented in this 2D simulations. The important selection factors implemented were the coefficient of performance and torque. Finally, the optimized VAWT was selected based on transient computational simulation. Most of the previous works used Fluent-based computational investigation and the common observations were: symmetrical aerofoil was used, k-omega-based turbulence model was used and the unstructured mesh was implemented. With these observations, the current numerical simulation are been started.

3 Methodology Used—CFD

3.1 Conceptual Design

The symmetrical airfoil NACA 0012 is used in this VAWT with a chord length of 300 mm, the span of 1150 mm and the three blades are in the radius of 2000 mm for analysis. The 3D models are generated with the help of CATIA and the models are analyzed in ANSYS Fluent. There are two different types of cuts that are generated in the leading edge (LE) of the blade. One is a saw tooth-based sharp cut and the other is a curvy cut. The radius of the curvy cut is 6 and 4 mm based on the teeth. The VAWT is taken to be having three blades for the analysis purpose [10–13].

The conceptual design of all the eight VAWT models is revealed in Figs. 1, 2, 3, 4, 5, 6, 7, and 8, wherein Figs. 1, 2, 3, and 4 corresponds for saw tooth cuts loaded

Fig. 1 Blade with five saw tooth cuts at LE

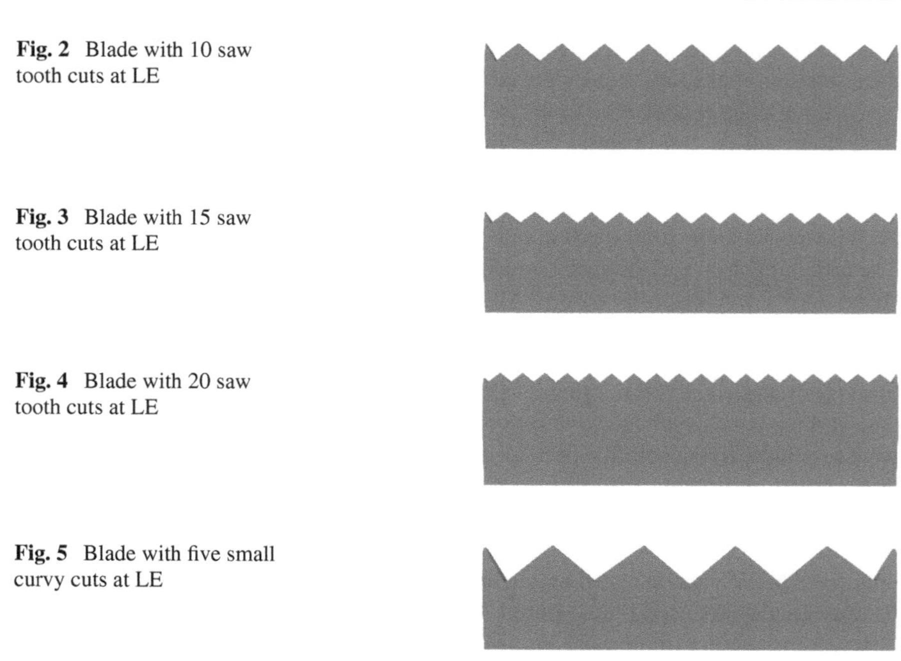

Fig. 2 Blade with 10 saw tooth cuts at LE

Fig. 3 Blade with 15 saw tooth cuts at LE

Fig. 4 Blade with 20 saw tooth cuts at LE

Fig. 5 Blade with five small curvy cuts at LE

on LE of VAWT and Figs. 5, 6, 7 and 8 corresponds for curvey cuts loaded on LE of VAWT [14–17].

3.2 *Computational Aerodynamic Analysis on Various VAWT Model*

The rectangular shape-based external control volume is surrounded by the VAWT, wherein the dimensions are 10 times greater than the length of VAWT. The main control volume of the VAWT is converted into small cells to get fine fluid properties in the required places in and around the VAWT. In cells, the nodes and centroids are

Fig. 6 Blade with 10 small curvy cuts at LE

Fig. 7 Blade with 15 small curvy cuts at LE

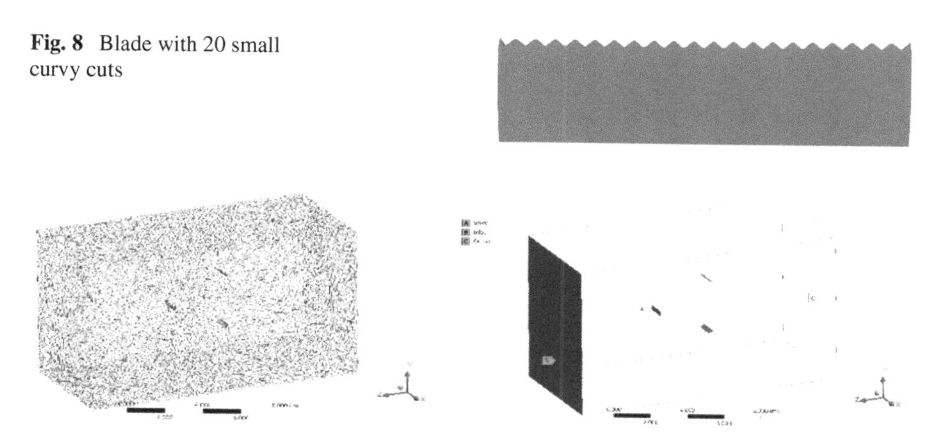

Fig. 8 Blade with 20 small curvy cuts

Fig. 9 Discretized structure of various VAWT with control volumes

having the capacity to get the fluid properties with the help of CFD tool. Figure 9 is revealed the discretized models of the saw tooth and curvy cuts equipped on the LE of the VAWT. In the mesh, proximity and curvature setups are incorporated and thereby the fine unstructural cum tetrahedral elements are used for the construction of the mesh. The grid convergence test is computed for 20 curvy cuts based VAWT, in which Mesh Case 4 is selected as an optimized mesh facility to provide reliable outcomes with the mesh computing process. The comprehensive mesh test report is shown in Fig. 10. The same picked mesh is used in all nine simulation cases.

3.3 Boundary Conditions and Flow Behaviors

The flowing nature around the VAWT undergoes the effect of incompressible flow condition. The variation of fluid density in and around the VAWT is an inconsiderable one so it is assumed as constant, which has been directly provided through known magnitude. Thus, the considerable fluid properties for this investigation are pressure variation and velocity variations in three directions. These two flow parameters-based energy variations have been mainly relayed on the formation of eddies so it needs to be predicted effectively. Therefore, the pressure-based solver with k-epsilon turbulence model and SIMPLE scheme-oriented pressure correction technique are selected as a computational platform. The fluid velocity is given as 10 m/s, the operating pressure is given as 101,325 Pa, which was extracted through fieldworks [18–22].

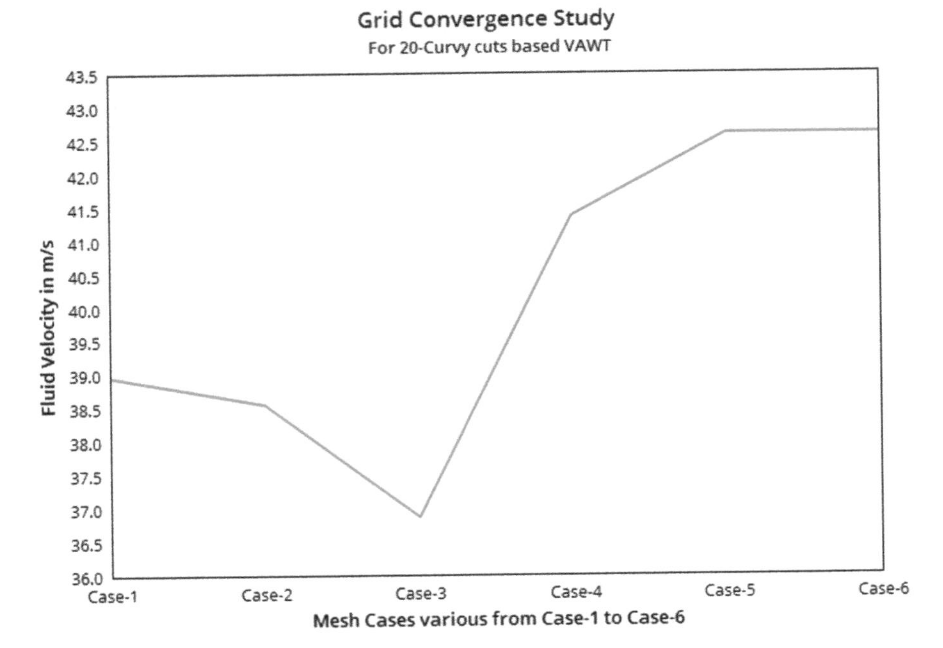

Fig. 10 Test of grid convergence for VAWT-curvy cut

4 Results and Discussions

As per the flow nature-based boundary conditions, the CFD simulation is carried out for nine VAWT models. The base VAWT model is included in this comparative analysis for reference purpose, wherein the CFD predictions for base VAWT are revealed in Figs. 10 and 11.

4.1 Base

See Figs. 11 and 12.

4.2 Five Saw Tooth Cuts at the Leading Edge of VAWT

See Figs. 13 and 14.

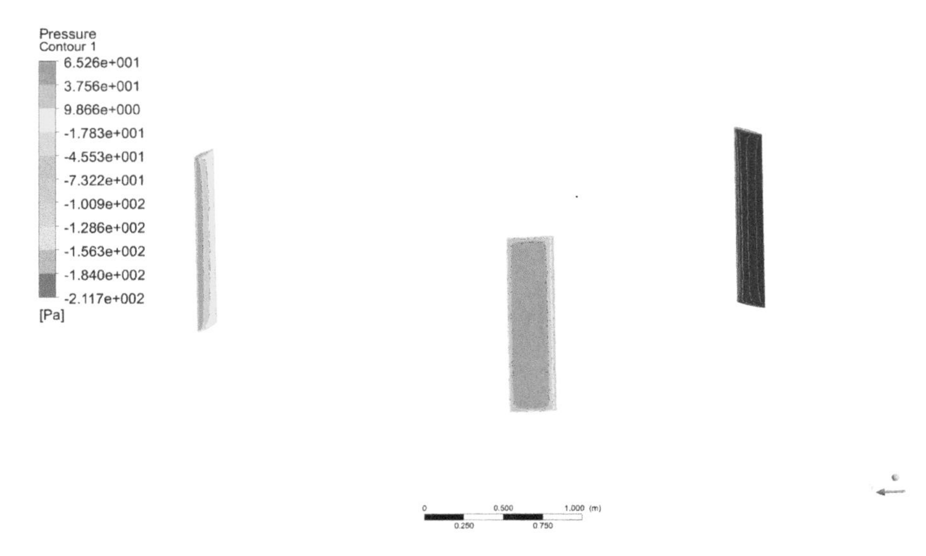

Fig. 11 Pressure distributions on-base VAWT model

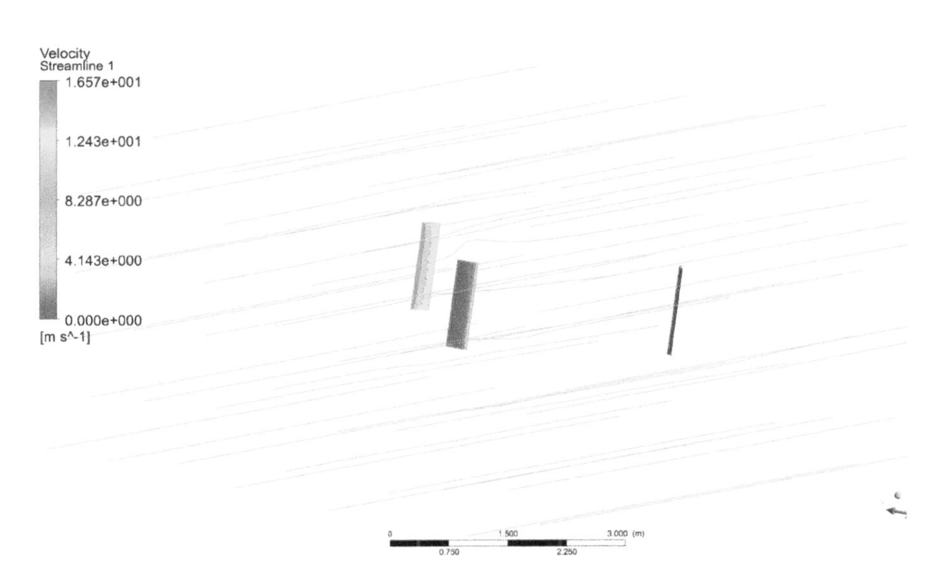

Fig. 12 Velocity variations on base VAWT model

4.3 Ten Saw Tooth Cuts at the Leading Edge of VAWT

See Figs. 15 and 16.

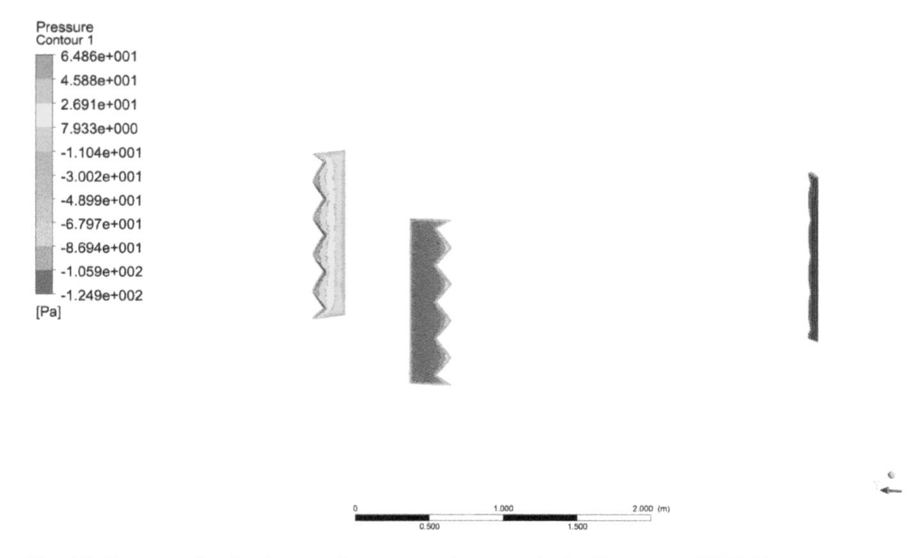

Fig. 13 Pressure distributions on five saw tooth cuts at the leading edge of VAWT

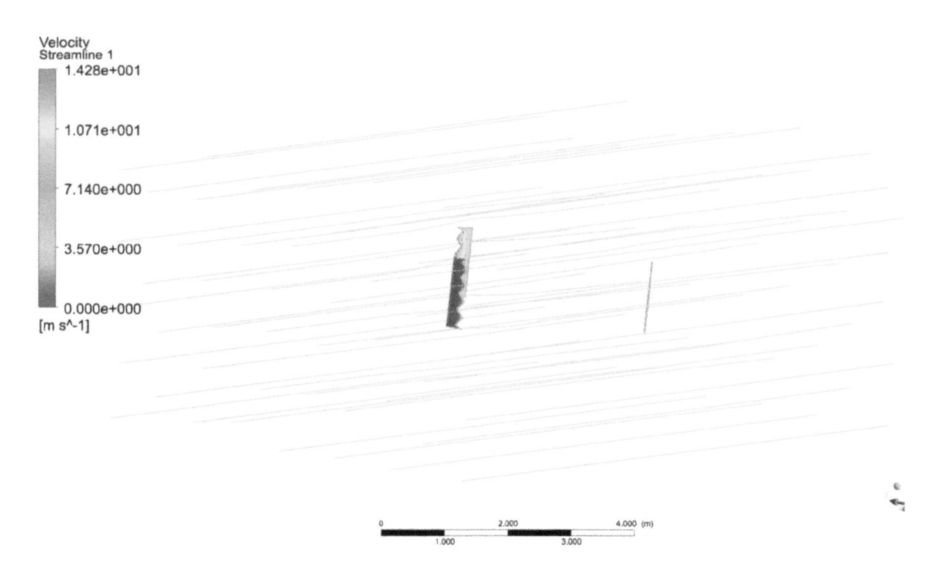

Fig. 14 Velocity variations on five saw tooth cuts at the leading edge of VAWT

4.4 Fifteen Saw Tooth Cuts at the Leading Edge of VAWT

See Figs. 17 and 18.

In the first phase, the CFD analysis for the VAWT with saw tooth cuts are performed. The predominant fluid properties of pressure and velocity are revealed

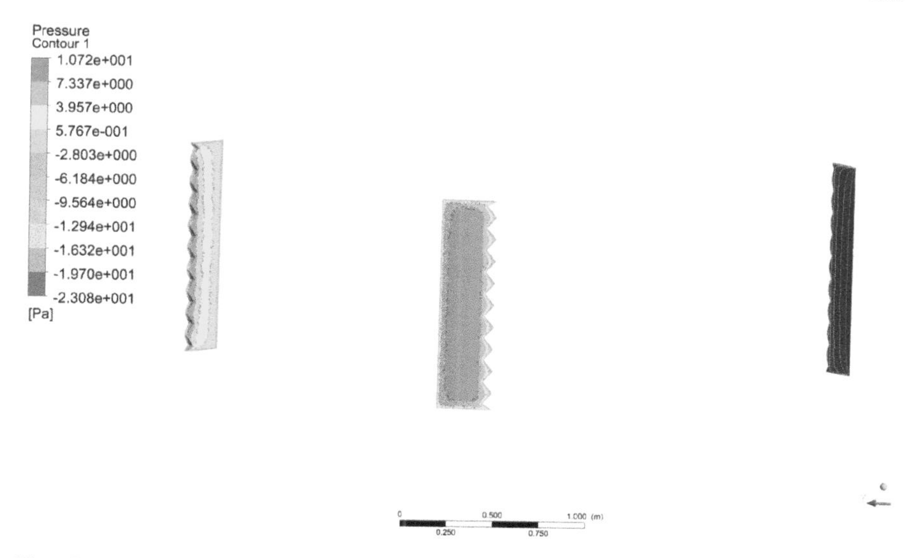

Fig. 15 Pressure distributions on ten saw tooth cuts at the leading edge of VAWT

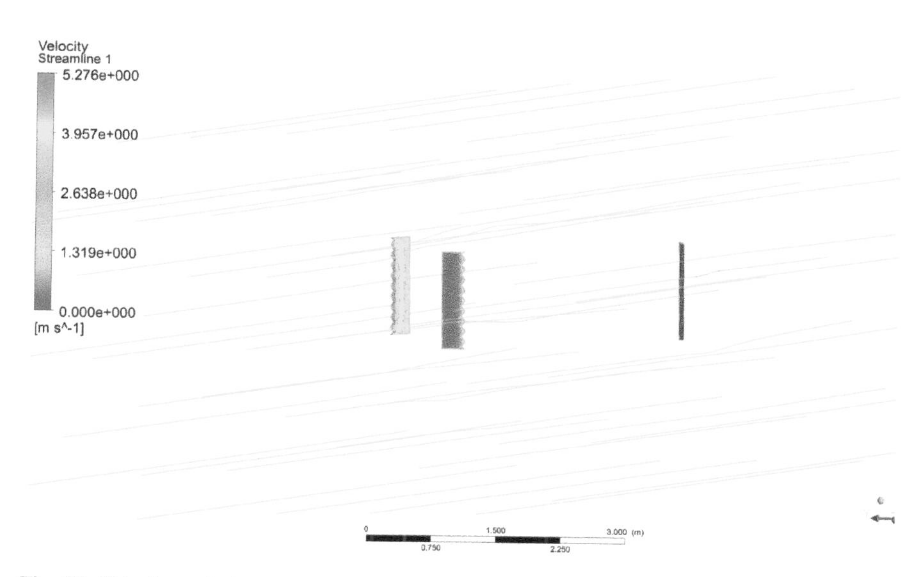

Fig. 16 Velocity variations on ten saw tooth cuts at the leading edge of VAWT

from Figs. 12, 13, 14, 15, 16, 17, 18 and 19, in which Figs. 12 and 13 correspond for five saw tooth cuts at LE of VAWT, Figs. 14 and 15 correspond for 10 saw tooth cuts at LE of VAWT, Figs. 16 and 17 corresponds to 15 saw tooth cuts at LE of VAWT, and Figs. 18 and 19 correspond for 20 saw tooth cuts at LE of VAWT.

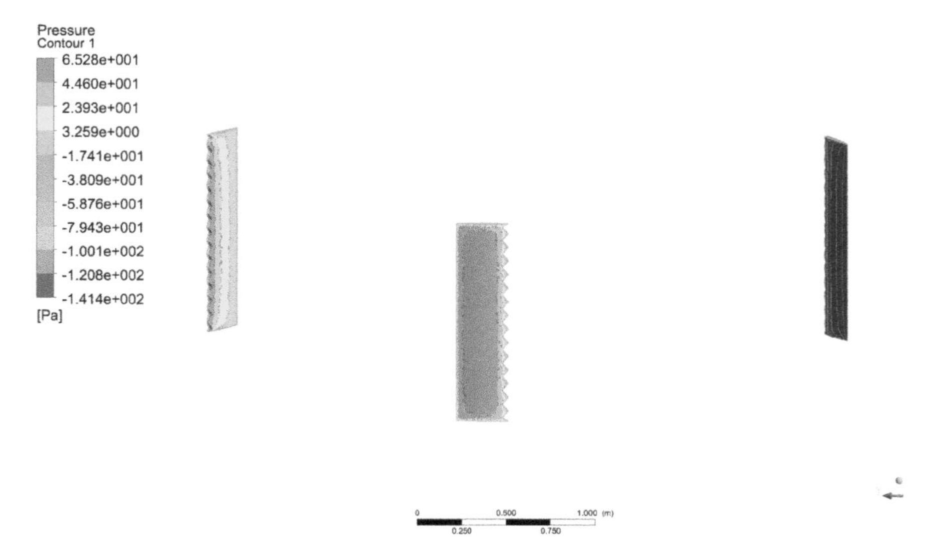

Fig. 17 Pressure distributions on fifteen saw tooth cuts at the leading edge of VAWT

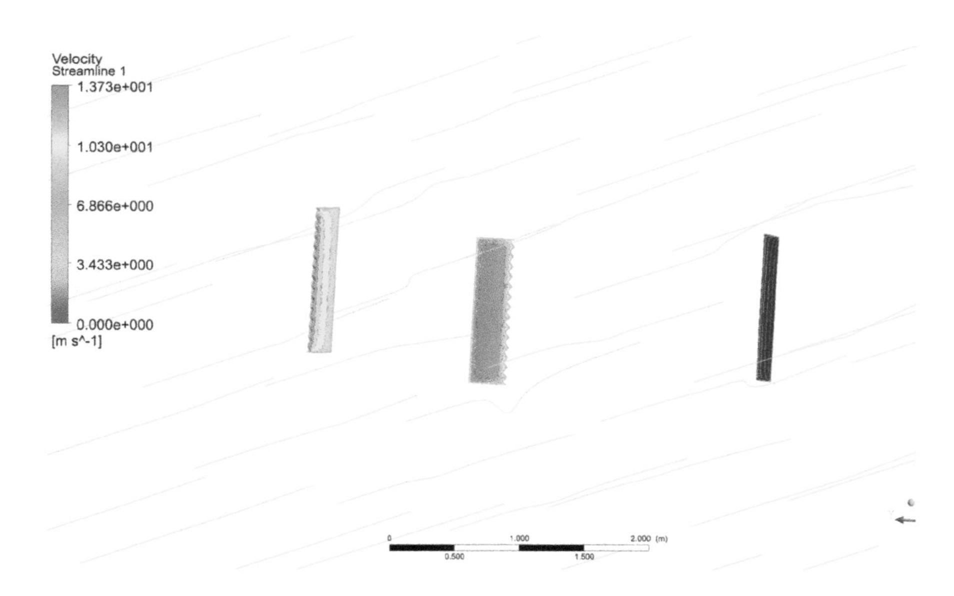

Fig. 18 Velocity variations on fifteen saw tooth cuts at the leading edge of VAWT

4.5 Twenty Saw Tooth Cuts at the Leading Edge of VAWT

See Figs. 19 and 20

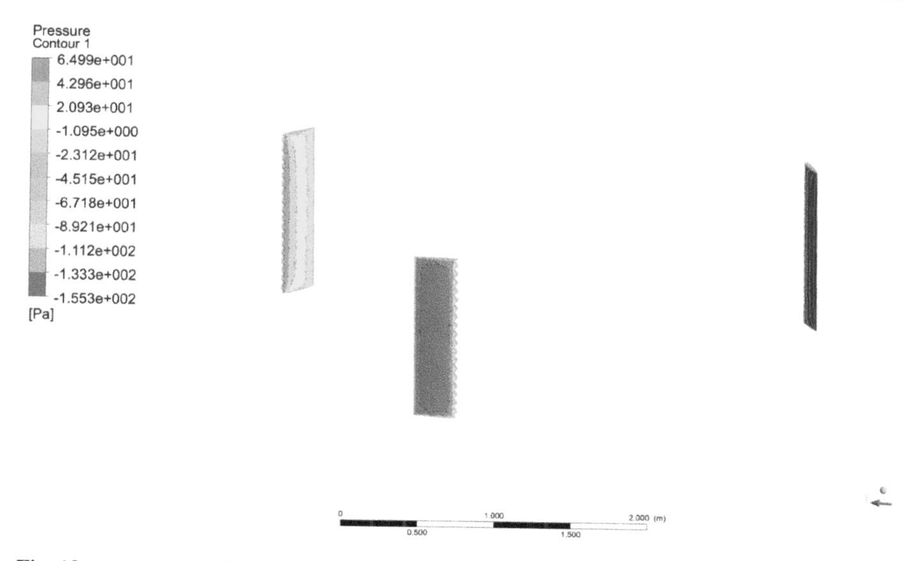

Fig. 19 Pressure distributions on twenty saw tooth cuts at the leading edge of VAWT

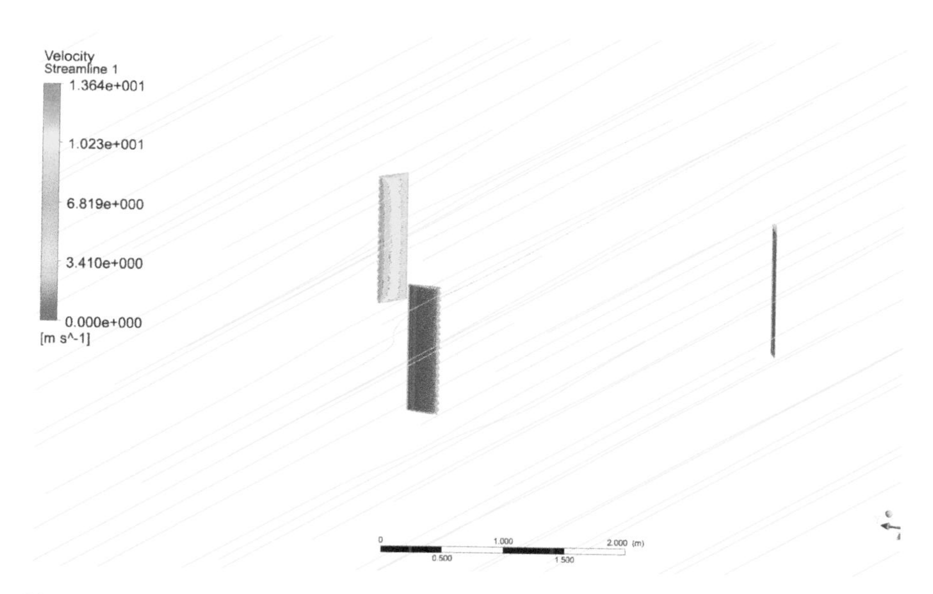

Fig. 20 Velocity variations on twenty saw tooth cuts at the leading edge of VAWT

4.6 Five Curvy Cuts at the Leading Edge of VAWT

See Figs. 21 and 22

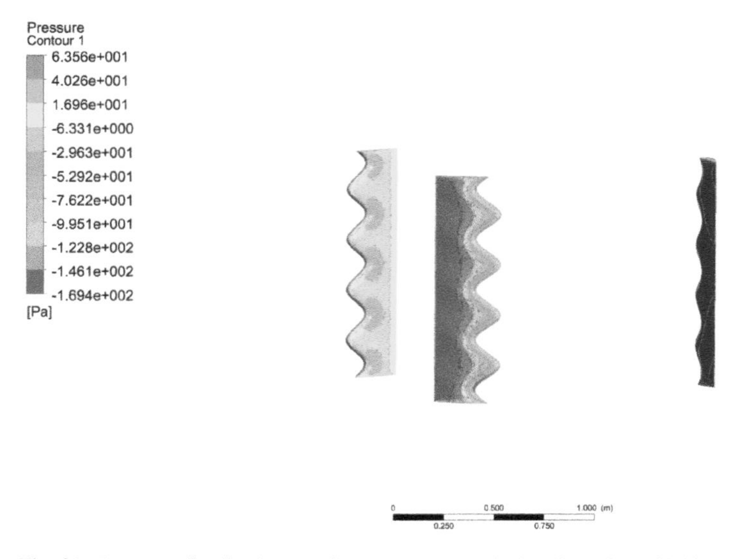

Fig. 21 Pressure distributions on five curvy cuts at the leading edge of VAWT

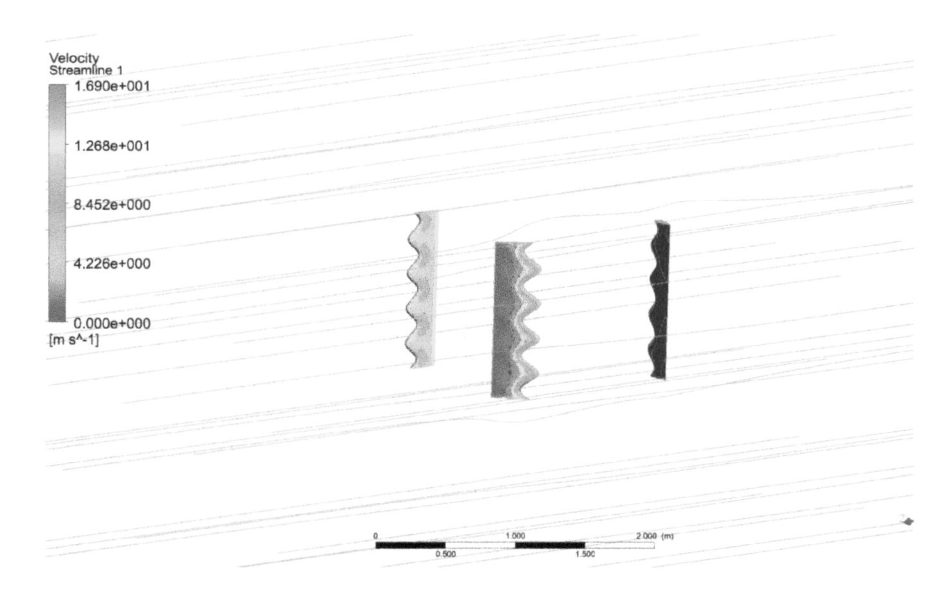

Fig. 22 Velocity variations on five curvy cuts at the leading edge of VAWT

4.7 Fifteen Curvy Cuts at the Leading Edge of VAWT

See Figs. 23 and 24.

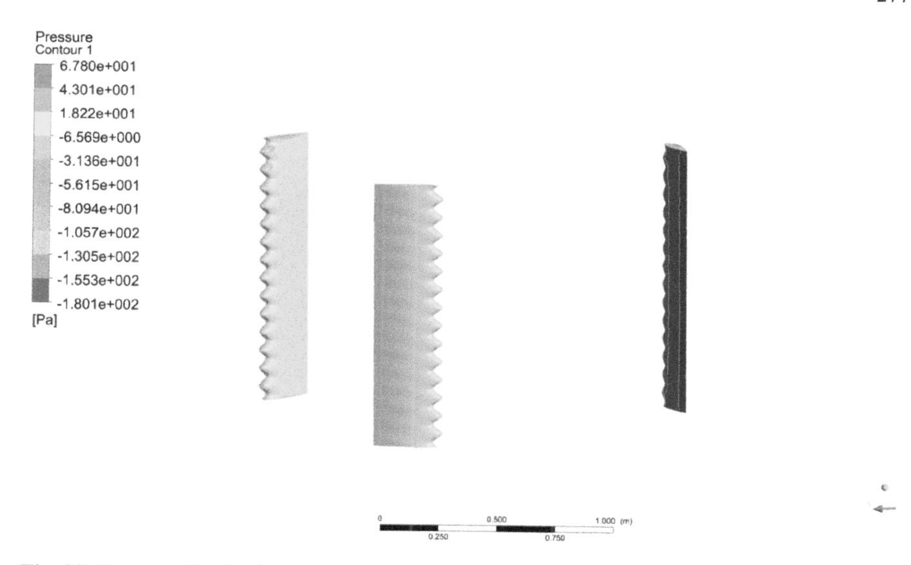

Fig. 23 Pressure distributions on fifteen curvy cuts at the leading edge of VAWT

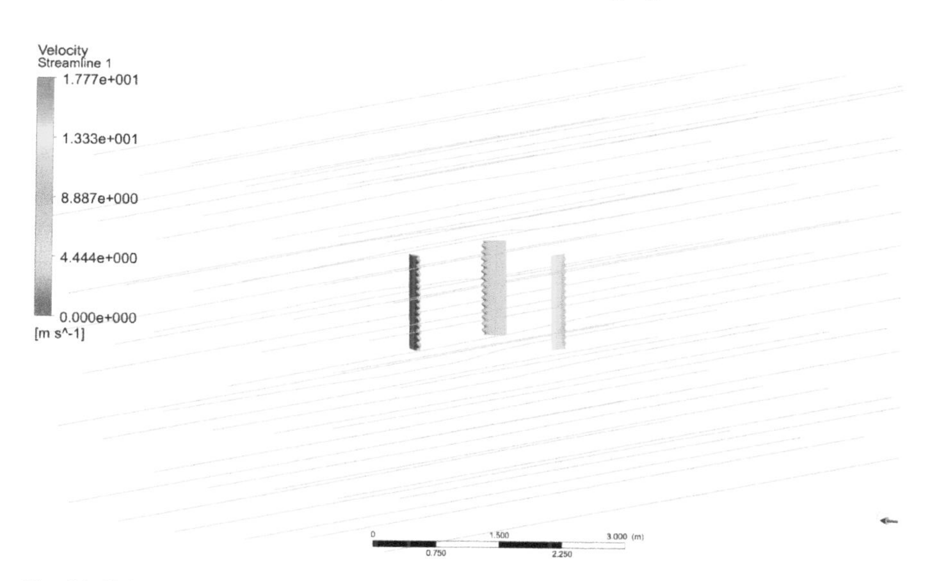

Fig. 24 Velocity variations on fifteen curvy cuts at the leading edge of VAWT

In the second comparative phase, the curvy cut-based VAWT models were analyzed. The major fluid properties of pressure and velocity are shown in Figs. 20, 21, 22, 23, 24 and 25, in which Figs. 20 and 21 correspond for five curvy cuts at LE of VAWT, Figs. 22 and 23 correspond to fifteen curvy cuts at LE of VAWT, and Figs. 24 and 25 correspond for twenty curvy cuts at LE of VAWT.

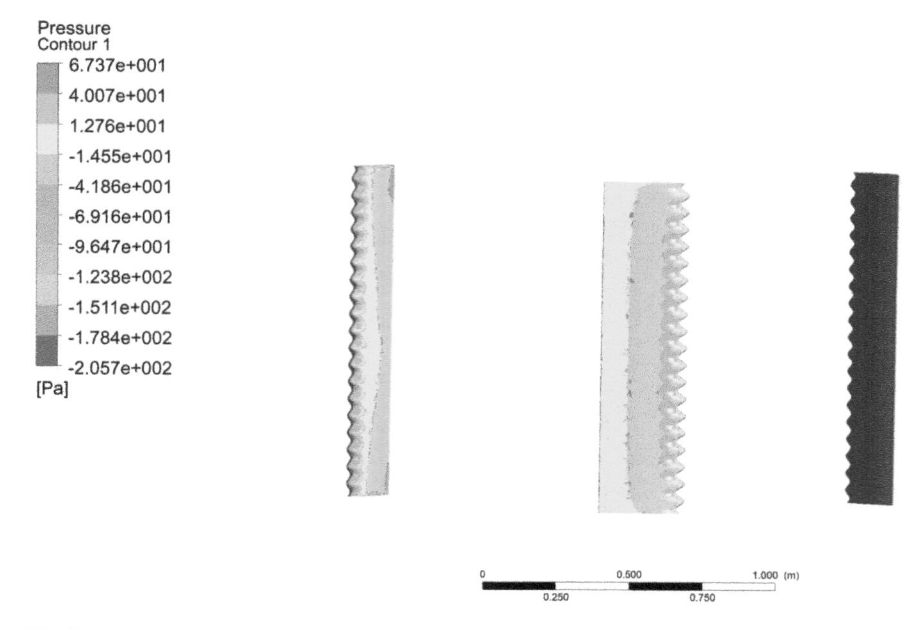

Fig. 25 Pressure distributions on twenty curvy cuts at the leading edge of VAWT

4.8 Twenty Curvy Cuts at the Leading Edge of VAWT

See Figs. 25 and 26.

4.9 Comparative Analysis

The comparative aerodynamic performance results are provided in Table 1, in which the 20 teeth curvy cuts have generated the higher rotational force, which is 42.5888 N. The eddy formations are quite higher in the presence of curvy cuts, which increased the aerodynamic performance of VAWT than the base VAWT model.

5 Conclusions

The wind turbine is an emerging renewable energy device, as VAWT is more flexible and suitable for implementation, and in this work, VAWT with various design configurations was analyzed. Two different families are VAWT equipped with various saw tooth cut and VAWT equipped with various curvy cuts and base VAWT model. In each family, four different modifications are executed, which are the leading edge

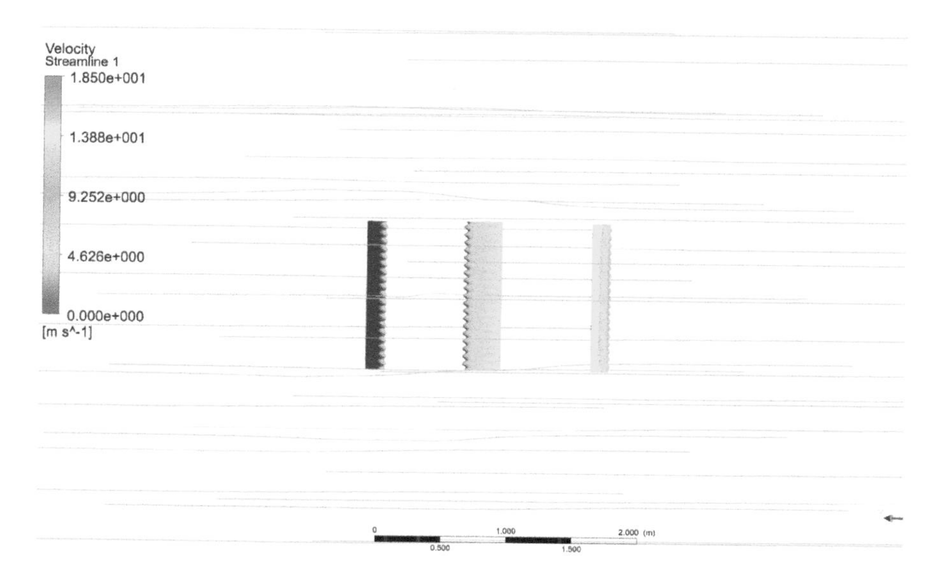

Fig. 26 Velocity variations on twenty curvy cuts at the leading edge of VAWT

Table 1 Aerodynamic performance analysis on various VAWT models

S. No.	VAWT—types	Exit velocity (m/s)	Rotational force (N)
1	Base	16.57	38.9578
2	5-saw tooth cut	14.28	38.5439
3	10-saw tooth cut	13.28	36.8640
4	15-saw tooth cut	13.73	41.9691
5	20-saw tooth cut	13.64	39.5411
6	5-curvy cut	14.35	39.0037
7	10-curvy cut	14.10	41.3711
8	15-curvy cut	14.17	42.2696
9	20-curvy cut	13.74	42.5888

modified with 5 cuts, 10 cuts, 15 cuts, and 20 cuts. CATIA is the design tool, which is used for the construction of all nine different VAWT models. An advanced CFD tool, i.e., ANSYS Fluent is used for the aerodynamic performance parameter studies on various models. From the analysis, it is found that the blade with 20 teeth curvy cut has higher rotational force about 42.5888 N. The probability for the increase in the rotational force maybe because of the cuts that channel the flow over the blade into narrower streamline producing higher velocities and reduction of flow over the tip of the blades. This combined action results in the reduction of parasitic drag and vortices at the tip of the blades. From this analysis, it is suggested that the wing with 20 teeth curvy cut is suitable for the vertical axis wind turbine.

References

1. Vijayanandh, R., Senthil Kumar, M., Naveenkumar, K., Raj Kumar, G., Naveen Kumar, R.: Design optimization of advanced multi-rotor unmanned aircraft system using FSI (Chapter No. 28). In: Innovative Design, Analysis and Development Practices in Aerospace and Automotive Engineering. Lecture Notes in Mechanical Engineering, pp. 299–310 (2019). https://doi.org/10.1007/978-981-13-2718-6. eBook ISBN: 978-981-13-2718-6
2. Balaji, S., Prabhagaran, P., Vijayanandh, R., Senthil Kumar, M., Rajkumar, R.: Comparative computational analysis on high stable hexacopter for long range applications (Chapter No. 31). In: Unmanned Aerial System in Geomatics, Lecture Notes in Civil Engineering, pp. 369–391 (2020). https://doi.org/10.1007/978-3-030-37393-1_31. ISBN: 978-3-030-7393-1
3. Vijayanandh, R., Senthil Kumar, M., Rahul, S., Thamizhanbu, E., Durai Isaac Jafferson, M.: Conceptual design and comparative CFD analyses on unmanned amphibious vehicle for crack detection (Chapter No. 14). In: Unmanned Aerial System in Geomatics. Lecture Notes in Civil Engineering, pp. 133–150 (2020). https://doi.org/10.1007/978-3-030-37393-1_14.eBook ISBN: 978-3-030-7393-1
4. Vijayanandh, R., Kiran, P., Indira Prasanth, S., Raj Kumar, G., Balaji, S.: Conceptual design and optimization of flexible landing gear for tilt-hexacopter using CFD (Chapter No. 15). In: Unmanned Aerial System in Geomatics. Lecture Notes in Civil Engineering, pp. 151–174 (2020). https://doi.org/10.1007/978-3-030-37393-1_15. eBook ISBN: 978-3-030-7393-1
5. Vijayanandh, R., Ramesh, M., Raj Kumar, G., Thianesh, U.K., Venkatesan, K., Senthil Kumar, M.: Research of noise in the unmanned aerial vehicle's propeller using CFD. Int. J. Eng. Adv. Technol. **8**(6S), 145–150 (2019). https://doi.org/10.35940/ijeat.F1031.0886S19. ISSN: 2249-8958
6. Alaimo, A., Esposito, A., Messineo, A., Orlando, C., Tumino, D.: 3D CFD analysis of a vertical axis wind turbine. Energies **8**, 3013–3033 (2015). https://doi.org/10.3390/en8043013
7. Bai, C.-J., Lin, Y.-Y., Lin, S.-Y., Wang, W.-C.: Computational fluid dynamics analysis of the vertical axis wind turbine blade with tubercle leading edge. J. Renew. Sustain. Energy **7**(033124), 1–14 (2015). https://doi.org/10.1063/1.4922192
8. Suffer, K., Usubamatov, R., Quadir, G., Ismail, K.: Modeling and numerical simulation of a vertical axis wind turbine having cavity vanes. In: Fifth International Conference on Intelligent Systems, Modelling and Simulation. IEEE Xplore, 15507935, pp. 479–484 (2014). https://doi.org/10.1109/ISMS.2014.88
9. Mothilal, T., Harish Krishna, P., Jagadeesh Babu, G., Suresh, A., Baskar, K., Kaliappan, S., Rajkamal, M.D.: CFD analysis of different blades in vertical axis wind turbine. Int. J. Pure Appl. Math. **119**(12), 13545–13551 (2018)
10. Arul Prakash, R., Sarath Kumar, R., Vijayanandh, R., Raja Sekar, K., Ananda Krishnan, C.: Design optimization of convergent—divergent nozzle using computational fluid dynamics approach. Int. J. Mech. Prod. Eng. Res. Dev. **9**(1), 220–232 (2019). ISSN:2249-8001
11. Senthil Kumar, M., Vijayanandh, R., Srinivas, R., Arun Karthik, D., Tamil Mani, M.: Acoustic analysis and comparison of chevron nozzle using numerical simulation. Int. J. Mech. Prod. Eng. Res. Dev. **8**(7), 1089–1103 (2018). ISSN(E): 2249-8001
12. Senthil Kumar, M., Vijayanandh, R.: Vibrational fatigue analysis of NACA 63215 small horizontal axis wind turbine blade. Mater. Today Proc. **5**(2), Part 2, 6665–6674 (2018). https://doi.org/10.1016/j.matpr.2017.11.323
13. Vijayanandh, R., Senthil Kumar, M., Sanjeev Kumar, B., Akshaya, V., Nishanth, B., Sindhu, K.: Numerical study on drag effect of waste collector attachment in the train. Int. J. Mech. Prod. Eng. Res. Dev. **8**(7), 1060–1078 (2018). ISSN(E): 2249-8001
14. Senthil Kumar, M., Vijayanandh, R., Gopi, B.: Numerical investigation on vibration reduction in helicopter main rotor using air blown blades. Int. J. Mech. Prod. Eng. Res. Dev. **8**(7), 152–164 (2018). ISSN (E): 2249-8001
15. Senthil Kumar, M., Vijayanandh, R., et al.: Conceptual design and comparative computational analysis of secondary inlet of rotary-wing aircraft engine. J. Adv. Res. Dyn. Control Syst. **9**(Sp–14), 1189–1209 (2017)

16. Naveen Kumar, K., Vijayanandh, R., Ramesh, M.: Design optimization of nozzle and second throat diffuser system for high altitude test using CFD. IOP J. Phys.: Conf. Ser. 1355 012012 (2019). https://doi.org/10.1088/1742-6596/1355/1/012012
17. Vijayanandh, R., Ramesh, M., Venkatesan, K., Raj Kumar, G., Senthil Kumar, M., Rajkumar, R.: Comparative acoustic analysis of modified unmanned aerial vehicle's propeller (Chap. 45). In: Advances in IC engines and combustion technology. Lecture Notes in Mechanical Engineering, pp. 557–571 (2021). https://doi.org/10.1007/978-981-15-5996-9_45
18. Sonaimuthu, B., Panchalingam, P., Raja, V.: Comparative analysis of propulsive system in multi-rotor unmanned aerial vehicle. In: Proceedings of the ASME 2019, Gas Turbine India Conference—GTINDIA2019, vol 2, 8 p. https://doi.org/10.1115/GTINDIA2019-2429, ISBN: 978-0-7918-8353-2, V002T08A004
19. Murugesan, R., Raja, V.: Acoustic investigation on unmanned aerial vehicle's rotor using CFD-MRF approach. In: Proceedings of the ASME 2019, Gas Turbine India Conference—GTINDIA2019. vol. 2, 7 p. https://doi.org/10.1115/GTINDIA2019-2430. ISBN: 978-0-7918-8353-2, V002T08A005
20. Naveen Kumar, K., Vijayanandh, R., Bruce Ralphin Rose, J., Swathi, V., Narmatha, R., Venkatesan, K.: Research on structural behavior of composite materials on different cantilever structures using FSI. Int. J. Eng. Adv. Technol. **8**(6S3):1075–1086 (2019). https://doi.org/10.35940/ijeat.F1178.0986S319
21. Jagadeeshwaran, P., Natarajan, V., Vijayanandh, R., Senthil Kumar, M., Raj Kumar, G.: Numerical estimation of ultimate specification of advanced multi-rotor unmanned aerial vehicle. Int. J. Sci. Technol. Res. **9**(01), 3681–3687 (2020). ISSN 2277-8616
22. Vijayanandh, R., Venkatesan, K., Senthil Kumar, M., Raj Kumar, G., Jagadeeshwaran, P., Raj Kumar, R.; Comparative fatigue life estimations of marine propeller by using FSI, IOP. J. Phys. Conf. Ser. **1473**, 012018, 1–8 (2020), https://doi.org/10.1088/1742-6596/1473/1/012018

A Detailed Analysis on Carbon Negative Educational Institute for Sustainable Environment

Krishna Pai and Tejas Doshi

Abstract Carbon dioxide is the central ozone-harming substance that consequences from human workout routines and reasons a hazardous atmospheric deviation and climatic change. The ingesting of the herbal substances in non-renewable electricity sources produces vitality and discharges carbon dioxide and one-of-a-kind mixes into the world's climate. Ozone-harming materials can be radiated through vehicle, land freedom, introduction and utilization of nourishment, fills, made products, materials, wood, streets, constructions and administrations and so forth. A carbon footprint is the percentage of the measure of ozone-depleting substances, estimated in devices of carbon dioxide, delivered through human exercises. An automobile footprint shall be estimated of every individual or an association, and is many times given in massive amounts of carbon dioxide equally every year. Hence, this paper manages the examination on outflow of carbon dioxide from special emanation inventories in any or in a particular educational institute. The investigation has been embraced in the university campus so as to check the measure of carbon dioxide and to recommend the therapeutic measures which can be used to decrease the discharges as a piece of social responsibility.

Keywords Carbon footprint · GHGs · Ecological footprint · LCA · CO_2 · Absorb

1 Introduction

The impact of carbon footprint on the environment and its changes is due to the effect of our exercises. It identifies and estimates the measure of ozone-depleting materials, i.e. greenhouse gases (GHGs) through burning of petroleum derivatives

K. Pai (✉)
Department of Electrical and Electronics Engineering, KLE Dr. M. S. Sheshgiri College of Engineering and Technology, Belagavi, India
e-mail: krishnapai271999@gmail.com

T. Doshi
Department of Civil Engineering, KLE Dr. M. S. Sheshgiri College of Engineering and Technology, Belagavi, India

P. Muthukumar et al. (eds.), *Innovations in Sustainable Energy and Technology*, Advances in Sustainability Science and Technology,
https://doi.org/10.1007/978-981-16-1119-3_25

for power, cooking and transportation that has tons (or kilogram) of CO_2 equivalent. A carbon footprint is the entire arrangement of ozone-depleting substances introduced by way of an organization, occasion, object or character through vehicle, emission, land, creation and its utilization of the nourishment, merchandise, materials, timbers, structures and administrations. The carbon footprint has two sections; i.e. the primary footprint is a share of immediate emission of CO_2 due to the burning of petroleum merchandises such as family vitality utilization and transportation [1]. The discharges can be controlled directly in this section. The secondary footprint is the proportion of the backhanded CO_2 outflows of the items that are used during entire lifecycle related to their assembling and inevitable breakdown. The concept of carbon footprint begins from the ecological footprint discussion leading to the concept of life cycle assessment (LCA) [2].

2 Survey

Suitable outflow inventories are to be chosen to wreck down the carbon footprint from the campus. These inventory evaluations are to be finished for one scholarly year. The selected inventories are transportation, human activities, power, solid waste, food nourishment and manufacturing, LPG and structures. The data obtained through distinct attendant are accumulated and perceived [3].

2.1 Human Activities

Carbon dioxide discharged by means of a person for every day is important and is proportional to the outflow of a single car in a 5 km stretch. People emanate 26 gigatons of CO_2, while CO_2 in the air is ascending at 15 gigatons for each year. Only for breathing, each human discharges 1140 g of CO_2 per day, assuming the consumption of 2800 kcal of mean diet. The considered CO_2 emission is in highest population which includes teaching, non-teaching staff and the students of the institute.

2.2 Transports and Logistics

The most used fuel for transportation is fossil fuels. Worldwide, 13% of GHG emission is contributed by fossil fuels. Different fuels emit different amounts of CO_2; for instance, petrol emits 2.3 kg whereas diesel emits 2.7 kg of CO_2 per L. The transportation details for the institution include various types of vehicles based on the fuel used, the covered distance and the fuel consumption. A distance of 1000 km travelled by a car can emit 200–230 kg, whereas a bus can emit 1075 kg of CO_2 into the atmosphere [4].

2.3 Power

The major emission inventory contributing to the carbon footprint of the institution is electricity. The survey reveals that electricity also generates 1.297 lbs. The average generation of CO_2 per kW due to electricity usage results in heating of buildings.

2.4 Solid Waste

One of the least emission inventories considered is solid waste. Solid waste generation is based on the income of the country. For instance, solid waste generation is 1.1–5 kg, 0.45–0.89 kg and 0.52–1 kg per capita per day for high-income, low-income and middle-income countries, respectively. The surveyed details include the solid waste collected from canteen and hostels of the institution.

2.5 Food Nourishment and Manufacturing

The European Commission (2006) study mentions that the food consumption has an impact of 20–30% on the environment. Fourteen percentage of GHG emission is contributed by agriculture worldwide. The factors affecting the GHG production are food and food type, for instance, food such as fruits and vegetables produces less carbon footprint than food such as meat. In our survey, we have investigated food consumptions for different zones.

2.6 Natural Gases and LPG

The utilization of 1 L of liquid petroleum gas can discharge 1.5 kg of carbon dioxide to the air. Burning of 250 kg of wood adds 33 kg of CO_2. The usage of LPG and natural gas in canteens and hotels of the institute is investigated.

2.7 Structures

Structures contribute to carbon footprint considerably. About 12% of CO_2 is emitted into the atmosphere due to the materials used to construct the structure. The study says that about 28 kg of CO_2 is produced by square metre of brick work and around 0.0001867 tons per square metre per year. Forty percentage of carbon emission is due to the structures and buildings.

3 Methodology

Two phases are formulated to determine the carbon footprint, for example, characterizing the carbon footprint and evaluating the carbon footprint [5].

3.1 *Phase I: Define the Carbon Footprints*

Each inventory is bounded by certain decision to abide the protocol of GHG. Decisions include:

1. Defining the location of project: Select an instructive institution area as the website for calculating carbon impression.
2. Defining the length of duration: Select the academic year, say (2019–'20).
3. Defining suitable and appropriate zones of the campus.

3.1.1 Defining GHG Inventory Parameters

1. Selection of appropriate emission inventory at each location.
2. Selection of the data loggers at each emission inventory.

3.2 *Phase II: Quantification of the Carbon Footprints*

3.2.1 Source Data Collection and Data Gaps Determination

1. The tedious process of collecting emission source data from identified data loggers.
2. Gather the collected information.
3. From the collected data, identify/determine the data gaps.

3.2.2 Estimate and Model Missing Data

The second request of the data holders is initiated if there is a gap in the information after uploading a database.

3.2.3 Conduct Procedural Quality Assurance

The quality and the accuracy of the project are assured by cross-checking and verifying the data.

3.2.4 GHG Emission Calculations

1. Unit conversion factors and emission factors are collected for each inventory data base.
2. To obtain the compatibility of the available emission factor, it is required to convert consumption quantities to a common unit.

3.2.5 Inventory Management Plan and Summarize Results

1. The database is used at different levels of granularity and in different formats to report the emission data.
2. For the college campus, the carbon footprint is calculated zone-wise.
3. Determination of total carbon footprint.

3.2.6 Suggestive Measures/Remedies to Reduce Carbon Footprint

1. The inventories contributing to the high degree of emission are to be identified.
2. Recommend optimal measures to reduce CO_2 emission in the identified inventories.

4 Carbon Footprint Analysis

The motivation of this investigation is to make campus carbon negative by performing the analysis of carbon footprint, which is one of the major contributors to the global warming. KLE Dr. M. S. Sheshgiri College of Engineering and Technology campus, Belagavi, was considered for survey study. This paper details the various carbon emission inventories affecting the carbon footprint in the campus.

4.1 Carbon Emission Within Campus

The identified emission inventories have been investigated for the analysis of carbon footprint within the campus as shown in Figs. 1 and 2 and Tables 1 and 2.

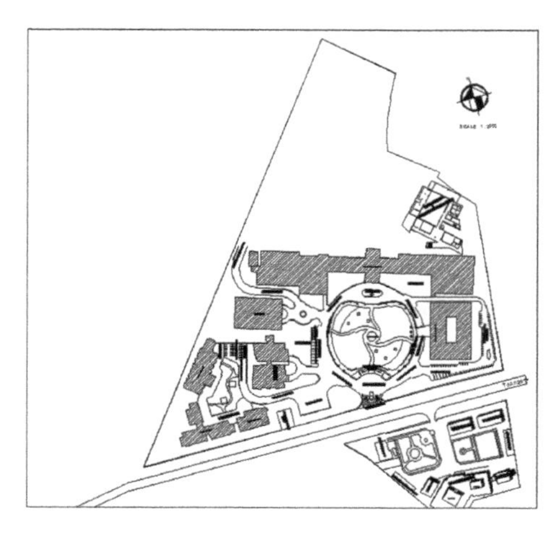

Fig. 1 Layout of KLEMSSCET campus

Fig. 2 Layout of parking KLEMSSCET

Table 1 Various factors of emission of carbon in campus [4]

Emission inventory	Carbon emitted (kg/month)
Human activates	37,620
Transports and logistic	5272.8
Power	46,723.20
Solid waste	7185
Food nourishment and manufacturing	902.7
Natural gases and LPG	2670
Structures	1,110,352.32
Electrical equipments	10,498,122
Water usage	4,512,000
Total inventories	6,772,538.02

Table 2 Calculation of emission of carbon in campus by various factors [4]

Emission inventory	Calculation of carbon emitted
Human activates	Total population of campus was 2800. Considering carbon emission factor to be 1.14 kg/person/day. Total amount of CO_2 emitted by humans was 37,620 kg/month (considering 8 h working for day scholars and staff; 24 h for hostelite)
Transports and logistic	A survey on vehicle count was conducted for a period of 7 days for a month. The average number of two wheelers, cars and buses is 672, 32 and 8, respectively. Therefore, amount of petrol consumed by the two wheelers and cars was 0.1 L/day and 0.2 L/day, respectively. Amount of diesel consumed by the bus was 0.3 L/day. Carbon emission factor for petrol as 2.3 kg/L and for diesel as 2.7 kg/L. Total amount of carbon emitted due to transports and logistics was 5272.8 kg/month
Power	Amount of average electricity utilized in a month is 67,758 kWh. Carbon emission factor considered was 0.68956 kg/kWh. Amount of carbon emitted due to power was 46,723.20 kg/month
Solid waste	Considering population of campus to be 2800. Carbon emission factor for hostel was 0.125 kg/person/day and for day scholars and staff was 0.06 kg/person/day. Total carbon emitted by solid waste was calculated to be 7185 kg/month
Food nourishment and manufacturing	A survey was conducted to calculate total amount of food production in the campus. As per the survey, it was that amount of food production in the campus was 117 kg. Carbon emission factor considered to be 17% of total food production. So, amount of carbon emitted by food nourishment and manufacturing was 902.7 kg/month
Natural gases and LPG	Amount of average natural gases and LPG consumed per month is 1780 kg. Carbon emission factor considered was 1.5 kg/month of natural gases and LPG. Therefore, amount of carbon emitted due to natural gases and LPG was 2670 kg/month
Structures	Total area of campus is 68,797 m^2. Total built-up area of plinth is 11,978.87 m^2, and total built-up area was found to be 39,655.44 m^2. Considering carbon emission factor as 28 kg/m^2 of brick work. Total carbon emitted by all the structures is 1,110,352.32 kg/month
Electrical equipments	Total amount of carbon emitted by electrical equipments in the campus is 10,498,122 kg/month. This number is considering all the electrical and electronics equipments in the campus used for 24 h
Water usage	By considering the usage of water under all the above-mentioned inventories. The carbon emission by water usage is found to be 4,512,000 kg/month

(continued)

Table 2 (continued)

Emission inventory	Calculation of carbon emitted
Total inventories	6772.53802 tons/month

Table 3 Carbon absorbed within campus [4]

Carbon absorption	Carbon quantity
Quantity of carbon absorbed via greenery of the campus	4480 tons/month
Total quantity of carbon absorbed with the aid of grass	43.212 kg/month
Total quantity of carbon decreased via campus greenery	4480.043 tons/month

4.2 Carbon Absorbed Within Campus

The investigation of amount of absorption of carbon is considered for suggesting remedial measures, and considering the greenery of the campus and the area of grass grown as discussed in Table 3 plays an important role in the analysis [6].

Amount of carbon yet to be reduced = Total carbon emitted − Total carbon absorbed = **2292.5 tons/month**.

5 Results

Carbon negative campus can be achieved through various systems. There is a necessity for making the structure and its surroundings carbon free. There are different types of measures which can be adopted to make campus carbon negative from carbon positive. Some remedial measures are discussed below.

5.1 Rainwater Gathering or Harvesting System

Rainwater gathering or harvesting is an innovation used to gather, omit on and keep downpour water for later use from commonly smooth surfaces, for example, a rooftop, land floor or rock catchment. RWH is the technique of gathering water from rooftop, filtering and inserting away for extra employments. Water harvesting is a fundamental approach of getting and keeping water the place where it falls. It is viable that we can shop it in tanks for extra utilization or we can make use of it to revive groundwater relying on the circumstance as shown in Figs. 3 and 4. The framework constituted for RWH in the college offers stable and well structured well-springs to have a high quality, low wastage and one of the kind water storage

Fig. 3 RWH system at hostel

Fig. 4 RWH system at main building

system. RWH framework is monetarily much less pricey in improvement contrasted with exclusive sources, for instance, indeed, channel, dam, streams, redirection and so on.

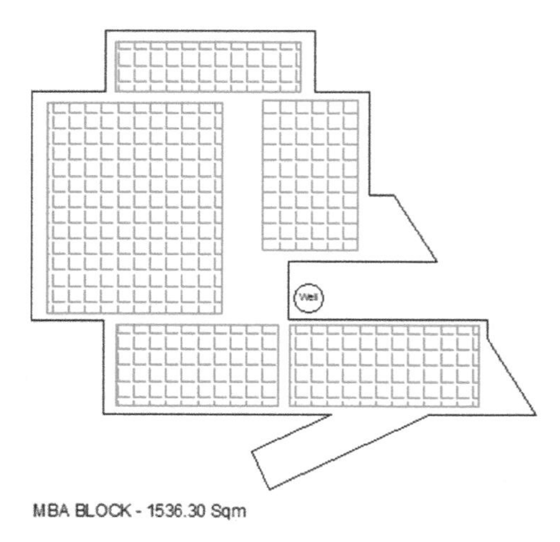

Fig. 5 Layout of MBA Building

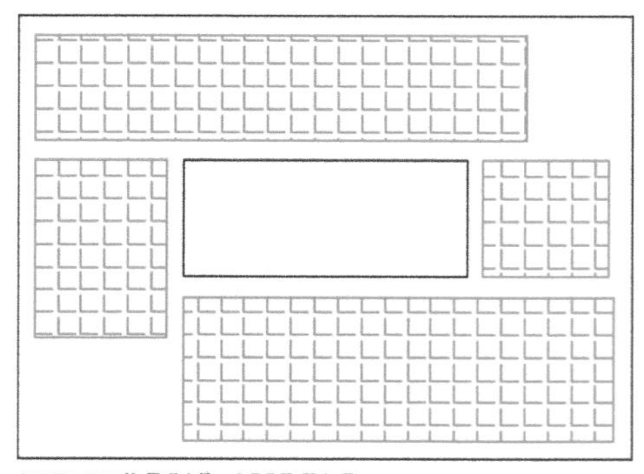

Fig. 6 Layout of civil department

5.2 Solar Energy System

Solar energy system is transformation of sunlight hours into electricity using photovoltaic (PV) cell. The photovoltaic cell converts the sunlight energy into usable electrical energy by three basic steps such as the light is absorbed and emits the electrons, the free electrons constitute the electrical current and the current is collected at the load [7] (Figs. 5 and 6).

5.3 Terrace Gardening

Terrace gardening not only absorbs the carbon from the atmosphere, but also contributes to cooling effect of the building [8] (Fig. 7).

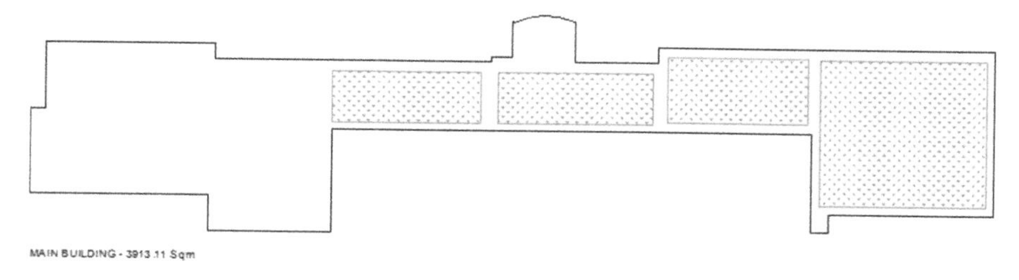

Fig. 7 Layout of main building

5.4 *Other Measures*

The other suggested measures that can be effective in reducing the amount of carbon emission by different identified inventories include planting more tress in the campus, avoiding rampant consumerism, use of efficient fuels for transportation and adhering to emission norms encouraging bicycles and electric vehicles in the campus such that there is reduction in usage of petroleum products, using electricity of each building effectively, adopting appropriate waste management techniques, following the do's and don'ts of the food consumption within the campus, efficient usage of LPG, encouraging the usage of natural gas and eco-friendly building construction materials in an attempt to achieve the objective of carbon negative campus [9] (Tables 4 and 5).

All the emitted carbon is now absorbed by making use of the remedial measures. Thus, campus or so-called campus is made carbon negative.

Table 4 Carbon reduction

Remedial measures	Amount of carbon reduced (KG/month)
Rainwater harvesting	165,745.61
Solar energy system	46,723.20
Green cover	2,252,000
Terrace gardening	162.24
Total	2,464,631.05

Table 5 Comparative results of carbon absorption and emission

Carbon emitted	Carbon absorbed by present green cover	Carbon absorbed by measures	Carbon positive/negative (carbon emitted—total carbon absorbed)
6772.538 tons	4480.043 tons	2464.631 tons	−172.13 tons

6 Conclusions

With the presented study of analysis, it can be concluded that it is a need of the hour to make our ecosystem free from positive carbon and move towards negative carbon. The analysis carried out in the campus proved that we can obtain a negative carbon campus. The suitable inventories were identified, and the optimal recommendations were suggested at the end of the study. The study concludes that rainwater harvesting system reduces carbon by 165,745.61 k kg/month, the solar energy system installed on MBA Building, Civil Department, and Electronics and Communications Department absorbed carbon by 46,723.20 kg/month. The greenery and the grass in the campus absorbed carbon by 2,252,000 kg/month. The terrace gardening on Mechanical Block, hostel and canteen absorbed carbon by 162.24 kg/month. The above recommended systems demonstrated an effect of—172.13 tons/month, contributing to carbon negative environment.

Acknowledgements Authors would like to acknowledge Sonali Muddebihal, Jaisal Dalal and Jyesta Walikar for their time and efforts in completion of this study. We also extend our gratitude to KLE Dr. M. S. Sheshgiri College of Engineering and Technology, Belagavi, for the support and encouragement.

References

1. Schwartz, Y., Raslan, R., Mumovic, D.: The life cycle carbon footprint of refurbished and new buildings a systematic review of case studies. Renew. Sustain. Energy Rev. **81**, 231–241 (2018)
2. Dossche, C., Boel, V., De Corte, W.: Use of life cycle assessments in the construction sector: critical review. Proc. Eng. **171**, 302–311 (2017)
3. Allen, et al.: Warming caused by cumulative carbon emission: the trillionth tone. Nature **458**(7242), 11631166 (2009)
4. Sudarsan, J.S., Jyesta, W., Jaisal, D.: Carbon footprint estimation of education building a drive towards sustainable development. Int. J. Eng. Adv. Technol. (IJEAT). **9**(1S4) (2019). ISSN: 2249–8958
5. Tanger, K.: How to calculate carbon footprint-part 2. EHS J (2010)
6. Chau, C.K., Leung, T.M., Ng, W.Y.: A review on life cycle assessment, life cycle energy assessment and life cycle carbon emissions assessment on buildings. Appl. Energy **143**, 395–413 (2015)
7. Sadeghsaberi, J., et al.: Passive solar building design. J Novel Appl. Sci. (2013). ©2013JNASJournal-2013-2-S4/1178-1188ISSN 2322-5149©2013JNAS
8. Terra Green: Green challenges—Passive cooling for comfortable indoor temperature (2017)
9. Buyle, M., Braet, J., Audenaert, A.: Life cycle assessment in the construction sector: a review. Renew. Sustain Energy Rev. **26**, 379–388 (2013)

Performance Evaluation of Adsorption Refrigeration System Using Different Working Pairs

Abhishek Dasore, Ramakrishna Konijeti, Bukke Kiran Naik, and Surya Prakash Rao Annam

Abstract Adsorption refrigeration system has attained intrigue in the field of refrigeration and air-conditioning due to its potential in applying available low-grade energy (waste heat from tail pipes of automobiles and solar heat) leading to savings in energy requirements and no ozone depletion potential (ODP) and global warming potential (GWP) emissions. The objective of the present work is to assess the performance of various adsorbents like activated carbon, silica gel and zeolite with methanol or water as refrigerants in an adsorption refrigeration system with respect to cycle time. Also, a mathematical model is proposed from first principles to probe the impact of adsorption and desorption temperatures on the adsorption cycle performance. This work enables us to determine the optimal adsorbent and refrigerant working pair based on the available waste heat temperature and refrigeration capacity theoretically. The results obtained are contrasted with the empirical data from the issued reports and are found to be in a reasonable agreement.

Keywords Low-grade energy sources · Adsorption refrigeration · Adsorbent–refrigerant working pairs · Coefficient of performance (COP) · Specific cooling power (SCP)

A. Dasore (✉)
School of Mechanical Engineering, RGM College of Engineering and Technology, Nandyal 518501, India
e-mail: dasoreabhishek@gmail.com

R. Konijeti · S. P. R. Annam
Department of Mechanical Engineering, Koneru Lakshmaiah Education Foundation, Vaddeswaram, Guntur 522502, India

B. K. Naik
School of Mechatronic Systems Engineering, Simon Fraser University, Surrey, BC V3T0A3, Canada

Department of Mechanical Engineering, National Institute of Technology Rourkela, Odisha 769008, India

P. Muthukumar et al. (eds.), *Innovations in Sustainable Energy and Technology*, Advances in Sustainability Science and Technology,
https://doi.org/10.1007/978-981-16-1119-3_26

1 Introduction

The applications of refrigerators and air conditioners in homes and offices around the world are growing by leaps and bounds. Due to this growing demand, of all the electricity produced worldwide more than 17% is being consumed by refrigeration and air-conditioning sector and out of this 72% of energy is lost as waste heat [1]. In addition to this, working medium used in these apparatuses also imparts to emission of greenhouse gases and depletion of stratospheric ozone layer [2]. In the course of recent decades, researchers have been trying to develop a refrigeration system that has the potential to overcome these problems and found sorption cooling systems to be a better substitute [3]. Sorption system includes both adsorption and absorption cooling systems. But, the main difference between them lies in sorbents (beds) which are generally liquids in absorption and granular or compact solids in adsorption. Sorption cooling technique does not contain rotating parts as that of mechanical vapor compression system, which makes them maintenance free, no vibrations, lower running costs and more reliable. Moreover, sorption systems effectively utilize available waste thermal energy like exhaust heat from tailpipe of automobiles, waste heat from industries, etc., and also sources of renewable energy like sun's energy [4].

Among sorption system, adsorption cooling system do not encounter the problems of corrosion and crystallization as that of absorption systems. Thus, the adsorption systems, which provide desired effect at low temperature levels, will be an optimistic alternative for intuitive energy management and for a sustainable development of refrigeration and cold production systems. In spite of its specific advantages, adsorption system has not been established as a commercially available technology. The main reason for this can be attributed to the fact that adsorption systems usually have lower thermal COP and lower SCP. However, these constraints can be surpassed by upgrading the adsorption characteristics of the working pairs by inflating adsorbent heat and mass transfer properties and by better heat governance during the adsorption cycle. Hence, the most research on adsorption refrigeration system is concentrated in enriching its performance through evaluating thermodynamic properties of the working pairs [5–8].

The basic adsorption refrigeration system is the single-bed cycle which mainly comprises a condenser, a refrigerant storage tank, an evaporator, a throttling device and an adsorption reactor as appeared in Fig. 1a [9]. The adsorption bed is made up of sort of solid element which substantially adsorbs and desorbs the refrigerant vapor. The cycle comprises the following four steps: pressurized heating, desorbed condensation, depressurized cooling and adsorbed evaporation. Adsorption refrigeration system working principle is illustrated in Fig. 1b [9].

Adsorbent–refrigerant working pairs play a requisite role in an adsorption refrigeration system. The adsorbent ought to have attributes like large adsorption capacity, good rapport with refrigerant and flat desorption isotherm. The characteristics of a refrigerant are analogous to that in VCR systems, like right freezing point, no toxicity, good thermal stability, large latent heat per volume, correct saturation vapor pressure, no corrosion, nonflammable, etc. But practically, there are no single working pairs to

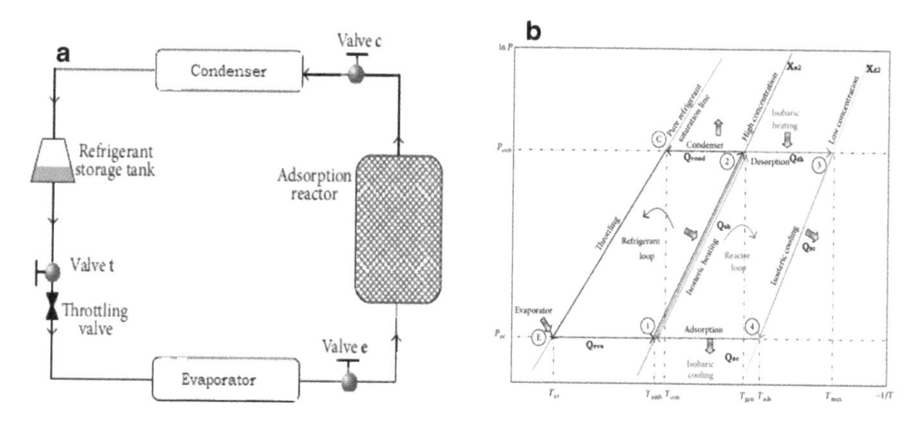

Fig. 1 **a** Schematic diagram of single-bed adsorption refrigeration system. **b** Clapeyron diagram of single-bed adsorption refrigeration cycle

totally meet all these necessities. But researchers found activated carbon–methanol [10, 11], zeolite–water [12, 13] and silica gel–water [14, 15] working pairs closely met these requirements and hence there are most regularly utilized working sets in adsorption refrigeration systems.

As far as to our knowledge, meager studies are dealt with the variation of adsorption capacity of different adsorbent–refrigerant pairs with respect to time. Hence, the present work aims at study of adsorption capacities of ACM, SGW and ZW at various temperatures with respect to time theoretically. Also, COP and SCP variations of these working pairs with time in adsorption refrigeration system are estimated.

2 Mathematical Modeling

Aggregated variable technique based mathematical model is adopted in the present work and following assumptions have been made in formulating it.

1. Uniform dispersal of temperature and vapor pressure inside the adsorbent.
2. Refrigerant adsorbed is uniformly distributed inside bed and is in liquid state.
3. Pressure drop between bed, condenser and evaporator is ignored.
4. No heat loss from or to the ambient.
5. No accumulation of refrigerant and cooling and hot water in the pipes.

2.1 Thermodynamic Properties

Adsorption Isotherms. The adsorption equilibrium of silica gel–water is [16]

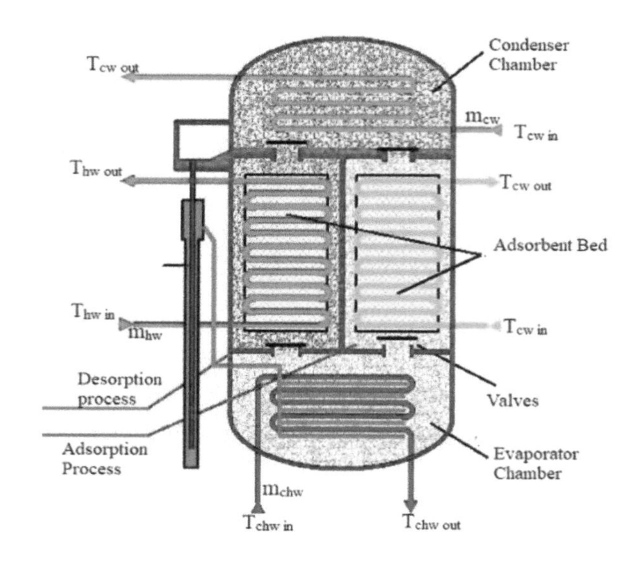

Fig. 2 Cross-sectional view of single-bed adsorption refrigeration system

$$x = x_0 \left[\frac{p(T_w)}{p(T_s)} \right]^{1/n} \tag{1}$$

For silica gel–water working pair [16], $x_0 = 0.346$ and $n = 1.6$

The simplified Dubinin–Astakhov equation [4] is employed to calculate adsorbed quantity of activated carbon and zeolite at different temperatures.

$$x = x_0 \exp\left[-K \left(\frac{T}{T_s} - 1 \right)^n \right] \tag{2}$$

For adsorption, $T_s = T_e$, and for desorption, $T_s = T_c$.

For activated carbon–methanol working pair, $x_0 = 0.682$, $K = 10.84$ and $n = 1.21$.

For zeolite–water working pair, $x_0 = 0.331$, $K = 2.99$ and $n = 2$ [3]

Rate of adsorption and desorption. Difference in adsorbed amounts is the driving force for rate of adsorption or adsorption velocity [3].

$$\frac{dx}{dt} = K_s a_p (x^* - x) \tag{3}$$

where

$$K_s a_p = \frac{15 D_{so}}{R_p^2} \exp\left(\frac{-E_a}{RT} \right) \tag{4}$$

Energy balance equation for desorption bed. During desorption process, bed is heated by exterior heating fluid, transferring heat to the bed. Refrigerant adsorbed by bed is desorbed and is condensed in condenser releasing heat to the surroundings or to the cooling fluid. The energy balance of the bed in desorption is denoted by the following equation:

$$\frac{\mathrm{d}}{\mathrm{d}t}\{[M_a(C_a + C_{ref}X) + C_{cu}M_{cu} + C_{al}M_{al}]T_{des}\} = M_a H_{ads}\frac{\mathrm{d}x_{des}}{\mathrm{d}t} + m_{hw}C_{hw}(T_{hw,in} - T_{hw,out}) \tag{5}$$

Hot water outlet temperature coming from desorption bed is given by

$$T_{hw,out} = T_{des} + (T_{hw,in} - T_{des})\exp\left(\frac{-U_{des}A_{des}}{m_{hw}C_{hw}}\right) \tag{6}$$

Energy balance equation for evaporator. Condensed refrigerant from condenser is collected in evaporator, and heat from the refrigerating space is used to evaporate the refrigerant which is adsorbed by bed.

Energy balance of evaporator is given by

$$\frac{\mathrm{d}}{\mathrm{d}t}\{[C_w M_{evpw} + C_M M_M]T_{evp}\} = \left[-LM_a\frac{\mathrm{d}x_{ads}}{\mathrm{d}t} + m_{cw}C_{cw}(T_{cw,in} - T_{cw,out})\right] \tag{7}$$

Chilled water outlet temperature from evaporator is

$$T_{cw,out} = T_{evp} + (T_{cw,in} - T_{evp})\exp\left(\frac{-U_{evp}A_{evp}}{m_{cw}C_{cw}}\right) \tag{8}$$

2.2 Performance Evaluation of Adsorption Refrigeration System

Adsorption kinetics can be used to determine the rate of adsorption and time needed to reach equilibrium and to calculate adsorbed amount of adsorbent over period of time. In order to calculate adsorption kinetics of SGW, ACM and ZW, data mentioned in Table 1 is used. For analyzing the system performance of SGW, ACM and ZW adsorption refrigeration system, heating power and refrigerating effect of the system are to be calculated.

Expressions for cooling and heating power

Table 1 Thermodynamic properties of adsorbents [17, 18]

Properties	Silica gel	Zeolite	Activated carbon
$D_{so}(m^2/s)$	2.54×10^{-4}	3.92×10^{-6}	
$R_p(m)$	0.35×10^{-3}	1×10^{-6}	
$15D_{so}/R_p^2$ (1/s)			7.35×10^{-3}
E_a (J/mol)	4.2×10^4	2.8035×10^4	8.13×10^3

$$\text{Cooling power} = Q_{ref} = \frac{\int_0^{t_{cyc}} C_w m_{w,e}\left(T_{chill,in} - T_{chill,out}\right)dt}{t_{cyc}} \tag{9}$$

$$\text{Heating power} = Q_h = \frac{\int_0^{t_{cyc}} C_w m_{w,h}\left(T_{h,in} - T_{h,out}\right)dt}{t_{cyc}} \tag{10}$$

Expressions for COP and SCP

COP and SCP are the two important parameters to analyze the performance of adsorption refrigeration system. SCP of adsorption refrigeration system is demarcated as cooling power generated per unit mass of adsorbate

$$\text{SCP} = \frac{\int_0^{t_{cyc}} m_{ch} C_w\left(T_{chw,in} - T_{chw,out}\right)dt}{M_a t_{cyc}} \tag{11}$$

COP of adsorption refrigeration system is demarcated as the ratio of cooling power produced to the heat given to the system

$$\text{COP} = \frac{\int_0^{t_{cyc}} m_{ch} C_w\left(T_{chw,in} - T_{chw,out}\right)dt}{\int_0^{t_{cyc}} m_{hw} C_w\left(T_{hw,in} - T_{hw,out}\right)dt} \tag{12}$$

3 Results and Discussion

From Fig. 3, it is obvious that rate of adsorption is very high in the beginning and is gradually decreasing with increase in time. This is because, initially as there is no adsorption, adsorption pores are free. As the adsorption process starts, adsorbate starts filling into the process and as there is no resistance to adsorbate to get adsorbed, rate of adsorption is high. Gradually, all the pores are filled with adsorbate and there is no enough space to be filled. Hence, rate of adsorption decreases with increase in time and reaches zero when equilibrium is obtained. Adsorption capacity of various working pairs at different adsorption temperatures for same cycle time is presented in Table 2.

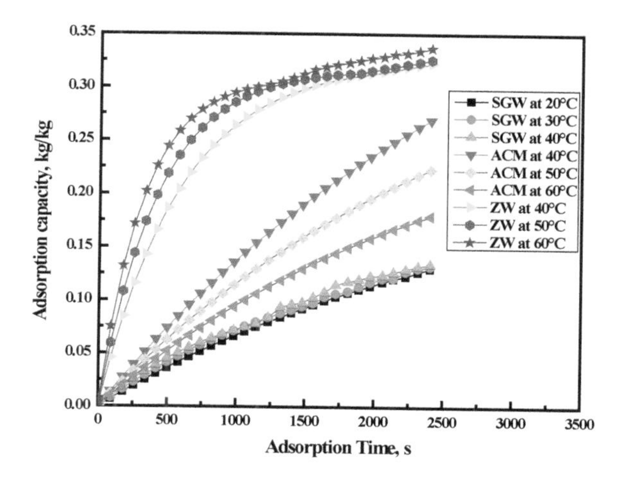

Fig. 3 Adsorption capacities of ACM, SGW and ZW at different adsorption temperatures

Table 2 Adsorbed quantity of working pairs at various adsorption temperatures

Working pair	Temperature (°C)	Adsorbed quantity (kg/kg)
ACM	40	0.2626
	50	0.2212
	60	0.1792
SGW	20	0.1299
	30	0.1221
	40	0.1032
ZW	40	0.2498
	50	0.285
	60	0.3034

From Table 2, it is noticed that adsorption capacity of adsorbate declines with increase in adsorption temperature. This is because, as the adsorption temperature increases, kinetic energy of gas molecules increases. This results in desorption of previously adsorbed molecules. Increase in kinetic energy does not give enough time for gaseous molecules to settle down. As desorption temperature increases, more amount of heat is provided to adsorbent bed and amount of desorption increases. Table 3 provides results obtained by evaluating refrigeration capacity of different adsorption working pairs at various adsorption and desorption temperatures. Refrigeration capacity may vary with mass of adsorbent and adsorption/desorption time.

Desorption temperature of silica gel cannot be greater than 100 °C because hydroxyl bonds present in silica gel break at higher temperature. Desorption temperatures of ACM pair cannot be greater than 120 °C because decomposition of methanol occurs from that temperature. Adsorption heat of ZW is higher than SGW system and activated carbon system. ZW is stable at higher temperatures. Due to large adsorption

Table 3 Refrigeration capacity of different working pairs

Adsorbent–refrigerant working pair	Saturation pressure (bar)	Adsorption temperature (°C)	Desorption temperature (°C)	Refrigeration capacity (kW)
ACM	0.403744	40	90	0.4573
			100	0.895
			110	1.26
	0.574825	50	90	0.1
			100	0.525
			110	0.987
	0.943054	60	90	–
			100	0.178
			110	0.520
SGW	0.0233	20	70	0.686
			80	1.203
			90	1.5806
	0.0423	30	70	0.510
			80	1.02739
			90	1.404
	0.0736	40	70	0.083
			80	0.600
			90	0.977
ZW	0.0736	40	180	3.079
			210	4.096
			240	4.985
	0.1231	50	180	3.021
			210	4.037
			240	4.926
	0.1987	60	180	2.849
			210	3.865
			240	4.755

heat and higher desorption temperatures, performance of zeolite is poorer than silica gel and activated carbon at lower temperatures than 150 °C.

Variation of COP and SCP of ACM, SGW and ZW system with cycle time is presented in Figs. 4, 5 and 6, respectively. SCP along with cycle time increases initially and reaches maximum value when the system reaches equilibrium. Further with increase in cycle time, SCP diminishes as the adsorption rate of adsorbate decreases with rise in time. On the other hand, COP escalates with cycle time as it is well acknowledged that increasing cycle time enhances adsorption/desorption

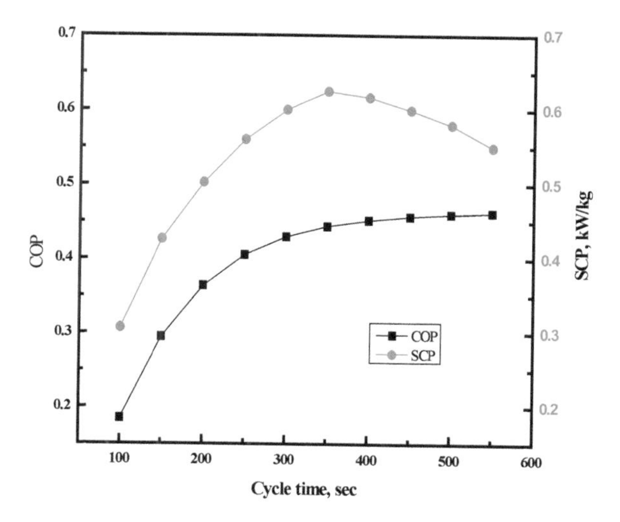

Fig. 4 COP and SCP of ACM system

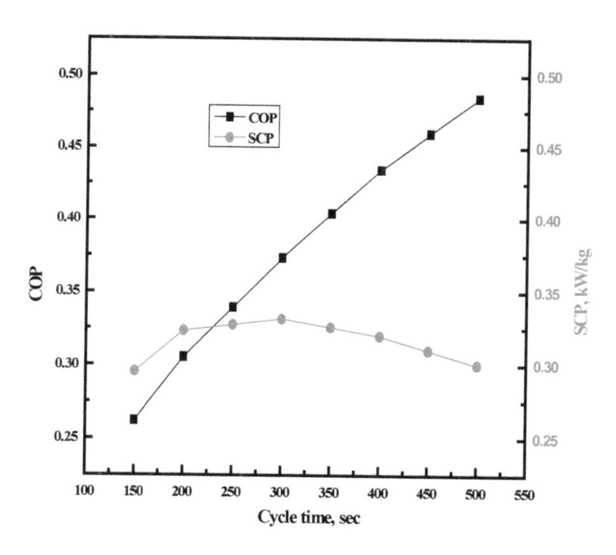

Fig. 5 COP and SCP of SGW system

process. COP and SCP of SGW system are compared with the available data [18] in Figs. 7 and 8 and found to be in good agreement.

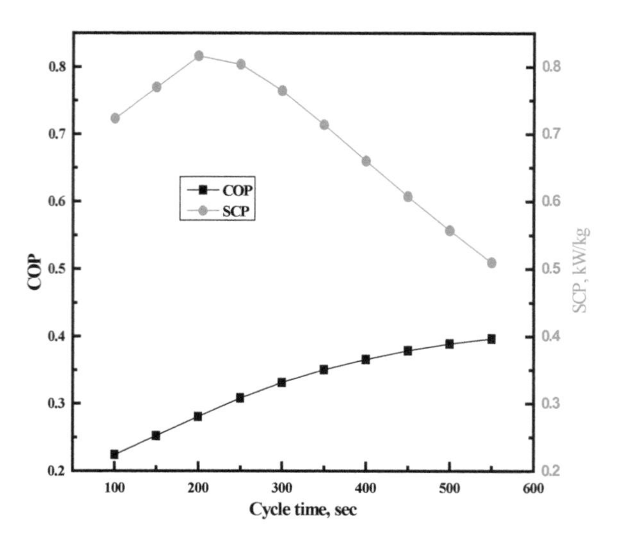

Fig. 6 COP and SCP of GW system

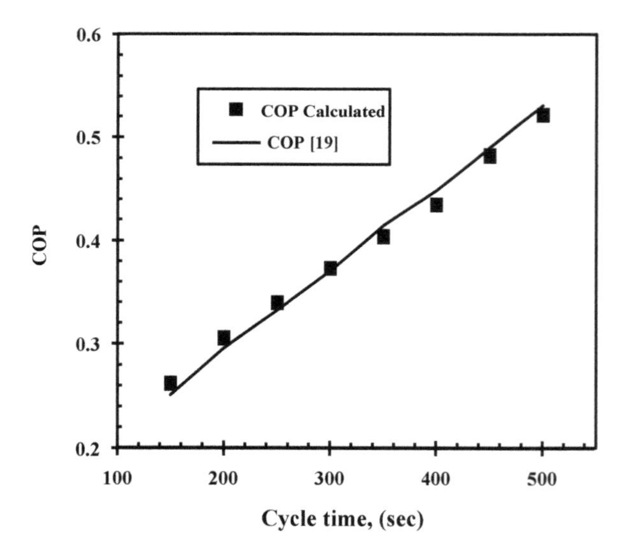

Fig. 7 COP comparison of SGW working pair with reference [19]

4 Conclusions

The following inferences can be drawn

1. In the present work, a simple thermodynamic performance assessment of various working pairs for an adsorption refrigeration system is introduced.
2. The analysis is according to energy conservation principle and DA equation.
3. Adsorption capacities of ACM, SGW and ZW adsorption refrigeration system with respect to cycle time are calculated and presented.

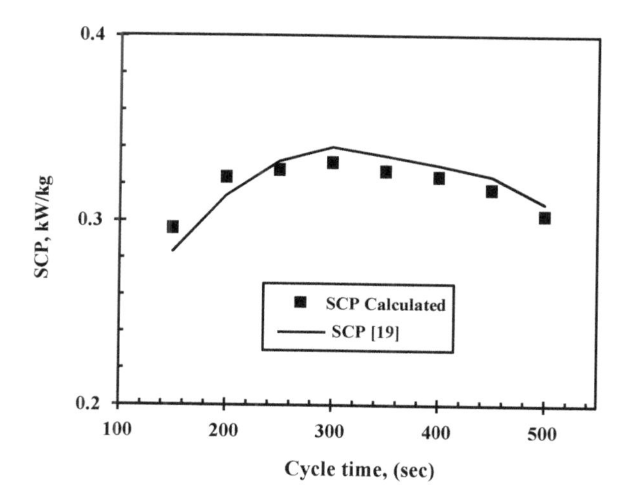

Fig. 8 SCP comparison of SGW working pair with reference [19]

4. As adsorption temperature increases, adsorption capacity of working pairs decreases, and as desorption temperature increases, desorption capacity of working pairs increases.
5. ZW system is best suitable for adsorption refrigeration systems utilizing waste energy available at high temperatures.
6. SGW system is ideal for systems utilizing waste energy available at lower temperatures.
7. Change of COP and SCP with cycle time is presented graphically and found that with increase in cycle time COP increases and SCP decreases.

References

1. Sachin, G., Narasimha, K.K., Ramakrishna, K., Abhishek, D.: Thermodynamic analysis and effects of replacing HFC by fourth—generation refrigerants in VCR systems. Int. J. Air-Condition. Refriger. **26**(1), 1850013-1–1850013-12 (2018)
2. El-Sayed, A.R., El Morsi, M., Mahmoud, N.A.: Experimental investigation of a walk-in refrigerator performance using R290 as a retrofit for R22, Int. J. Air-Condition. Refriger. **26**(4), 1850029-1–1850029-14 (2018)
3. Wang, R., Wang, L., Wu, J.: Adsorption Refrigeration Technology Theory and Application. Wiley, Singapore (2014)
4. Choudhury, B., Saha, B.B., Chatterjee, P.K., Sarkar, J.P.: An overview of developments in adsorption refrigeration systems towards a sustainable way of cooling. Appl. Energy **104**, 554–567 (2013)
5. Returi, M.C., Konijeti, R., Dasore, A.: Heat transfer enhancement using hybrid nanofluids in spiral plate heat exchangers. Heat Transf. Asian Res. **48**(7), 3128–3143 (2019)
6. Faizan, S., Muhammad, S., Yasir, N., Muhammad, U., Sobhy, M.I., Yongqiang, F., Kiran, N.B., Abdul, N., Imran, A.: Steady-state investigation of carbon-based adsorbent-adsorbate pairs for heat transformation application. Sustainability **12**(17), 7040 (2020)

7. Naik, B.K., Muthukumar, P.: Energy, entransy and exergy analyses of a liquid desiccant regenerator. Int. J. Ref. **105**, 80–91 (2019)
8. Wang, K., Vineyard, E.A.: Adsorption refrigeration: new opportunities for solar. ASHRAE J., **53**(9), 14–22 (2011)
9. Hassan, H.Z.: Energy analysis and performance evaluation of the adsorption refrigeration system. ISRN Mech Eng (Article ID 704340), 1–14 (2013)
10. Restuccia, G., Cacciola. G.: Performances of adsorption systems for ambient heating and air conditioning. Int. J. Refriger. **22**(1), 18–26 (1999)
11. Hu Eric, J.: A study of thermal decomposition of methanol in solar powdered adsorption refrigeration machines. Sol. Energy **62**(5), 325–329 (1998)
12. Zhang, L.Z.: Design and testing of an automobile waste heat adsorption cooling system. Appl. Therm. Eng. **20**(1), 103–114 (2000)
13. Tather, M., Erdem-Senatalar, A.: When do thin zeolite layers and a large void volume in the adsorber limit the performance of adsorption heat pumps? Microporous Mesoporous Mater. **54**(1–2), 89–96 (2002)
14. Chua, H.T., Ng, K.C., Chakraborthy, A., Oo, N.M., Othman, M.A.: Adsorption characteristics of silica gel +water systems. J. Chem. Eng. Data **47**(5), 1177–1181 (2002)
15. Alam, K.C.A., Saha, B.B., Kang, Y.T., Akisawa, A., Kashiwagi, T.: Heat exchanger design effect on the system performance of silica gel adsorption refrigeration machines. Int. J. Heat Mass Transf. **43**(24), 4419–4431 (2000)
16. Rezk, A.R.M., Al-Dadah, R.K.: Physical and operating conditions effects on silica gel/water adsorption chiller performance. Appl. Energy **89**(1), 142–149 (2012)
17. Ghilen, N., Gabsi, S., Benelmir, R., El-Ganaoui, M.: Performance simulation of two-bed adsorption refrigeration chiller with mass recovery. J. Fundam. Renew. Energy Appl. **7**(3), 1–7 (2017)
18. Ko, D., Wardane, R.S., Biegler, L.T.: Optimization of a pressure-swing adsorption process using zeolite 13X for CO2 sequestration. Ind. Eng. Chem. Res. **42**(2), 339–348 (2003)
19. Gado, M.G., Elgendy, E., Elsayed, K., Fatouh, M.: Parametric study of an adsorption refrigeration system using different working pairs, In: 17th International Proceedings on Aerospace Sciences and Aviation Technology (ASAT-17), pp. 1–15. Cairo (2017)

Forecasting Electricity Demand Using Statistical Technique for the State of Assam

Kakoli Goswami and Aditya Bihar Kandali

Abstract Accurately predicting the rise or fall in the electrical load with satisfactory result requires a good forecasting model. Proper planning of power companies needs a definite and accurate prediction of load. This in turn helps to supply energy to the consumers in a reliable manner. Statistical forecasting scheme has been used to study load for the state of Assam. Real-time load data of three years has been used in this study. Daily electricity load of 10 a.m. for three consecutive years has been divided into different data set for training and testing. The entire dataset has been provided by the Assam State Load Dispatch Center. The analysis is carried out using regression model. Performance of the different models has been evaluated using different accuracy measures. The final outcome helps to compare the regression models thereby assisting in selecting the best model for forecasting electricity demand. The entire analysis was carried out using MATLAB R2016a.

Keywords Forecasting electricity demand · Statistical method · Regression model

1 Introduction

Load forecasting forms an indispensable part of the operation and control of power system and is a method of predicting accurately the rise or fall in electricity demand [1]. Load forecasting can be either long-term, medium-term or short-term. Any form of expansion of power utility leads to a complicated and distressed power system. On the other hand, a distressed power system is at a high risk to cascade outages [2]. A mechanism that enables regular supply of electricity without any interruption ensures rapid economic and industrial growth of any region. A good reliable forecasting model has become indispensable for the states situated in the north-eastern part of India. In spite of numerous literatures on load forecasting, research on recent

K. Goswami (✉) · A. B. Kandali
Electrical Engineering Department, Jorhat Engineering College, Dibrugarh University, Assam, India
e-mail: kakoligoswami2009@gmail.com

P. Muthukumar et al. (eds.), *Innovations in Sustainable Energy and Technology*, Advances in Sustainability Science and Technology,
https://doi.org/10.1007/978-981-16-1119-3_27

forecasting work on Assam is limited, thereby providing ample scope of research work.

Load predicting techniques are in general divided as traditional methods and artificial intelligence (AI) methods [3]. Differences between these two methods however have been greatly reduced as a result of merging of different disciplines. Classical forecasting techniques include statistical methods, regression methods and exponential smoothing. Extensive literature work is available on forecasting electricity demand [3–11].

2 Technique Used

The study has been carried out using statistical regression models namely autoregressive integrated moving average ARIMA) and seasonal autoregressive integrated moving average model (SARIMA).

ARIMA is considered to be one of the simplest and most popular statistical methods. Its popularity can be attributed to Box and Jenkins [12]. ARIMA model consist of two forms of linear regressions, the autoregressive (AR) and the moving average (MA) [13]. The model in its simplest form is written as

$$y(t) = c + a_1 y_{t-1} + a_2 y_{t-2} \ldots + a_p y_{t-p} + e_t \tag{1}$$

$$y(t) = \mu + u_t + m_1 u_{t-1} + m_2 u_{t-2} \ldots + m_q u_{t-q} \tag{2}$$

$a_1, a_2, a_3, \ldots, a_p$, and $m_1, m_2, m_3, \ldots, m_q$ refer to the parameters while the subscripts p, q refer to the order of the respective AR and MA portions with the constant represented by the term c and the white noise denoted by the e_t. u_t, u_{t-1}, u_{t-2}, …, u_{t-q} refer to the white noise terms or the error term. μ in Eq. (2) denotes the expectation. The two regression models can be combined into a single regression model ARIMA(p, q) as

$$y(t) = c + a_1 y_{t-1} + \ldots + a_p y_{t-p} + u_t + m_1 u_{t-1} + \ldots + m_q u_{t-q} \tag{3}$$

In Eq. (3) "p" refers to the AR or autoregressive terms used to represent the number of observations from the past data taken to forecast a future data. "q" refers to the MA or moving average terms representing the lagged values of the error terms. The first move to develop this model lies in determining if the time series (TS) data is stationary in terms of statistics and to look for any seasonality which can be used to develop a statistical model. In case of data that is not stationary, differencing technique can be used. This results in the ARIMA (p, d, q) model with d as the required degree of differencing required to makes the time series (TS) stationary. The SARIMA model can be used to deal with the seasonal changes that lead to periodic oscillations in electrical load. This model is written as SARIMA (p, d, q)(P, D, Q)S where p,q and

d have their usual meaning as in the ARIMA model. They refer to the AR order, MA order and differencing order, respectively. The uppercase letters P, D, Q and S convey the seasonality involved in the model.

3 Steps Adopted

There are four steps to identify the most appropriate model for forecasting. These are test to check stationary followed by identification of the regression model, model fitting and evaluating the performance of the model [14]. In this paper, we have tried to analyze two types of statistical models using electrical load demand data for three consecutive years.

This study used 24 hourly electrical loads. The 24 hourly load data corresponds to 1095 different electrical load demand points. These load data has been presented as TS in Fig. 1. Statistical measures of load data were calculated. Figure 1 clearly shows periodic oscillations that result from seasonal changes. The ACF and PACF plots of the time series are presented in Fig. 2. ACF and PACF plots assist in finding the statistical correlation that might be present in the data. The ACF of white noise is basically an impulse. So, while building the forecasting model, the left over residuals should not have any predictability left and as such its ACF should be that of an impulse. This is done using residual test.

3.1 Stationary Test

The elemental step to develop the regression this model is to make the time series (TS) stationary such that the statistical characteristics do not change over time. In this study, the differencing degree d for ARIMA and D for SARIMA has been assessed

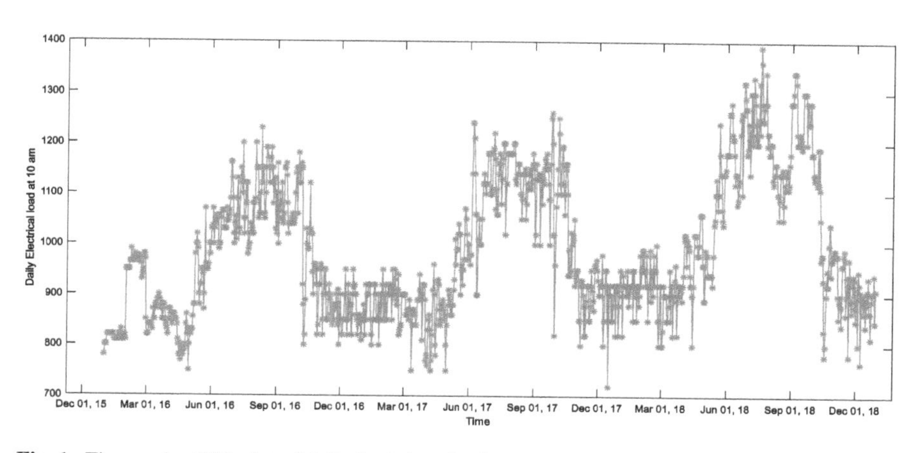

Fig. 1 Time series (TS) plot of daily load data for three consecutive years

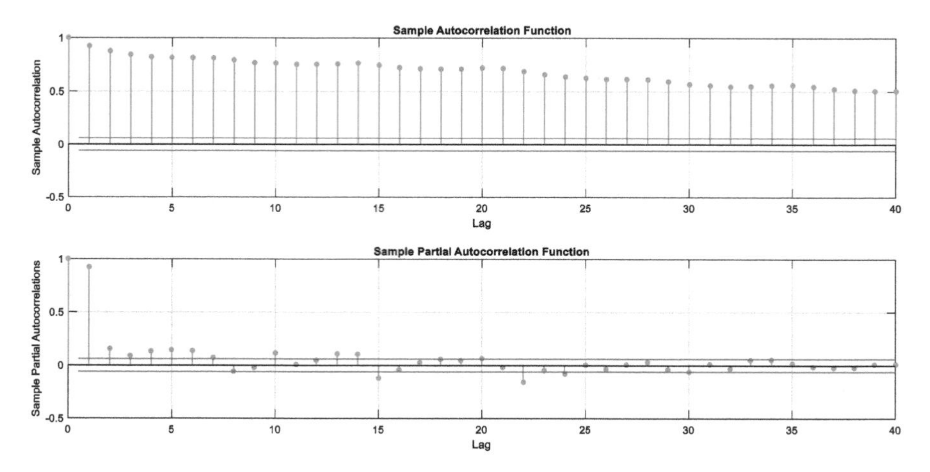

Fig. 2 ACF plot and PACF plot of load data of Fig. 1

using ACF and PACF plots and supported by the use of augmented Dickey–Fuller test (ADF) test using "adftest" command. The three plots of the first column in Fig. 3 show the original load data, its first-order differencing and second-order differencing, respectively, together with their corresponding correlation function plots. It can be observed from the figures that differencing the TS with non-stationary data leads to a TS with stationary data. Figure 2 indicates there are significant spikes at multiples of 7. As a consequence, lag 7 seasonality may be used in forming the SARIMA model. Figure 4 shows plot of lag 7 seasonality of load. This plot seems stationary when compared with original load data TS plot.

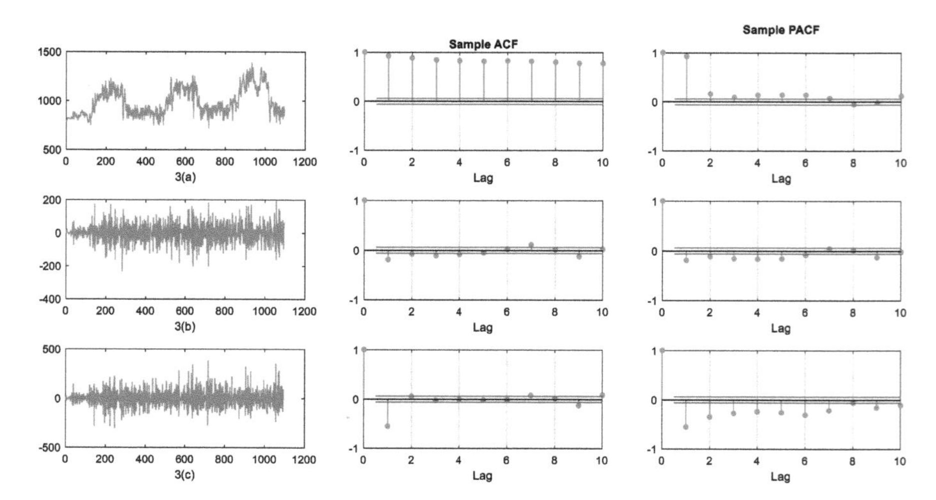

Fig. 3 TS plot along with ACF and PACF of load data followed by the first-order differencing and second-order difference

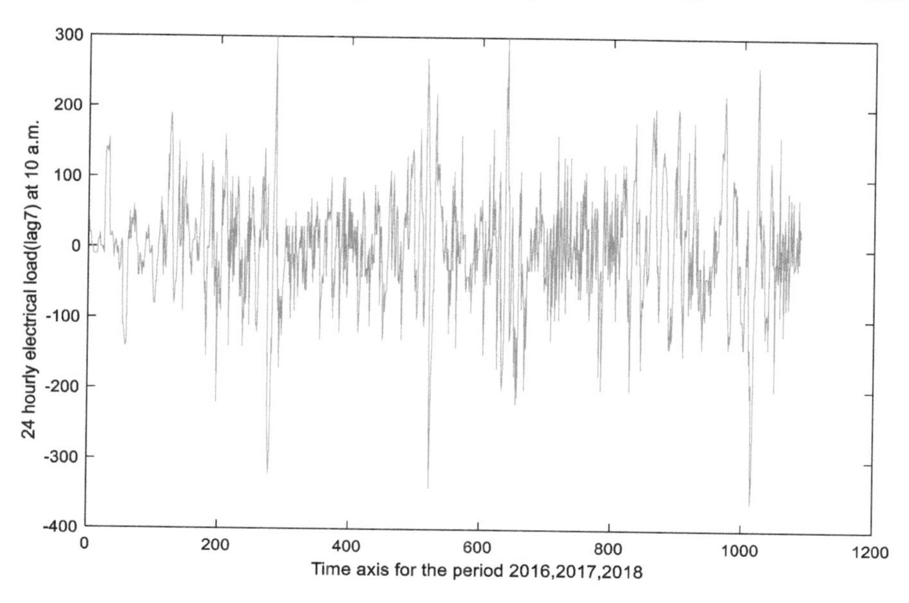

Fig. 4 Plot of lag 7 load data

3.2 *Model Identification*

The next step after addressing the stationarity and seasonality is to determine the parameter p and q (P and Q) for a model. This is referred to as model identification. Different techniques can be found in literature for identification of model. This study used the ACF and PACF plots for determining the parameters. The value of Akaike information criterion (AIC), Bayesian information criterion (BIC) and log-likelihood function ($\log L$) was used in determining the best possible model; the model with the minimum AIC, BIC values. Different ARIMA model configurations were formed by varying the values of the parameters p *and* q [14, 15].

The ACF in Fig. 5 indicates a diminishing pattern indicating AR of order 1. PACF shows the first prominent spike at lag 1. The next prominent spike is visible at lag 7. In case of ACF, there is a diminishing pattern in multiples of 7. These information can assist in building the SARIMA model with SMA(1) and $S = 7$.

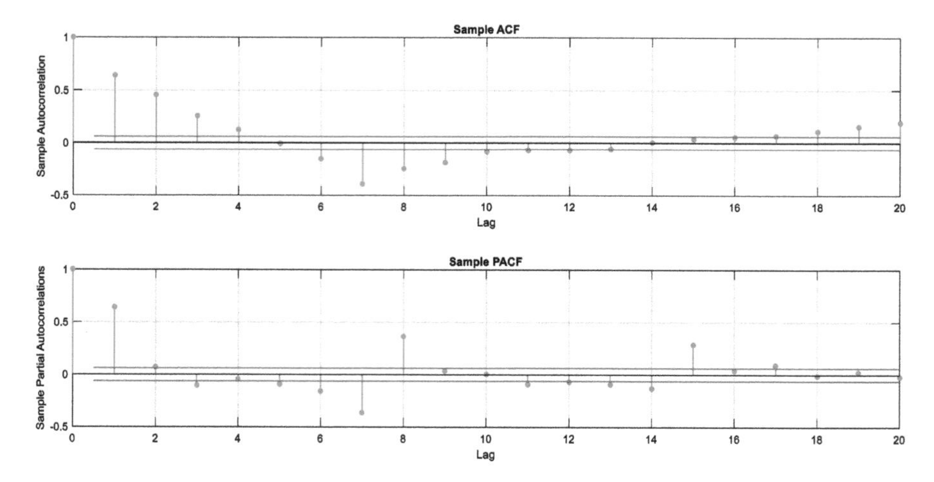

Fig. 5 ACF and PACF of lag 7 load data

3.3 *Model Fitting*

The entire load data set has been partitioned into training data and testing data as shown in Fig. 6. The training data was used to train the various models. Table 1 represents the estimated parameter for each of the regression model. Any residual autocorrelation was verified using the Ljung–Box Q-test present in MATLAB. If residual ACF of a model is within the 95% significant level as in Fig. 7, it indicates there is no correlation left in the residuals. However, if the residual ACF is not within the significant level as shown in Fig. 8, it indicates that some correlation are still left in the residuals.

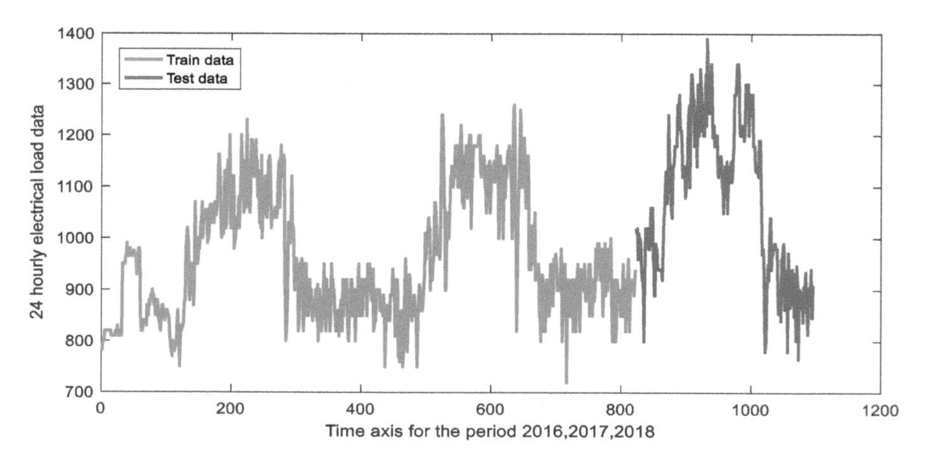

Fig. 6 Load data divided into training and testing data

Table 1 Estimated parameters and calculated error of ARIMA and SARIMA models

Forecasting Model	Log Likelihood	AI Criterion	BI Criterion	RMSE	MAE	R	MAPE
1.ARIMA(0,1,2)	-2.41E+03	9.82E+03	9.83E+03	732	665	0.52	63
2.ARIMA(1,2,1)	-4.37E+03	8.63E+03	8.66E+03	192	159	0.75	15
3.ARIMA(3,2,2)	-4.37E+03	8.79E+03	8.79E+03	206	164	0.68	17
4.ARIMA(3,2,1)	-4.32E+03	8.71E+03	8.74E+03	352	327	0.47	32
5.SARIMA (0 1 1)(0 1 2)7	-5.78E+03	1.15E+04	1.15E+04	198	124	0.89	10
6.SARIMA (0 1 2)(0 1 3)7	-4.59E+03	1.46E+04	1.46E+04	215	148	0.78	13
7.SARIMA (1 0 2)(0 1 2)7	-4.67E+03	1.28E+04	1.28E+04	205	139	0.86	19
8.SARIMA (2 1 1)(0 1 2)7	-4.78E+03	1.18E+04	1.18E+04	215	165	0.65	18

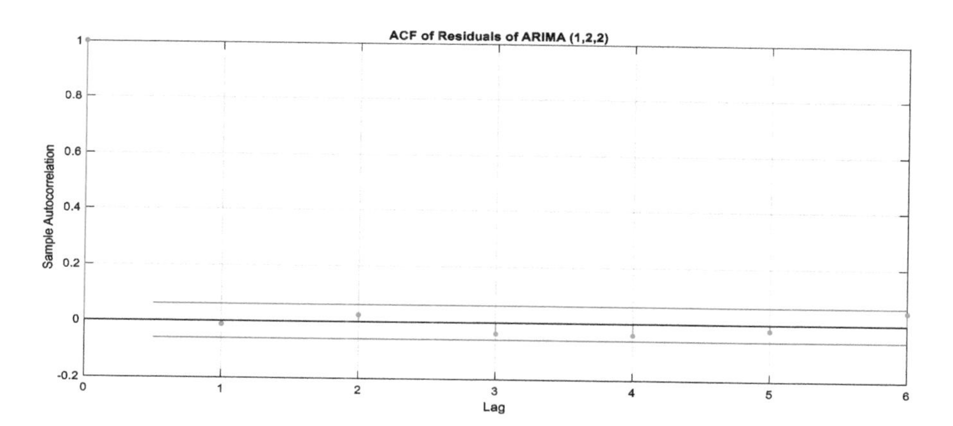

Fig. 7 ACF plot for the residuals of ARIMA model ARIMA(1,2,2) [16]

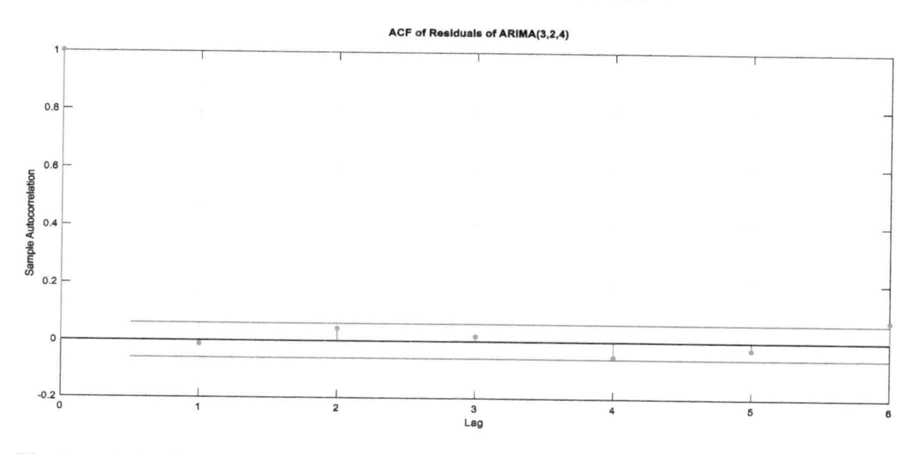

Fig. 8 ACF plot for the residuals of ARIMA model ARIMA(3,2,4) [16]

3.4 Performance of the Model

The precision of all models has been checked using three different accuracy measures namely root mean square error (RMSE), mean absolute error (MAE) and correlation coefficient *R*. Table 1 gives a summary of the results. Between the two types of regression model used in our study, the results obtained using SARIMA have least error with a better fit, thereby proving the superiority of SARIMA over ARIMA in this study. Thus, it can be said that both types of regression models can be used for predicting electricity load demand. However, for time series data, having seasonal trend SARIMA offers a better forecast.

4 Results and Analysis

The results of Table 1 show SARIMA (0 1 1)(0 1 2)7 as the most deserving forecasting model among the various model. This model was selected to forecast the electrical load using the test data. Figure 9 shows that the predicted load follows the observed load. This model may be used for long-term forecasting. SARIMA model has the ability of extracting the seasonal characteristic of the time series data which is not available with the ARIMA model.

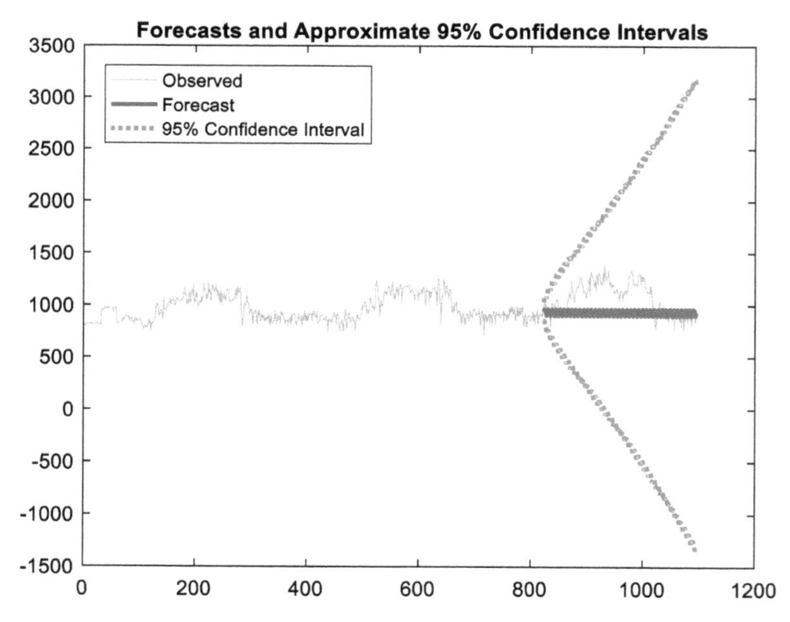

Fig. 9 Forecasted load of SARIMA model

5 Conclusion

In this study, statistical regression models were used to predict the electrical load for Assam, a north-eastern state in India. This work helped in understanding the advantages of the SARIMA model compared to the ARIMA model for periodic time series data. SARIMA model takes into account the periodic variations of electrical load and is thus advantageous for time series data that has a periodic nature. This facilitated in obtaining better forecasting results. As a result, SARIMA outperformed the ARIMA with least error and better fit. SARIMA model when used along with AI models may further show better prediction. The authors intend to use hybrid model in their next forecasting model.

References

1. Hagan, M.T., Behr, S.M.: The time series approach to short term load forecasting. IEEE Trans. Power Syst. **2**(3) (1987)
2. Saxena, D., Singh, S.N., Verma, K.S.: Application of computational intelligence in emerging power systems. Int. J. Eng. Sci. Technol. **2**(3)
3. Wang, Q., Li, S., Li, R.: Forecasting energy demand in China and India: using single-linear, hybrid-linear, and non-linear time series forecast techniques. Energy Elsevier **161**(C), 821–831 (2018). https://doi.org/10.1016/j.energy.2018.07.168
4. Zhang, X., Wang, J.: A novel decomposition-ensemble model for forecasting short-term load-time series with multiple seasonal patterns. J. Appl. Soft Comput. **65**(C), 478–494 (2018)
5. Saleh, A.I., Rabie, A.H., Abo-Al-Ez, K.M.: A data mining based load forecasting strategy for smart electrical grids. Adv. Eng. Inf. 30, 422–448 (2016)
6. Hyndman, R.J., George, A.: 8.9 Seasonal ARIMA Models. Forecasting: Principles and Practice. Texts. Retrieved 19 May 2015
7. Chikobvu, D., Sigauke, C.: Regression-SARIMA modeling of daily peak electricity demand in South Africa. J. Energy South. Afr. **23**(3), 23–30 (2012)
8. Song, K.B., Baek, Y.S., Hong, D.H., Jang, G.: Short-term load forecasting for the holidays using fuzzy linear regression method. IEEE Trans. Power Syst. **20**(1), 96–101 (2005)
9. Beccali, M., Cellura, M., Brano, V.L., Marvuglia, A.: Forecasting daily urban electric load profiles using artificial neural networks. Energy Conver. Manage. **45**(18–19), 2879–2900 (2004)
10. Lv, G., Wang, X., Jin, Y.: Short-term load forecasting in power system using least squares support vector machine. In: Proceedings of the 2006 International Conference on Computational Intelligence, Theory and Applications, 9th Fuzzy Days, pp 117–126 (2006)
11. Yu, K.W., Hsu, C.H., Yang, S.M.: A model integrating ARIMA and ANN with seasonal and periodic characteristics for forecasting electricity load dynamics in a state. In: 2019 IEEE 6th International Conference on Energy Smart Systems (ESS), Kyiv, Ukraine, pp. 19–24 (2019)
12. Box, G.E.P., Jenkins, G.M.: Time series analysis: forecasting and control. J. Am. Stat. Assoc. **68**(342), 199–201 (1970)
13. Yang, Y., Zheng, H., Zhang, R.: Prediction and analysis of aircraft failure rate based on SARIMA model. In: 2017 2nd IEEE International conference on Computational Intelligence and Applications
14. Al-Musaylha, M.S., Deoa, R.C., Adamowskic, J.F., Lia, Y.: Short-term electricity demand forecasting with MARS, SVR and ARIMA models using aggregated demand data in Queensland, Australia. Adv. Eng. Inf. **35**, 1–16 (2018)

15. Park, S., Han, S., Son, Y.: Demand power forecasting with data mining method in smart grid. In: 2017 IEEE Innovative Smart Grid Technologies—Asia (ISGT-Asia)
16. Goswami, K., Kandali, A.B.: Electricity demand prediction using data driven forecasting scheme: ARIMA and SARIMA for real-time load data of Assam. In: 2020 International Conference on Computational Performance Evaluation (ComPE), Shillong, India, pp. 570–574 (2020). https://doi.org/10.1109/ComPE49325.2020.9200031

Analysis and Prediction of Air Pollution in Assam Using ARIMA/SARIMA and Machine Learning

Th. Shanta Kumar, Himanish S. Das, Upasana Choudhary, Prayakhi E. Dutta, Debarati Guha, and Yeasmin Laskar

Abstract The classical methods like autoregressive integrated moving average (ARIMA) have been playing a vital role in solving time series problems. It is also seen that the machine learning approach is gaining popularity in solving various problems including time series. In this paper, we compare ARIMA and machine learning algorithms—linear regression, neural network regression, Lasso, ElasticNet, decision forest regression, extra trees regression, decision tree, AdaBoost, XGBoost. Our primary focus is to find how well each model forecasts the seventeen years data collected from Pollution Control Board Assam (PCBA). PCBA is monitoring air through nationwide programs. This paper aims to analyze the spread of the pollutants in the air focusing mainly on respirable suspended particulate matter (RSPM) and to predict for the most polluted district. We found that data has a strong seasonal structure, and the ARIMA model outperforms the machine learning algorithms as all the methods tend to perform poorly in forecasting.

Keywords Time series · ARIMA · SARIMA · Machine learning · Air pollution

1 Introduction

Air pollution is causing hazardous problems to human health, plants, and animals. In general, air quality is the state of the air around us, while ambient air quality is the quality of outdoor air in our surroundings. Certain substances present in the air, like particulate matter (PM10 and PM2.5), gases—nitrogen oxide, sulfur dioxide, and carbon monoxide, can have severe effects on all living beings. PM2.5 refers to atmospheric aerosol particles with a diameter of less than 2.5 μm. This particular pollutant is so tiny that it can be easily inhaled while breathing, thus causing severe

Th. S. Kumar (✉) · U. Choudhary · P. E. Dutta · D. Guha · Y. Laskar
Girijananda Chowdhury Institute of Management and Technology (GIMT), Assam, India
e-mail: skumar@gimt-guwahati.ac.in

H. S. Das
Cotton University, Assam, India

P. Muthukumar et al. (eds.), *Innovations in Sustainable Energy and Technology*, Advances in Sustainability Science and Technology,
https://doi.org/10.1007/978-981-16-1119-3_28

health hazards [1]. It has also been proved that air pollution also causes damages to buildings and sculptures [2]. These pollutants can cause dangerous health effects in human and other life forms if the permissible levels for each criteria pollutant are crossed.

It is also a fact that there is an increase in energy demand; India, in particular, faced a significant increase in fossil fuel consumption due to the growth in industrialization. Thus, the substantial increase in vehicular emission, industrial emission, and the burning of dried agricultural plants, the quality of air in rural and urban areas, is deteriorating. Since the production of vehicles has increased in recent years, the emissions thus produced have caused air pollution [3]. The result of this has affected the inhabitants, especially humans and domestic animals, causing allergies, asthma, cancers, etc. [4, 5].

The air quality status is represented in terms of air quality index (AQI) or air pollution index (API), which is used as an air quality model in monitoring and controlling air pollutants [6]. Over the years, a model that combines the AQI and other scales has been used to assess the effect of air pollution on humans [7]. The air quality indices (AQI) were first used by the US Environmental Protection Agency (USEPA) and used by many cities [8]. The approach helps to find the current status of air in any location. Since it is indexed with time, forecasting AQI is mostly based on time series models [9]. Overall, presently the RSPM level in the air is on an all-time high throughout the different parts of the world.

The research in the field of air quality has evolved to a high standard. Predicting air quality can help in identifying the industries whose emission is above the acceptable threshold. It also warns the public to take the appropriate steps when needed. It will also help in government to frame in administrative decisions.

2 Related Works

IBM is using machine learning techniques in monitoring the air quality data of Beijing. With this, they can make a real-time prediction as well as warn for the airborne pollutants. Using extreme learning machine is seen in [10], where a case study of Hong Kong has been performed to predict the concentration of air pollutants. The use of the principal component regression technique is being observed in [11] in predicting the air pollution of New Delhi concerning the daily air quality index [12] proposed a hybrid model of artificial neural networks (ANNs) and ARIMA for times series prediction. The application of a multilayer perceptron for predicting the pollution levels is observed in [13]. They used the pollutants SO_2, PM10, and CO. They also used meteorological features such as temperature, wind direction, pressure, humidity, and wind speed. They found higher errors for SO_2 than other pollutants.

Masih, A. proposes principles of machine learning techniques of ensemble learning algorithms (ELA) [14], neural network (NN), and support vector machine (SVM) algorithms are popular to forecast air pollution. Machine learning techniques mainly conducted in Europe and America. The paper discusses several studies based on machine learning tools undertaken in air quality modeling. Hourly prediction of air pollutants concentration like O_3, PM2.5, and SO_2 is seen in [15] using machine learning. This paper proposes methodologies to predict the hourly concentration of air pollution which is totally based on meteorological data of previous days by formulating the prediction over 24 h as a multi-task learning (MTL) problem.

Use of big data and machine learning-based techniques to forecast air quality is seen in [16]. Using big data, we can model the air system, which is considerably dynamic, spatially expensive, and behaviorally heterogeneous. They combined linear regression, neural network, dynamic aggregator, and inflection predictor while implementing big data. They also demonstrated the use of machine learning prediction models: ANN, genetic algorithm-ANN model, random forest, decision tree, and deep belief network.

The air quality index study discovers that the particulate matter (PM), mainly PM10, is the dominant pollutant in the index value [17]. PM10 is causing severe public health issues because of their synergetic action. So, we have to look after appropriate pollution control and management techniques like a plantation, green belt, etc. to betterment public and environmental health.

Huiping Peng focused on the air quality analysis and prediction of six stations in Canada of the air particulate like O_3, PM2.5, and NO_2 by using machine learning models like—the stepwise multiple linear regression(MLR), online-sequential multiple linear regression (OSMLR), multi-layer perceptron neural network (MLP NN), and online-sequential extreme learning machine (OSELM) [18].

Sak, H. et al. explored LSTM RNN architectures which is very good for sequence to sequence translation and can be used to model for speech recognition. In their work, they showed that LSTM RNNs outperforms DNNs and conventional RNNs for acoustic modeling. Their proposed approach makes great usage of model parameters by addressing the computational efficiency needed for training large networks.

3 Methods and Materials

3.1 Air Quality Data and Study Area

This study emphasizes the analysis and forecasting of ambient trends in the air pollution level of different districts of Assam, India. The dataset is collected from the Pollution Control Board Assam (PCBA), Guwahati, for 17 years between 2003 and 2019. Some of the features collected are location, sulfur dioxide (SO_2), nitrogen dioxide (NO_2), respirable suspended particulate matter (RSPN), date of collection.

3.2 Dataset Preprocessing

After removing the unwanted columns, the dataset contained some missing values that were imputed using interpolation techniques. The entries of both location and date are erroneous. Many of the date entries were not in *date* format, and hence while opening using pandas, we had to use special functions to handle it. The 'location' entries had a lot of misspelling, which had to be corrected. Also, trimming had to be done both at the start and end of the locations. At most, care was taken while preprocessing these features.

3.3 Methodology

The study's main objective is to analyze and predict this RSPM value by examining the efficiency of the ARIMA/SARIMA model and machine learning models, namely linear regression, neural network, Lasso, ElasticNet, decision forest, extra trees, and boosted decision tree.

3.4 Autoregressive Integrated Moving Average (ARIMA)

ARIMA is a popular statistical method for time series forecasting. The acronym itself describes the features of the technique. The components are autoregressive (AR), integrated component, and moving average (MA). Autoregression represents the effects of the previous observations. The integrated component represents trends and seasonality. The moving average represents the effects of previous random errors. The mathematical representation of the ARIMA model is given in Eq. (1) [15].

$$y_t' = c + \emptyset_1 y_{t-1}' + \ldots + \emptyset_1 y_{t-p}' + \theta_1 \varepsilon_{t-1} + \ldots + \theta_q \varepsilon_{t-q} + \theta_1 \varepsilon_t \quad (1)$$

where y_t' is the differenced series. The predictors in the right-hand side comprised of lagged values of y_t and lagged errors. The model takes the form ARIMA (p, d, q), where,

$p =$ order of the autoregressive part;

$d =$ degree of first differencing involved;

$q =$ order of the moving average part.

3.5 Linear Regression (LR)

Linear regression is a well-known and well-understood algorithm that belongs to both statistics and machine learning. It is defined as:

$$y_i = \beta_0 + \beta_1 x_{1i} + \ldots + \beta_k x_{ki} + \varepsilon_i \tag{2}$$

where $i = 1, 2, \ldots, n$, y is the dependent variable, β_i are the parameters, x_i are the k independent variables, ε_i are the errors, k is the number of features.

3.6 Regression Neural Network

Regression artificial neural networks predict an output variable given a function of inputs. These input features can be either categorical or numeric, while the dependent feature should be a numeric type. The weights are initialized randomly and are updated by backpropagation.

While modeling a regression neural network, gradient descent algorithms are used. One popular algorithm is stochastic gradient descent (SGD). In SGD, we update the weights in the negative direction of the slope leading to our goal. The error function is given by:

$$\begin{aligned} E(w) = \sum \big[&(w_0 + w_1 x_1 - y_1)^2 + (w_0 + w_1 x_2 - y_2)^2 \\ &+ \ldots + (w_0 + w_1 x_n - y_n)^2\big] \end{aligned} \tag{3}$$

where w's are the weights. If we differentiate the above equation w.r.t the weights, we get:

$$\begin{aligned} \partial E_0 &= 1, \\ \partial E_1 &= 2 \times (w_0 + w_1 x_1 - y_1) \times x_1 \end{aligned}$$

Update of the weights is done using the following equation:

$$w_n = w_n - \eta \times \frac{\partial E}{\partial w_n} \tag{4}$$

where η is the learning rate.

3.7 Lasso Regression

It stands for 'least absolute shrinkage selection operator.' It performs regularization, which adds a penalty equal to the absolute value of the magnitude of coefficients. If the dataset consists of many data inputs of higher value, it becomes computationally expensive. Moreover, the model derived from the training set, i.e., the output of the training set, becomes very complicated. To reduce this complexity, we have to shrink the magnitude of the regression coefficient. The penalty applied is given by:

$$+\lambda \sum_{j=0}^{p} |w_j| \tag{5}$$

A lambda value of zero is equivalent to the basic OLS equation, and larger the value of lambda, the more features shrink to zero.

3.8 ElasticNet

ElasticNet is a combination of both Ridge and Lasso regression. Regularization is the process of adding some penalty to the loss function. When there are tons of variables/parameters, we need some sort of regularization, and we use ElasticNet regression. Like Lasso and Ridge regression, ElasticNet regression starts with the least squares and then combines the Lasso regression penalty with the Ridge regression penalty. We also use cross-validation for Lasso and Ridge penalty to find the best values. When both the penalty of Lasso and Ridge is zero, then we get the original least square parameter estimates. When the penalty of Lasso is greater than zero and penalty of Ridge equals to zero, then we get Lasso regression and vice versa. When both are greater than zero, we set a hybrid of two. The ElasticNet is good at dealing with situations when there are co-relations between parameters. This is because Lasso tends to pick just one of the co-related terms and eliminates the other, whereas Ridge tends to shrink all the parameters together. The regularization is given by:

$$\frac{\sum_{i=1}^{n}\left(y_i - x_i^J \hat{\beta}\right)^2}{2n} + \lambda\left(\frac{1-\alpha}{2}\sum_{j=1}^{m}\hat{\beta}_j^2 + \alpha\sum_{j=1}^{m}\left|\hat{\beta}_j\right|\right) \tag{6}$$

3.9 Extra Tree (Extremely Randomized Trees)

Extra tree classifier is much faster than the random forest. Instead of locally optimal split for a feature combination, a random value is selected for the split of the extra tree. The point to split on is to randomly pick that leads to more diversified trees with fewer splitters to evaluate an extremely random forest. When extra tree classifier was tested with the readily available dataset, we observe that when we have a noisy feature in our dataset, extra tree classifier seems to outperform random forest. However, when all the features are relevant, and when we train both extra tree and random forest, both achieve the same performance, be it in terms of accuracy. The randomness is achieved from the random splits of all the observations and not from bootstrapping of data.

3.10 Decision Forest Regression

While applying decision trees when trees are grown too thick, they tend to form patterns that are irregular in nature. In most of the cases, the training sets suffer from overfitting problem. Random forests come as a solution taking the average of multiple deep decision trees that are trained on different forms using the same training set which reduces the variance [16]. This is done by applying the general technique of bootstrap aggregating or bagging. Given a training set $X = x_1, \ldots, x_n$ with the dependent variable, $Y = y_1, \ldots, y_n$, bagging repeatedly (B times) finds a random sample with replacement of the training set. For $b = 1, 2, \ldots B$ times, regression tree f_b has been trained as can be seen in Eq. (7). After the final training, the predictions for unseen sample x' can be calculated by taking the average of prediction values from all the individual regression trees on x'.

$$\hat{f} = \frac{1}{B}\sum_{b=1}^{B} f_b(x') \tag{7}$$

As a result, there is a slight increase in the bias and loss of interpretability. However, it greatly boosts the prediction accuracy.

3.11 AdaBoost

AdaBoost is another important ensemble classifier. It combines weak classifier algorithms forming strong ones. A single classifier may perform poorly. However, multiple classifiers are combined using a training set after assigning the right amount of weight, and a better accuracy score is obtained for the final prediction [16].

It is based on the following principle:

i. Retain the algorithm iteratively after choosing the training set depending on the accuracy of the previous training.
ii. The weightage of each trained classifier is updated based on the accuracy achieved.

A boosted classifier takes the form

$$F_T(x) = \sum_{t=1}^{T} f_t(x) \tag{8}$$

where each f_t being a weak learner which takes an object x as input and produces the class of the object.

Each weak learner results to hypothesis, $h(x_i)$, for every sample in the training set. At every iteration t, weak learning is chosen and assigned a coefficient α_t such that the sum training error, E_t, is minimized.

$$E_t = \sum_i E\left[F_{t-1}(x_i) + \alpha_t h(x_i)\right] \tag{9}$$

3.12 *XGBoost (eXtreme Gradient Boosting)*

XGBoost is another ensemble classifier implementing gradient boosting. The advantage of boosting method is that boosting creates trees with fewer splits. Since the trees are small and not deep, they are easy to interpret. The parameters such as a number of trees, learning rates, and depth of the tree can be optimally found using validation techniques.

Boosting consists of three simple steps:

i. An initial model F_0 is used to predict the target. It also produces a residual ($y - F_0$).
ii. A new model $h1$ is fitted to the residuals from the previous step.
iii. F_1, the boosted version of F_0, is formed by combining F_0 and $h1$.

After m iterations we get,

$$F_m(x) = F_{m-1}(x) + h_m(x) \tag{10}$$

3.13 Evaluation Metrics

The models' performance is measured using root-mean-square error (RMSE), which is given by:

$$\text{RMSE} = \sqrt{\frac{1}{n}\sum_{i=1}^{n}(y_i - \hat{y}_i)^2} \tag{11}$$

where y_i are the actual values and $\hat{y}_i$ are the predicted values for n observations.

4 Experiments and Results

4.1 Test for Stationarity

The air quality data taken from the Central Pollution Control Board (CPCB) is found to be stationary. Different tests were performed before we conclude. We divide the RSPM values into two halves and calculated mean and variance to check if it is constant. We obtained mean1 = 4.625, mean2 = 4.551, variance1 = 0.283, variance2 = 0.177. Though the mean looks to be constant, the variance is not. Then, we continued with the augmented Dickey–Fuller (ADF) test and Kwiatkowski–Phillips–Schmidt–Shin (KPSS) Test and obtained the result given in Table 1.

Table 1 shows the 'test statistics' values of both the tests—ADF and KPSS—are less than the critical values. Hence, we conclude that the series is stationary.

Table 1 Results for the test of stationarity

	Augmented Dickey–Fuller (ADF)	Kwiatkowski–Phillips–Schmidt–Shin (KPSS)
Test statistic	−3.356555	0.356935
p-value	0.012532	0.095717
#Lags used	15	15
Critical value (1%)	−3.464337	0.739
Critical value (5%)	−2.876479	0.463
Critical value (10%)	−2.574733	0.347

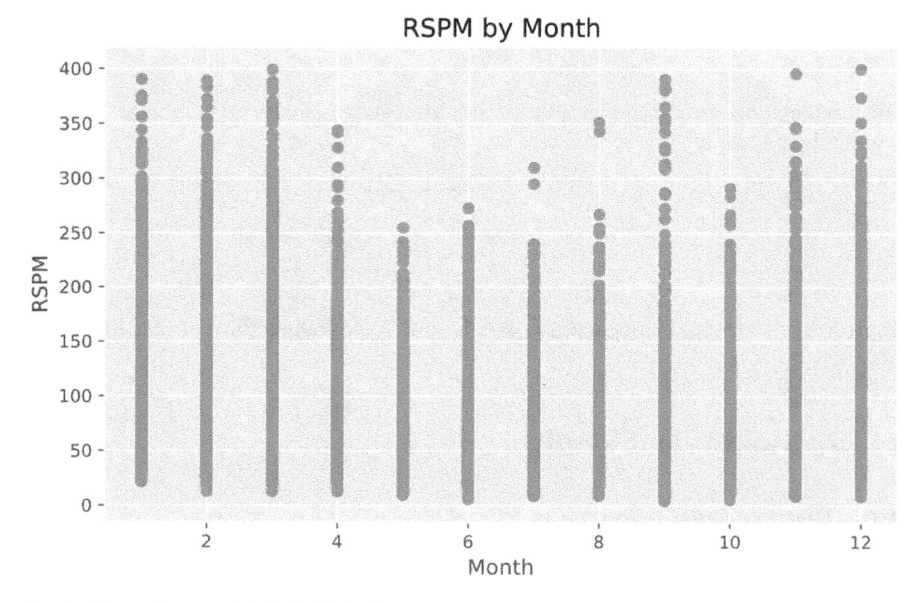

Fig. 1 Plot of the monthly RSPM level

4.2 Exploratory Data Analysis

When we plot the '*RSPM by month,*' we observed that the pollution level is high during the winter than the other seasons, as seen in Fig. 1. This may be because winter is dry and dusty.

The most polluted three districts of Assam are Guwahati (107.83), Nagaon (105.44), and Nalbari (94.52) as depicted in Fig. 2. The levels of RSPM indicated in the figure are the average of the 17 years' data.

4.3 Applying ARIMA/SARIMA

The moving average of RSPM with window size = 12 is found to be 91.6510 for Guwahati. The trend is depicted in Fig. 3.

As the RSPM value exhibits seasonality, we use SARIMA. SARIMA model is created for RSPM values with the order (p, d, q as 3, 1, 3) and (P, D, Q, s as 1, 1, 1, 12). The given order was found to be the best fit for the SARIMA model. Part of the '*state space model results*' is given in Table 2.

The criteria—Akaike information criterion (AIC), Bayesian information criterion (BIC), and Hannan–Quinn information criterion (HQC)—are used to estimate the coefficient of the models. These criteria depend upon maximum likelihood. We use that data to fit our model, and we would like to know how well our model generalizes

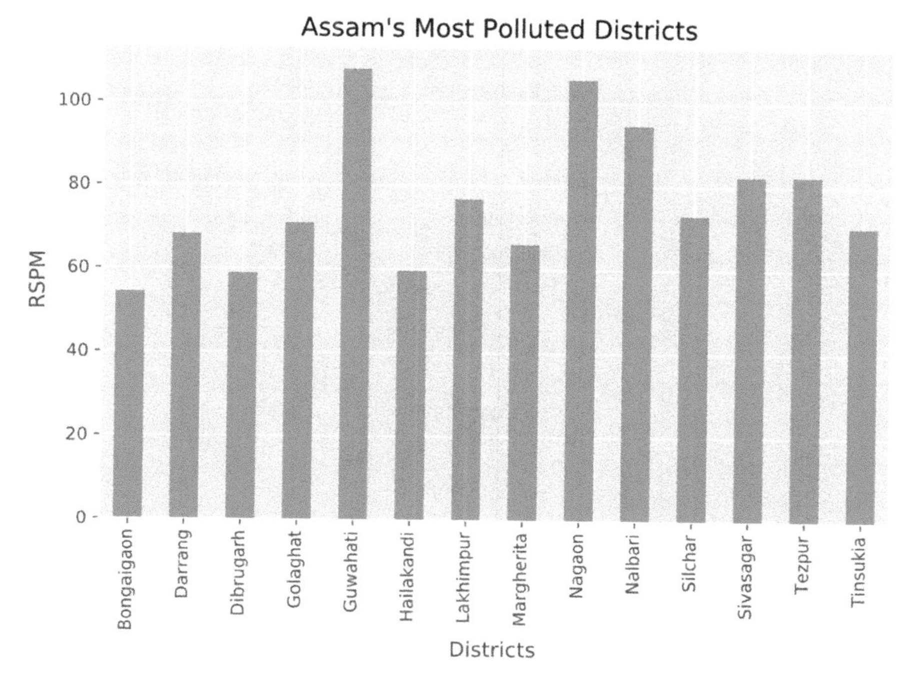

Fig. 2 Most polluted districts of Assam

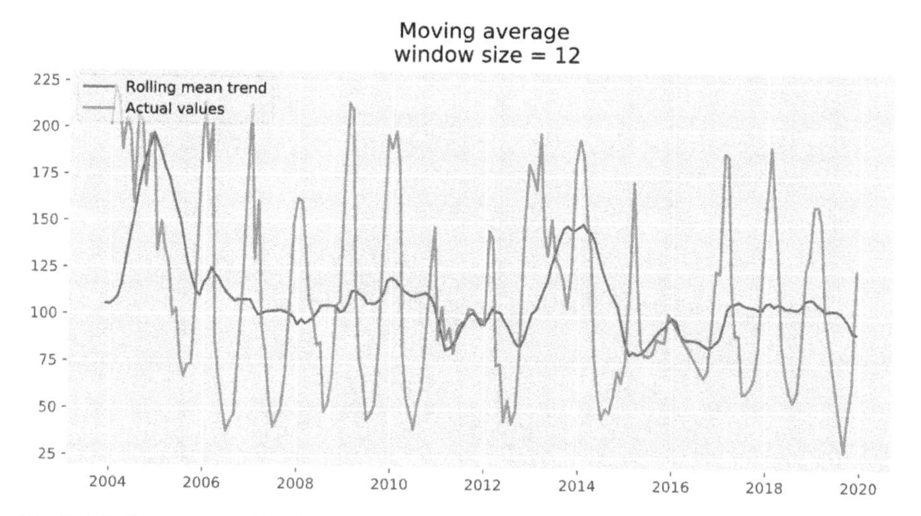

Fig. 3 Moving average with window size = 12

Table 2 Results from state space model

Heads	Values
Dependent variable	RSPM
Model	SARIMAX(3, 1, 3) × (1, 1, 1, 12)
Sample	01-01-2003 to 12-01-2019
No. observations	204
Log likelihood	−880.207
Akaike information criterion (AIC)	1778.413
Bayesian information criterion (BIC)	1807.684
Hannan–Quinn information criterion (HQC)	1790.269

out of sample predictions. We will fit our model using one sample of data and evaluate its predictive fit in other data. However, real-life data can be hard to combine, and so we do not have much data or independent datasets to evaluate the fit of the model. These criteria are used to fit the model accordingly and select the models with the least values.

The prediction has been made for four years: 2020 to 2024. The model was found to give root-mean-square error of 18.53 (Fig. 4).

The predicted value of RSPM value at Guwahati in 2020, as a sample, is given in Table 3.

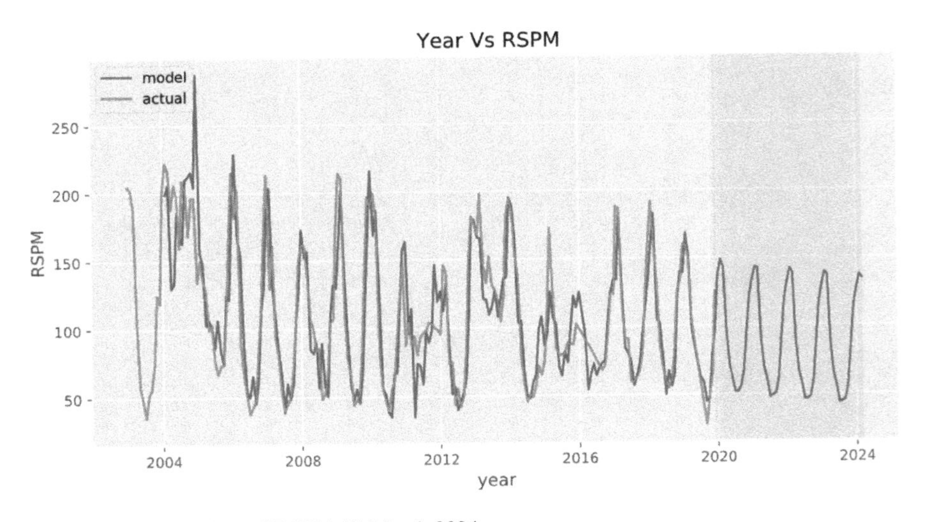

Fig. 4 Predicted values of RSPM till March 2024

Table 3 Predicted RSPM value of Guwahati for 2020

Date	RSPM	Date	RSPM
2020-01-01	136.915	2020-07-01	52.042
2020-02-01	148.377	2020-08-01	51.731
2020-03-01	142.774	2020-09-01	55.872
2020-04-01	95.053	2020-10-01	67.435
2020-05-01	71.173	2020-11-01	96.170
2020-06-01	61.464	2020-12-01	122.208

Table 4 Comparison of results from machine learning algorithms

	Score	RMSE
Linear regression	0.6982	34.35
Neural network regression	0.7052	33.95
Lasso regression	0.6983	34.35
ElasticNet regression	0.6983	34.35
Decision forest	0.7120	33.56
Extra trees	0.7237	32.87
Boosted decision tree	0.7105	33.65
XGBoost	0.7133	33.48

4.4 Applying Machine Learning Algorithms

The air pollution dataset of 'Guwahati' is applied to the machine learning algorithms. The result is shown in Table 4.

We can see that extra trees outperform all the selected machine learning algorithms with a score of 0.7237 and root-mean-square error of 32.87.

5 Conclusion

We have compared the performance of ARIMA and machine learning algorithms: Linear regression, neural network regression, Lasso, ElasticNet, decision forest regression, extra trees regression, and decision tree using the 17 years dataset obtained from PCBA. The ARIMA model outperforms the machine learning algorithms with RMSE of 18.53 while extra trees, the best among the machine learning algorithms, predicts RMSE of 32.87. It is to be noted that all the methods tend to perform poorly in forecasting because there is randomness in the data. Further research will be required on some techniques such as long short-term memory (LSTM).

Acknowledgements Our sincere and heartfelt gratitude to Pollution Control Board Assam (PCBA) for providing us with the required dataset, for letting us know the actual problems and how to handle it. We also would like to thank Assam Science and Technology University (ASTU) for funding the research work under the Collaborative Research Scheme through the TEQIP III program of MHRD funded by the World Bank.

References

1. Rao, C.S.: Environmental Pollution Control Engineering, 2nd edn. New Age International Publishers, India (2006)
2. Sarasamma, J.D., Narayanan, B.K.: Air quality assessment in the surroundings of KMML industrial area, Chavara. Aerosol Air Qual. Res. **14**, 1769–1778 (2014)
3. Mitra, A.P., Sharma, C.S.: Indian aerosols: present status. Chemosphere **49**, 1175–1190 (2002)
4. Lee, J.Y., Lee, S.B., Bae, G.N.: A review of the association between air pollutant exposure and allergic diseases in children. Atmos. Pollution Res. **5**, 616–629 (2014)
5. Daniel, V.: Fundamentals of Air Pollution, 5th edn. Academic Press, San Francisco (2014)
6. Kumar, A., Goyal, P.: Forecasting of daily air quality index in Delhi. Sci. Total Environ. **409**, 5517–5523 (2011)
7. Nur, H.A.R., Muhammad, H.L., Suhartono., Mohd, T.L.: Evaluation performance of time series approach for forecasting air pollution index in Johor, Malaysia. Sains Malaysiana **45**(11), 1625–1633 (2016)
8. Environmental Protection Agency: Air quality index: a guide to air quality and your health. In: Environmental Protection Agency Office of Air Quality Planning and Standards, Research Triangle Park (2003)
9. Argiriou, A.A.: Use of neural networks for tropospheric ozone time series approximation and forecasting—a review. Atmos. Chem. Phys. **7**, 5739–5767 (2007)
10. Zhang, J., Ding, W.: Prediction of air pollutants concentration based on an extreme learning machine: the case of Hong Kong. Int. J. Environ. Res. Public Health **14**, 114.
11. Kumar, A., Goyal, P.: Forecasting of air quality in Delhi using principal component regression technique. Atmos. Pollut. Res. **2**(4), 436–444 (2011)
12. Khashei, M., Bijari, M.: A novel hybridization of artificial neural networks and ARIMA models for time series forecasting. Appl. Soft Comput. **11**(2), 2664–2675 (2011)
13. Kurt, B., Gulbagci, F., Karaca, O.A.: An online air pollution forecasting system using neural networks. Environ. Int. **34**, 592–598 (2008)
14. Masih, A.: Machine learning algorithms in air quality modeling. Glob. J. Environ. Sci. Manag. (GJESM) **5**(4), 515–534 (2019)
15. Zhu, D., Cai, C., Yang, T., Zhou, X.: A machine learning approach for air quality prediction: model regularization and optimization. Big Data Cognit. Comput. **2**(1) (2018)
16. Kang, G.K., Gao, J.Z., Chiao, S., Lu, S., Xie, G.: Air quality prediction: big data and machine learning approaches. Int. J. Environ. Sci. Dev. **9**(1) (2018)
17. Nigam, S., Rao, B.P.S., Kumar, N., Mhaisalkar, V.A.: Air quality index—a comparative study for assessing the status of air quality. Res. J. Eng. Tech. **6**(2) (2015)
18. Peng, H.: Air quality prediction by machine learning methods. A thesis submitted in partial fulfillment of the requirements for the degree of Master of Science in The Faculty of Graduate and Postdoctoral Studies(Atmospheric Science), The University Of British Columbia (Vancouver) (2015)

Assessment of the Electric Rickshaws' Operation for a Sustainable Ecosystem

Sandeep Singh, B. Priyadharshni, Challa Prathyusha, and S. Moses Santhakumar

Abstract With changing institutional environments and higher mobility needs, electric vehicles have emerged as an alternative for commuting within an educational campus. Notably, electric rickshaws' operation inside an educational institution is considered an energy-efficient and eco-friendly mode of the para-transit system. This research paper presents a preliminary study conducted inside an educational institution to understand passengers' preference for electric rickshaw usage and examines commuters' travel patterns. An online-based questionnaire survey was conducted to collect the data concerning the dynamic aspects of usage and demand of the electric rickshaws. The study used descriptive analyses to assess the users' affinity toward green policy adoption. The preference for using of electric rickshaws for regular commuting was ranked the highest, thereby moving the campus to a sustainable transportation ecosystem. The user perception survey results reveal that there is a positive response to preference for implementation of the electric rickshaw services, which would have an impact on their daily usage of transport mode. The approach can be used to estimate trip makers demand in similar campuses of other universities and institutions towards adopting sustainable travel options like an electric rickshaw. The study also confirms that the electric rickshaws type of para-transit system unfolds unprecedented opportunities to address environmental sustainability issues.

Keywords Para-transit Electric vehicle (EV) · Electric rickshaw (E-rickshaw) · Passengers' preference · Sustainable ecosystem

---

S. Singh · B. Priyadharshni (✉) · S. Moses Santhakumar
National Institute of Technology Tiruchirappalli, Tiruchirappalli, India
e-mail: priyadharshni28195@gmail.com

C. Prathyusha
REVA University, Bengaluru, India

P. Muthukumar et al. (eds.), *Innovations in Sustainable Energy and Technology*, Advances in Sustainability Science and Technology,
https://doi.org/10.1007/978-981-16-1119-3_29

1 Introduction

The mobility patterns and travel behavior of the in-campus commuters inside an educational institution are distinct and complex compared to other commuters. University students form a social group, virtually autonomous in terms of mode choice decision making, and are gaining attention, as the completion of such everyday trips is rather complicated [1]. Hence, provision of a flexible mode of transport such as a para-transit system indicates the role in ensuring quick mobility, reliable accessibility, and excellent connectivity. The interest in renewable energy systems proliferates as they could replace polluting internal combustion engines [2]. Also, the feasibility of public transport in terms of connectivity, accessibility, and mobility has to be studied and analyzed in detail [3]. Electric vehicles are a crucial key element in the sustainable development of the transportation system as they could reduce energy consumption and the pollutants emitted from the vehicles.

Electric vehicles (EVs) have seen overgrowing growth rates over the last few decades. The EVs have even taken a substantial leap forward from early adopter technology toward becoming a mainstream consumer product. The transition of fossil fuel vehicles to EVs is unlikely to be achieved unless EV adoption barriers specific to electric rickshaws (e-rickshaws) are identified and overcome with immediate effect. The EVs are the sustainable, efficient, and economical public transportation systems, which reduces the depletion of resources and provides an environment-friendly transit system [4].

Nevertheless, research to date on the extent of usage of e-rickshaw and causes of the e-rickshaw adoption barriers has been limited and mostly confined to anecdotal shreds of evidence. In light of the above context, it makes a compelling objective to investigate and perform a detailed study on the first of its kind, implementing e-rickshaws inside the educational institution in the National Institute of Technology Tiruchirappalli (NITT) campus, India.

2 Literature Review and Background

Many studies have been conducted to analyze the acceptance of EVs and the influential antecedents on EV adoption.

Davison et al. [5], through their research, highlighted significant cultural differences and complexities in the travel behavior of students in the United Kingdom (UK) and Ireland. Das et al. [6] analyzed the travel behavior of university students resided on-campus and off-campus by studying the travel pattern and identifying transportation needs. Kotoulal et al. [1] examined various aspects of university students' travel behavior such as travel time, travel distance, safety, and comfort in the Xanthi, Greece. The authors' findings verify that distance and time are the most critical factors for both cases, while the use of public transport instead of walking increases the importance of economy and safety. In the latest study, Priye and Manoj [7] analyzed to understand passengers' perceptions of electric rickshaw safety in Patna, India, and developed an empirical model to explore the overall safety perceptions of electric rickshaw riders.

McLoughlin et al. [8] studied the effectiveness of electric bicycles and identified the barriers to bicycle use that can be overcome through public use of electric bicycles. One such barrier is the choice of an alternate mode of transport line EV and the public acceptance of its use, in terms of adoption by potential users.

Vazifeh et al. [9] proposed a novel methodology to perform data-driven optimization of EV charging station locations in the city of Boston, which provides computationally efficient solutions based on the genetic algorithm. The study results show that the genetic algorithm provides solutions that significantly reduce drivers' excess driving distance to charging stations, minimize overall energy overhead, and promote data-driven planning as a viable solution for optimal EV charging station location. Singh et al. [10, 11] used system dynamics simulation approach to model and predict the vehicular fuel consumptions and fuel emissions and fuel costs. They proposed sustainability based recommendations to reduce the fuel consumptions and fuel emissions and fuel costs for an urban city in India.

In Indian Institue of Technology, Roorkee, Rastogi and Doley [12] conducted a study to suggest a sustainable travel alternative and shift to e-rickshaws using logit analysis. This analysis used the random utility theory for the modeling of discrete data. These utilities were modeled based on multinomial logit (MNL) and nested logit (NL) specifications.

The literature reviewed in this study summarizes that the significant barrier to EVs is that consumers tend to resist new technologies that are considered new or unproven, availability of charging infrastructure at the shorter distance, the initial capital cost of the EV, and maintenance cost of the EV after purchase.

3 Study Methodology

In the present research work, an online-based user perception questionnaire survey using Google Forms was carried out among the daily commuters inside the institute campus for assessing the potential viability and the preference of the e-rickshaws. It was circulated online among the campus people to capture information on their travel patterns and preferences toward using the e-rickshaws. Two essential considerations were involved in the user survey design, such as socio-economic and perception-related information. In the year of study (2019), the students' strength was 6000 (including undergraduates, postgraduates, and PhDs), and faculties and employees' strength was 1000. Male students resided in 20 hostels, female students in 6 hostels, and faculties in 30 residential locations. All these residential locations were dispersed throughout the campus. The travel was initiated from these locations and directed to different destination locations (i.e., academic departments) in the daytime and other locations during day and evening times. Sample data was collected inside the campus through online questionnaire. About 10% out of the total population, which is around

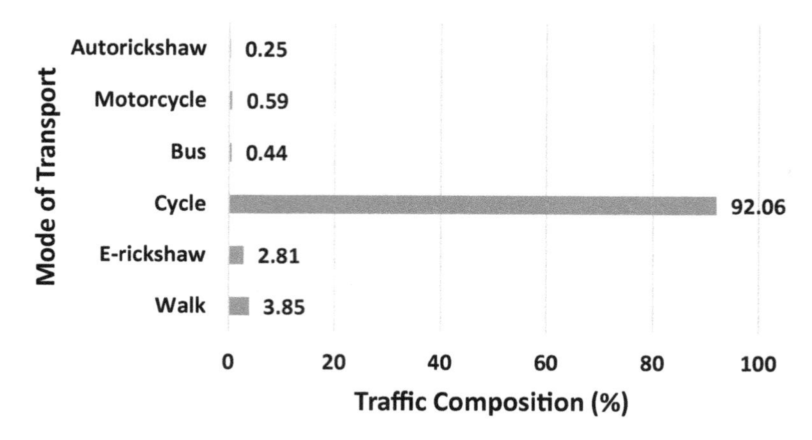

Fig. 1 Present choice of mode of transport inside the NITT campus

800, is taken for this study. The responses obtained from a Google Forms survey reveal users' willingness and perception inside the National Institute of Technology Tiruchirappalli (NITT) campus regarding the implementation of e-rickshaws.

4 Results and Discussions

4.1 Choice of the Mode of Transport

The choice of transport mode inside an educational institute is unique due to the availability of various modes of transport. Figure 1 depicts the choice of mode of transport by the commuters inside the campus. A look at the present mode of transport inside the NITT campus shows that most commuters choose to use the non-motorized mode of transport, such as the cycle. The higher percentage of choice of the cycle mode of transport is 92.06%. This is reasoned to be due to the presence of more number of students resided inside the NITT campus and also due to certain restrictions of the usage of the motorized mode of transport, such as a powered motorcycle. Though the share of e-rickshaws is significantly less compared to cycle and walk, it seems to be quite interesting to be in the third top mode of transport, which comes around 2.81%. This indicates that the commuters (walking) are more likely to shift to e-rickshaws and tend to shift more once and when implemented.

4.2 Trip Length-Frequency Distribution

Figure 2 depicts the trip length-frequency distribution of the commuters inside the

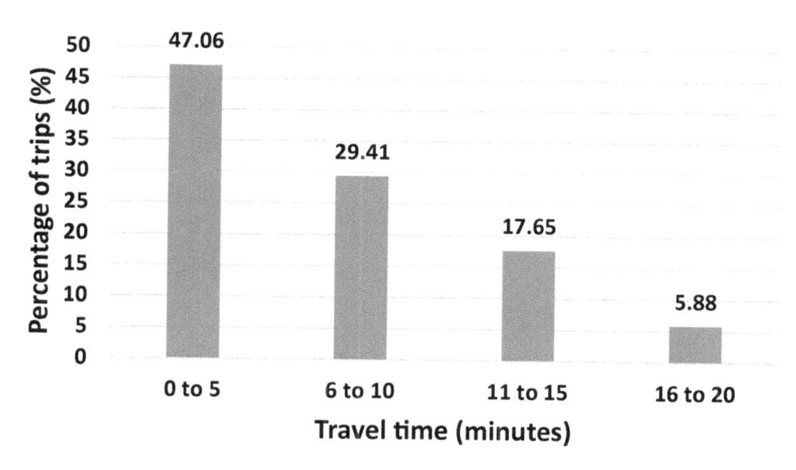

Fig. 2 Present frequency of trip length inside the NITT campus

campus. The commuters trip length-frequency survey reveals that up to 47.06% of the people prefer to keep the travel time of the trip shorter between 0 and 5 min. This is reasonably due to the low distance between the origin and destination of the trips made inside the NITT campus.

4.3 *Living and Social Status of the Respondents*

Figure 3 depicts the living status of the respondents. The preliminary surveyed data reveals that 97% of the respondents were residing inside the campus, clearly indicating the students of NITT.

Figure 4 depicts the gender distribution of the respondents. The questionnaire data reveals that the surveyed respondents were mostly male students with 78%, and

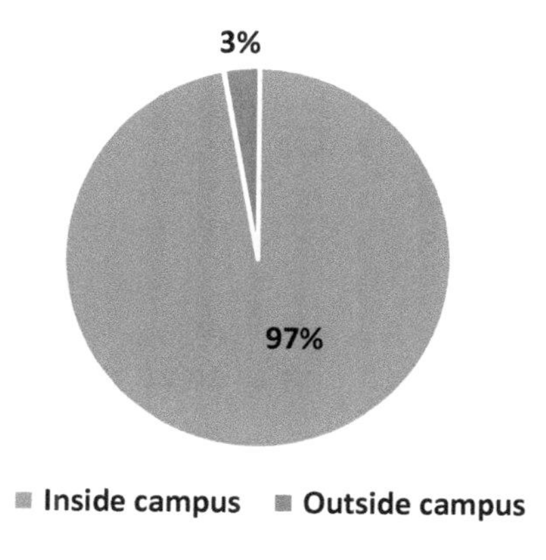

Fig. 3 Living status of the respondents

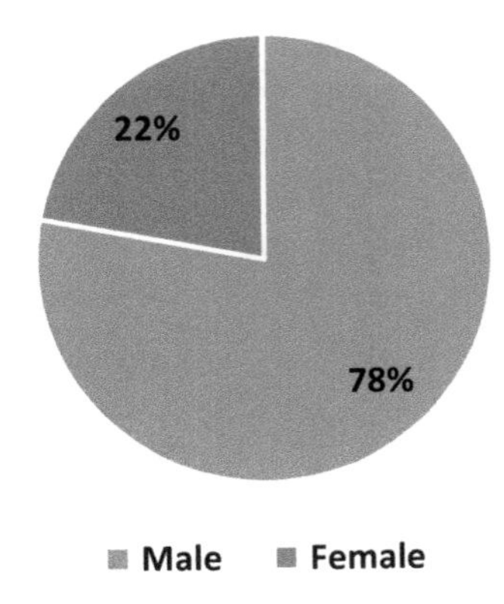

Fig. 4 Gender distribution of the respondents

the rest, females with 22%. The high male ratio confirms the more number of male students inside the campus.

4.4 Age Distribution of Respondents

Figure 5 depicts the age distribution of e-rickshaw users based on the collected data. The age distribution reveals that most of the users were in the age group of 18–22 years, belonging to undergraduates who are comparatively more in number than the other degree graduates.

It is evident from Fig. 5 that the proportion of male students is more compared to the female students in the age group of 18–22 years inside the institution. Also,

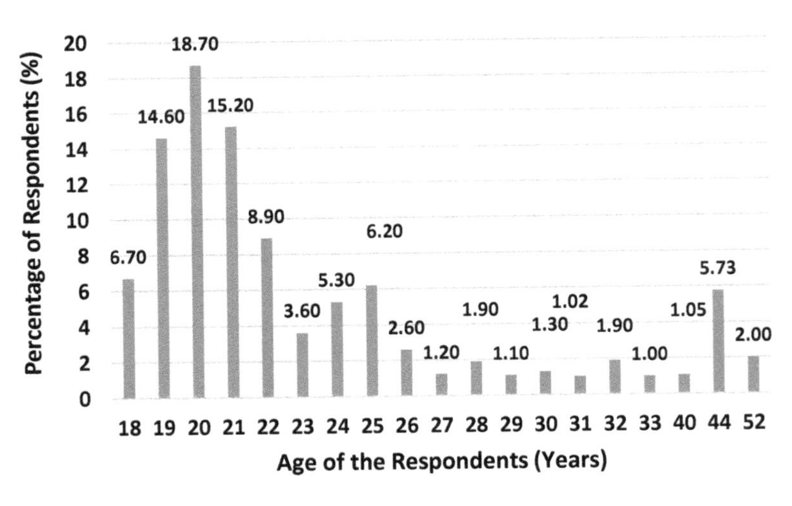

Fig. 5 Age distribution of the respondents

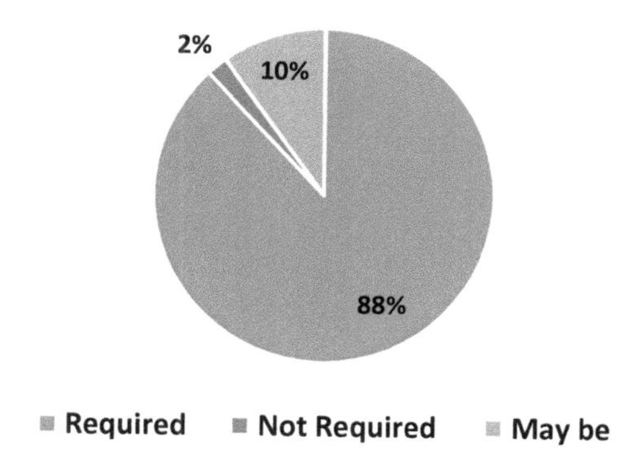

Fig. 6 Response to the importance of e-rickshaw

most of the respondents were found to be in the age group of 18–22 years, which cumulates to 57.4% out of the 100% respondents. People above the age of 30 years were also interviewed and were found to prefer using the e-rickshaws (contributing to 12.6%) considering their comfort and convenience compared to walking.

4.5 *Preference for E-rickshaw by the Respondents*

Figure 6 depicts the response to the importance of e-rickshaw by the respondents. The respondents were further queried regarding the importance of e-rickshaws and their cost to pay for a single trip. Almost 88% of the respondents indicated that they feel e-rickshaw is essential and immediately required for implementing inside the campus.

4.6 *Preference for the Willingness to Pay for E-rickshaw*

Figure 7 depicts the preference for the willingness to pay for e-rickshaw. One way of implementing non-proportional assignment is to incorporate equilibrium to the traffic flows that assign link cost functions, link travel cost, and path travel cost to the network a way to minimize travel costs using a target trip matrix. A look at the respondents' willingness to pay for e-rickshaws service showed that 42% of the people prefer to pay up to Rs. 5 per trip. However, it is surprising to note that nearly 25% of the respondents have agreed to pay double the latter, i.e., up to Rs. 10 per trip.

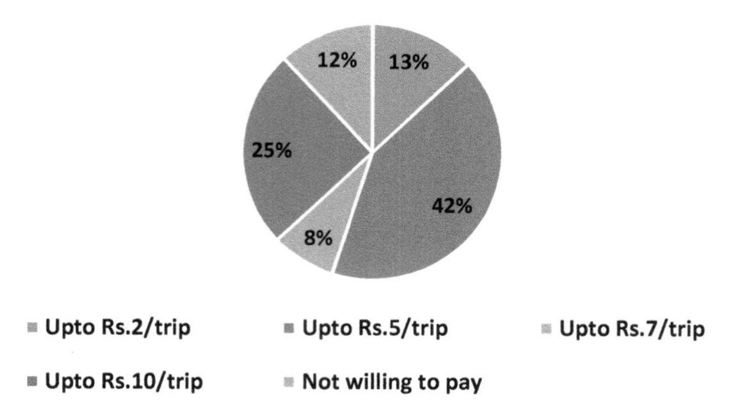

Fig. 7 Response to the willingness to pay for e-rickshaw

4.7 Preference for E-rickshaw Timing and Location

Figure 8 depicts the response to preferred timings for e-rickshaw service inside the campus. The respondents were requested to prefer their timings for the e-rickshaw service. The morning timing between 08:00 AM and 09:00 AM was the highest preferred time. This is because of the class hours of the students, which is around 8.30 AM, followed by the noon timing between 12:00 PM and 02:00 PM where they leave for lunch hours happened to be the highest, followed by 04:00 PM and 5.30 PM indicating students leaving back to their hostels.

Figure 9 depicts the response to the preferred location for transfer stations for e-rickshaw service inside the campus. Later, the respondents were asked to prefer the suitable location of the transfer stations for the e-rickshaw service according to

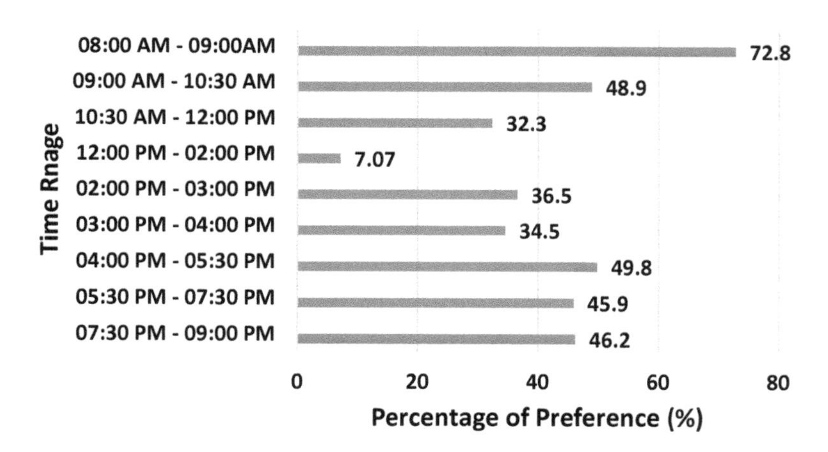

Fig. 8 Response to preferred timings for e-rickshaw service inside the campus

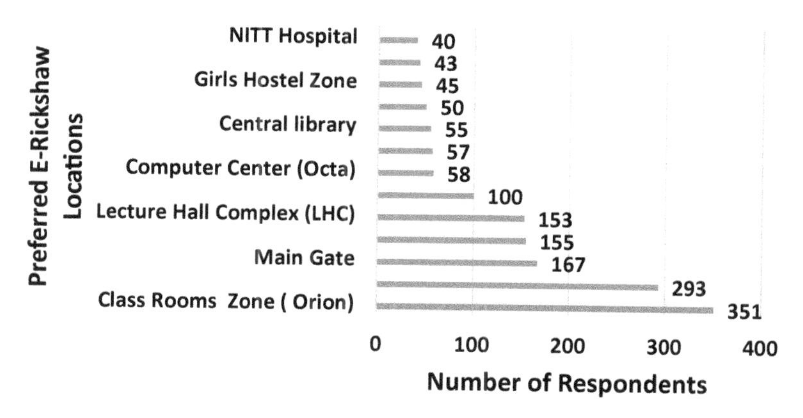

Fig. 9 Response to the preferred location for transfer stations for e-rickshaw service inside the campus

their travel pattern choice. Nearly, 351 respondents preferred to locate the transfer station at the Class Rooms Zone (Orion), ranking it as the most preferred location, followed by the transfer station at shopping complex and near main gate.

The efficient deployment of the network of public charging stations is an important matter, which plays a major role in further increasing EVs' market share in the near future [9].

4.8 Summary

The structured online-based questionnaire–interview survey conducted within the educational institute reveals that about 88% of daily commuters within the campus prefer to use the e-rickshaws, which would have a significant impact on the mode-shift. The service operation of e-rickshaws was found to be sustainable at Rs. 10 per trip; however, 42% of trip makers preferred Rs. 5. The study also further identified the best operational timing of the e-rickshaw inside the campus to be during the morning between 08:00 AM and 09:00 AM. The most preferred ideal locations for the transfer station for the efficient operation of the e-rickshaw were at the Class Rooms Zone (Orion).

Though the other vehicle ownership mode (motorcycle) strongly influences students' mode choice yet considering the weather and comfort conditions during summer and rainy seasons, students prefer to use the e-rickshaws. It can be noticed that the institute commuters, especially the students, tend to have unique and complex travel behavior since they move like a social group. The introduction of e-rickshaws provokes the commuters to prefer and shift toward a more sustainable transportation mode. Finally, this study interprets that the e-rickshaws can be a possible sustainable travel alternative that would majorly improve the mobility of the students at a low cost as well as make the NITT campus a sustainable place to live.

5 Conclusions

Students' mobility has a potentially significant impact on transport choices and travel behavior. The study is focused on assessing the operation of the e-rickshaws inside the NITT campus. The evaluation of the use of e-rickshaws through the online questionnaire survey provided many more insights that are relevant and useful for the assessment of the e-rickshaw service. The study found that the tendency of preference for the e-rickshaws was high among the users (students) who are residing inside the campus. This preference made a good case for implementing e-rickshaw operations within the campus. Also, the analyses from the study suggest that the introduction of e-rickshaws would lead to an increase in energy savings, a decrease in fossil fuel energy consumption, carbon emissions, and other cost externalities.

It positively influences university-related travel regarding distances, frequency, and mode: all of which have significant consequences for student travel sustainability and the environmental impacts of that travel. On this basis, it can be concluded that the e-rickshaws penetration could be advantageous for the energy-saving system since the carbon intensity would reduce to a considerable level, and the energy would be saved. Finally, it can be said that e-rickshaws can be considered to be a superior para-transit system than an internal combustion engine vehicle like a motorcycle or auto-rickshaw from an efficiency and environmental perspective.

Acknowledgements The authors would like to acknowledge and thank the Transport Section, National Institute of Technology Tiruchirappalli (NITT), India, for providing permission and support for conducting the data collection work within the institute campus.

References

1. Kotoula, K.M., Sialdas, A., Botzoris, G., Chaniotakis, E., Grau, J.M.S.: Exploring the effects of university campus decentralization to students' mode choice. Period. Polytechnica Transp. Eng. **46**(4), 207–214 (2018). https://doi.org/10.3311/PPtr.11641
2. Donateo, T., Licci, F., D'Elia, A., Colangelo, G., Laforgia, D., Ciancarelli, F.: Evaluation of emissions of CO2 and air pollutants from electric vehicles in Italian cities. Appl. Energy. **157**, 675–687 (2015)
3. Singh, S., Shukla, B.K.: A comparative study on the sustainability of public and private road transportation systems in an urban area: current and future scenarios. In: Ashish, D.K., de Brito, J., Sharma, S.K. (eds.) 3rd International Conference on Innovative Technologies for Clean and Sustainable Development. ITCSD 2020. RILEM Bookseries, vol. 29, pp. 173–189. Springer, Cham (2021). https://doi.org/10.1007/978-3-030-51485-3_12
4. Prathyusha, C., Singh, S., Shivananda, P.: Strategies for sustainable, efficient, and economic integration of public transportation systems. In: Jana, A., Banerji, P. (eds.) Urban Science and

Engineering. Lecture Notes in Civil Engineering. vol. 121, pp. 157–169. Springer, Singapore (2021). https://doi.org/10.1007/978-981-33-4114-2_13
5. Davison, L., Ahern, A., Hine, J.: Travel, transport, and energy implications of university-related student travel: a case study approach. Transp. Res. Part D Transp. Environ. **38**, 27–40 (2015). https://doi.org/10.1016/j.trd.2015.04.028
6. Das, R., Kumar, S.V., Prakash, B., Dharmik., Subbarao, S.S.V.: Analysis of University students travel behaviour: enroute to sustainable campus. Indian J. Sci. Technol. **9**(30), 1–6 (2016). https://doi.org/10.17485/ijst/2016/v9i30/99246
7. Priye, S., Manoj, M.: Passengers' perceptions of safety in para-transit in the context of three-wheeled electric rickshaws in urban India. Saf. Sci. **124**, 1–11 (2020). https://doi.org/10.1016/j.ssci.2019.104591
8. McLoughlin, I.V., Narendra, I.K., Koh, L.H., Nguyen, Q.H., Seshadri, B., Zeng, W., Yao, C.: Campus mobility for the future: the electric bicycle. J. Transp. Technol. **2**, 1–12 (2012). https://doi.org/10.4236/jtts.2012.21001
9. Vazifeh, M.M., Zhang, H., Santi, P., Ratti, C.: Optimizing the deployment of electric vehicle charging stations using pervasive mobility data. Transp. Res. Part A **121**, 75–91 (2019). https://doi.org/10.1016/j.tra.2019.01.002
10. Singh, S., Uma Devi G.: System dynamics simulation modeling of transportation engineering, energy, and economy interaction for sustainability. In: Arkatkar, S., Velmurugan, S., Verma, A. (eds.) Recent Advances in Traffic Engineering. Lecture Notes in Civil Engineering, vol. 69, pp. 379–401. Springer, Singapore (2020). https://doi.org/10.1007/978-981-15-3742-4_24
11. Singh, S., Prathyusha, C.: System dynamics approach for urban transportation system to reduce fuel consumption and fuel emissions. In: Jana, A., Banerji, P. (eds.) Urban Science and Engineering. Lecture Notes in Civil Engineering, vol. 121, pp. 385–399. Springer, Singapore (2021). https://doi.org/10.1007/978-981-33-4114-2_31
12. Rastogi, R., Doley, G.: Sustainable Transport option for University Campus. Research Gate (2015)

Location Potential Assessment and Feasibility Marking for Solar Power Generation on Basis of Source–Load Matching Strategies

Veeresh G. Balikai, Amit Shet, M. B. Gorawar, Rakesh Tapaskar, P. P. Revankar, and Vinayak H. Khatawate

Abstract The changes in living standards globally and rise in population have enhanced global energy consumption. The energy demand predictions for year 2050 present 'energy security' as a crucial parameter to development in majority of nations worldwide. The major policy matters linked to socio-economic progress now have their focal point on this domain. The growing energy demand has translated into irreparable ecology and environment damage. The global energy policy-makers strongly suggest a gradual, but confirmed switchover to renewable energy forms. The nature-based energy source being benevolent to environment offers technical challenges to transform as viable sources. The conversion devices to tap solar energy are more bulky than their coal or diesel fuelled counterparts. Thus, solutions to make renewable attractive need a two-point strategy that includes thorough resource assessment and effective equipment design. This article addresses these issues in relation to use of solar energy for electric power generation. The solar energy assessment at a location is based on use of established relations of 'solar radiation geometry,' to obtain fairly good estimates. The load prediction for an application can be obtained with a good accuracy for its use resource matching. There are also studies reported toward improving reliability of renewable energy extraction adopting concepts of maximum power point tracking. Thus, the second strategy of performance enhancement explores domains of better cooling of devices and use of novel energy storage materials. The SPV system to cater domestic load observed for typical urban household situated in three geographically distinct areas (A, B and C). The approach cited solar potential at these locations against load to formulate universal applicability of solar energy. The strategies of equipment selection and design to suit each case were unique. Site potential assessment formed an essential component of overall design viable function to deliver at least break-even energy output. The complete assessment included economic feasibility along with environmental benefit delivered by solar power installation. The location 'A' had fairly good solar potential, and hence, fixed

V. G. Balikai · A. Shet · M. B. Gorawar · R. Tapaskar · P. P. Revankar (✉)
Faculty, SME, KLE Technological University, Hubballi, India
e-mail: pp_revankar@kletech.ac.in

V. H. Khatawate
Faculty, Department of Mechanical Engineering, DJS College of Engineering, Mumbai, India

P. Muthukumar et al. (eds.), *Innovations in Sustainable Energy and Technology*,
Advances in Sustainability Science and Technology,
https://doi.org/10.1007/978-981-16-1119-3_30

panel was adequate to meet anticipated load. On the contrary, locations 'B' and 'C' situated in a different geographical location required tracked mode of installation due to paucity in solar potential. The study also compared various modes of tracking and inclination angle for solar collector.

Keywords Solar potential assessment · Load marking · Break-even generation · Green power

1 Introduction

Solar energy forms primary input to all ecosystems existent on the only living planet in the Universe. The photosynthesis driven by plants helps in fixation of carbon utilizing electromagnetic radiations. The plant matter forms as the perennial support to life on earth as food to all living organisms. Thus, solar energy is immutable support to all processes both natural and man-made. The developmental activities are capable of being supported by this abundant and clean energy. The energy liberated by nuclear fission reaction in the sun amounts to several trillion times magnitude than total energy consumed by entire human race. The huge magnitude of solar energy reaches earth in a time span of 8.5 min on a continuous basis, for several million years. Therefore, solar energy qualifies to be a reliable and eco-friendly means to meet all our energy needs. The technological today has made it possible to harness solar energy for both process heat and electric power generation. The quest lies in transformation to build more reliable and intelligent solar devices to match current energy scenario that emphasizes sustainability as bedrock of development.

The major hurdle associated with solar energy utilization roots form the variable and dilute nature of source that imposes challenges in selection of materials and conversion mechanisms. The SPV technology has made rapid progress to operate close to 20% efficiency and emerge as the largest renewable energy installations. The popularity of solar energy is also attributed to its modular nature, competitive pricing, minimum operating cost and ease in installation. The SPV system design and installation follows standardized procedure as reported by several researchers. The primary stage involves at least a year-long resource assessment at the site to capture diurnal and seasonal variations. The thorough assessment of barriers that lead to formation of shade on SPV modules helps in design of suitable orientation to maximize the annual input solar energy. The tracking of SPV modules can be proposed in case of locations with relatively poor solar insolation. The overarching factor that decides overall system design will be the economics of the system against existing power alternatives.

2 Overview of Solar Radiation Modeling and Site Implementation

The research findings revealed in this section can be broadly categorized into domains of site potential assessment and feasibility marking. The former group summarizes research findings that give insight into computational and experimental approaches to predict solar energy at a defined geographic location. The latter compilation explores adoption of site and situation specific strategies to make solar energy installations viable in techno-economic context.

2.1 Site Potential Assessment Studies

Ulgen and Hepbasli [1] investigated solar potential in city of Izmir located in western Turkey. The correlation to connect hourly diffuse radiation and monthly average global radiation with hourly clearness index was developed. The study gains prominence in the context of investigating atmospheric characteristics to assess solar potential on basis of the correlation.

Ricci et al. [2] investigated 'wind-solar' street light system to facilitate higher energy yield along with operational flexibility. The study explored various wind turbines to obtain high power while in operation in hybrid mode with PV controlled through maximum power point tracking.

Mousavi and Hizam [3] investigated solar energy assessment through sun-earth calculations and isotropic models (Liu-Jordan and Koronakis) for non-beam part of solar insolation. The Perez, Temps–Coulson, Klucher and Bugler anisotropic models provided good solar estimates.

Zhang et al. [4] reported on solar water heater and SPV/T system at Shanghai city for entire year of operation. The comparative measure revealed that conventional solar heater fared less attractive against SPV/T version in terms of overall efficiency estimates for the entire duration of year. The overall assessment indicated that PV/T system to competitive in auxiliary heating operation mode.

Harrison and Jiang [5] assessed prediction accuracy of simulation modeling tool against field tested date of energy output. The regulatory measures under Climate Change Act dictate mandatory use of renewable energy in new residential or commercial constructions. The solar potential assessment evaluated through simulation tools revealed a energy assessment gap of 8.6%.

Mills [6] explored 'Hybrid' tool for efficiency assessment of coal-based plant. The efficiency improvement could be realized by adopting hybrid mode operation with either solar or natural gas. The strategy leads to flexibility in operation and reduced emissions in generation.

Laseinde and Ramere [7] claimed solar power generation efficiency enhanced through low-cost controlled tracking systems. The percent rise of 23.95 in efficiency was reported by servo-based multiple-axis tracking.

2.2 Feasibility Marking Studies

Yibing [8] investigated concentrated PV modules at various concentration ratios and operating temperatures. The direct relationship of efficiency with concentration ratio was indicated at same operating temperature. On contrary, temperature had inverse relationship with efficiency for fixed concentration ratio that necessitated system cooling either using air or water.

Cuce et al. [9] reported on passive cooling for Si-cell efficiency improvement. The fins in PV cell cooling at different illumination and air temperature were studied. The cooling at 600 W/m^2 insolation was highest, and efficiency varied inversely with ambient temperature.

Rahman [10] has characterized SPV systems in MATLAB/Simulink. The charge controller for battery was designed to reduce duty cycle to prevent battery over-charge. The study aimed to regulate Inverter sine wave output for 220 V_{ac} load at low harmonic levels.

Mukherjee [11] studied embodied carbon footprint in silicon wafers manufacturing. The assessment of plant energy output and associated carbon credits earning was reported. The overall environmental impact study of solar power forms pivotal segment in the claim as green energy.

Ong [12] studied combined operation of heat pipe and thermo-electric module in solar generation of process heat and electricity. The unique feature to effectively transfer heat over large distance while operating with small temperature gradient made the passive heat pipes a suitable choice. The commercial availability of heat pipes and thermo-electric modules has made them viable in to low capacity, portable and off-grid process heat and electric power utilities.

Gao et al. [13] reported on economics of solar energy-based cooling through SPV, vapor absorption/compression on basis of economic indicators of annual cost. The study finds relevance in widening scope of solar energy into wider range of domestic applications.

Chen and Riffat [14] investigated combined heat and electricity through PV/T collector. The heat accumulated by cooling PV module was used for process utility, additionally leading to efficiency improvement of PV module. The PV/T concept can be an effective structure for shading on adoption of suitable material to meet the dual functional needs.

Maleki [15] investigated diesel generator to group with renewable energy sources for off-grid power through optimization of system components and overall cost of energy. The studies indicated that DG set operating in single-power source mode functioned better compared to hybrid mode. However from environmental degradation perspective, hybrid renewable power was preferred.

Baljit et al. [16] investigated multi-fluid characteristics for PV-T solar collector using Fresnel lenses for four operating modes. The air and water media were analyzed in terms of flow and incident insolation to quantify absorber plate temperature and operating efficiency.

3 Computational Analysis in System Design

Simulation is the art and science of predicting system behavior through a conglomeration of governing physical laws and experimental data sets. Physical device has operating characteristics that can be transformed into mathematical expressions to connect input and output variables. The common tools used for mathematical analysis include MATLAB and spread sheets. These tools have strong math-based libraries that can transform the task of decision making much easier and thereby help in building effective systems. This section highlights simulation studies on load and resource assessment coupled to characterizing various strategies and overall economics.

3.1 Load and Resource Assessment Studies

The load constitutes important factor in SPV system sizing to optimize performance and cost of power generation through prevention of oversize or undersize system components. The load curves help match resource with demand to evolve strategies for reliable and cost-effective power generation. The site load presented in Fig. 1 indicates power rating and net energy consumed during operation.

The typical household application has been cited for feasibility of adoption of SPV system [17].

The other loads include may include water lifting requiring high-capacity SPV installations to cater to water need of 50 L/day/person drawn through an average head of 8 m. The accounted load for SPV system takes up worst-case scenario to ensure adequate system reliability with desired autonomy. The higher duration of autonomy dictates larger storage capacity and escalated project cost. The simulation requires input data pertaining to specifications of devices that include inverter and

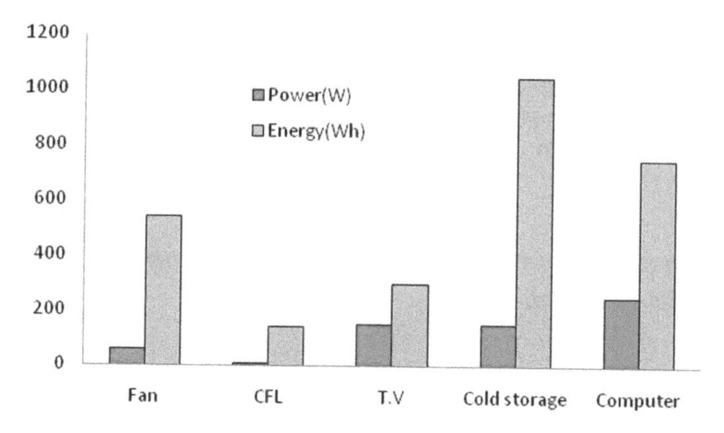

Fig. 1 Load used for the study

storage batteries to assess component sizing.

$$I_{\text{ext}} = I_{\text{SC}}\left[1 + 0.033\cos\left(\frac{360n}{365}\right)\right] \tag{1}$$

$$\text{LST} = \text{Local time} \pm 4\,\text{min}(L_{\text{std}} \pm L_{\text{Location}}) + \text{Eq. of time correction.} \tag{2}$$

$$\text{Eq. of time correction} = 9.87\sin(2B) - 7.53\cos(B) - 1.5\sin(B) \tag{2A}$$

$$B = \left(\frac{360}{365}(n - 81)\right) \tag{2B}$$

$$\text{Declination}(\delta) = 23.45 \times \sin\left(\frac{360}{365}(284 + n)\right) \tag{3}$$

$$\text{Hour angle } (\omega) = 15^{\circ}(\text{LST} - 12) \tag{4}$$

$$\begin{aligned}\text{Incidence angle } C\theta = \big[&(C\varphi C\beta + S\varphi S\beta C\gamma)C\delta C\omega + C\delta S\beta S\omega S\gamma \\ &+ S\delta(S\varphi C\beta - C\varphi S\beta C\gamma)\big]\end{aligned} \tag{5}$$

$$\text{Zenith angle } (\theta_Z) = C^{-1}[(C\varphi C\delta C\omega + S\delta S\varphi)] \tag{6}$$

$$\text{Sunset hour angle } (\omega_s) = \text{Cos}^{-1}[-(\tan\varphi\tan\delta)] \tag{7}$$

$$\text{No of sunshine hours} = \frac{2}{15} \times \text{Cos}^{-1}[-(\tan\varphi\tan\delta)] \tag{8}$$

The solar resource availability for power generation depends on geographic latitude of location, day of the year (n) other atmospheric conditions. The Eqs. 1–8 present solar radiation estimation. The relative position of sun and earth determines the magnitude of solar insolation that undergoes a periodic variation as depicted through sun-path diagram.

3.2 Characterization of Solar Energy Conversion Device and Its Economics

The SPV module works on the principle of converting electromagnetic radiation into dc power. The process is driven by magnitude of incident sunbeam as well as the operating temperature. The physics of this conversion process being established through I-V characteristics of a single solar cell provides a complete dynamics of electricity generation. The MATLAB software has been extensively used in studies to depict the behavior of PV module using the input cell characteristics: 15 W, V_{mp} = 18 V, I_{mp} = 0.83 A, Cell Area = 0.06 m^2 and operating temperature of 30 °C. The solar intensity on tilted module and its influence on efficiency at various levels of irradiation and temperature were investigated.

The economics of any energy conversion device sets a benchmark of its competitiveness with existing alternatives. The study was based on a thorough analysis for defined capital cost, operating cost and useful project life. The study based on Eqs. 9–12 evaluates parameters of life cycle cost and energy cost.

$$f = \frac{(1+i)^n - 1}{i(1+i)^n} \tag{9}$$

$$\text{LCC} = C_o + (C_m \times f) + C_r\left[\frac{1}{(1+i)^n}\right] - S\left[\frac{1}{(1+i)^n}\right] \tag{10}$$

$$\text{CR} = \text{LCC} \times \left(\frac{1}{f}\right) \tag{11}$$

$$\text{Cost per kWh} = \left(\frac{\text{CR}}{365 \times P}\right) \tag{12}$$

The economics of SPV power installation has been considered useful life of 20 year, initial cost of 12×10^3 Rs, annual maintenance of Rs. 200, one-time replacement cost (C_r) of Rs. 3×10^3 and annual rate of return on investment of 11% as economics data. The evaluated output parameters included life cycle cost of Rs. 14,326.2, capital recovery of Rs. 1799 and cost per kWh of Rs. 1.232 as per the estimates made for location A.

4 Results and Discussions

This section highlights the results obtained with respect to a SPV installation designed for a typical domestic load Fig. 1 that highlights commonly used electric devices.

4.1 *PV System Design for Estimated Domestic Load*

As studied a typical household application was considered for design and estimation of daily energy requirement; further calculations were done using this computational tool. The data presented in Table 1 gives specifications related to Inverter energy supply, total Ah-rating of battery for zero and defined power autonomy cases, number of batteries in series and parallel combinations, No. of SPV panels in series and

Table 1 Specifications of designed SPV power system

Energy supplied to inverter (Wh)	4224.194
Daily energy need (kWh)	5.5218
Total Ah of battery	176.0081
No.of batteries/series/parallel	5/2/2.5 ~ 3
Total Ah with Autonomy/No. of batteries	528.0242/ 15
PV energy output (Wh)	5521.822
Power rating/Voltage/ current W/V /A	75/15/5
Series/ parallel SPV panels	1.6 ~ 2/7.66 ~ 8

parallel configurations. The state of autonomy defines the number of days the system can support the load without any output delivered by SPV system. The higher range of autonomy defined for the power system demands a large Ah-rating of the support storage device. The quantity of autonomy for a installed SPV system depends upon the availability of solar energy at the site f installation. A very sunny location with adequate sunlight requires a fewer hours of autonomy as compared to the location with poor availability of sunlight. The study cites three different locations A (5.36° N, 75.12° E), B (28.61° N, 77.2° E) and C (53.55° N, 9.99° E) for investigations with respect to SPV system for identical load. The study revealed that location 'A' lying in the close vicinity of equatorial region had a fairly good magnitude of solar insolation as compared to locations B and C, situated beyond the tropic of cancer (23.5° N latitude).

Figure 2a, b represents defined load in congruence to available insolation (horizontal and inclined to latitude angle) with respect to locations A and B respectively. The Eqs. 13A, and 13B and 14A, 14B respectively show Insolation on horizontal and inclined surfaces for two locations. The common load shared by both locations A and B is indicated by Eq. 15 gave their comparison.

$$I(\text{hor})_{\text{A}} = -0.005x^6 + 0.3715x^5 - 10.789x^4 + 152.79x^3 - 1109.7x^2 + 4144.1x - 6659.7 \quad (13\text{A})$$

$$I(\text{hor})_{\text{B}} = 0.0019x^6 - 0.1442x^5 + 4.7196x^4 - 84.982x^3 + 843.17x^2 - 4001.1x + 6827.9 \quad (13\text{B})$$

$$I(\text{inc})_{\text{A}} = -0.0049x^6 + 0.3676x^5 - 10.68x^4 + 151.43x^3 - 1101.3x^2 + 4107.6x - 6608.4 \quad (14\text{A})$$

$$I(\text{hor})_{\text{A}} = 0.0036x^6 - 0.2718x^5 + 8.7021x^4 - 150.79x^3 + 1450.3x^2 - 6966.3x + 12766 \quad (14\text{B})$$

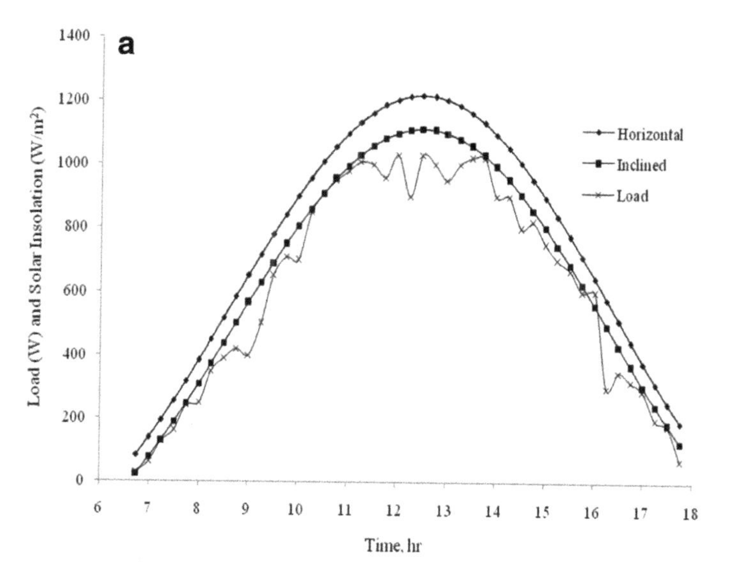

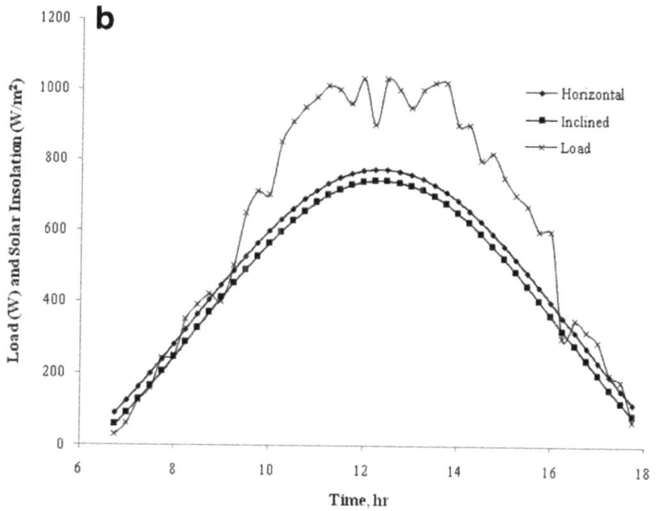

Fig. 2 **a** Variation of Load and Insolation for location A. **b** Variation of Load and Insolation for location B

$$\text{Load} = 0.0054x^6 - 0.4233x^5 + 14.09x^4 - 254.21x^3 + 2559.5x^2 - 13158x + 26603 \quad (15)$$

The energy gain on 30/8/2019 for SPV panel at 27° incline, facing due south at location A was 19,713.57 and 31,084.05 kJ/m^2-day respectively for horizontal and Inclined for 5.512 kWh/day.

4.2 *Sun Tracking Modes to Improve Input Irradiation Magnitude*

The location relative to equatorial region decides the intensity of solar energy availability. The variations in solar energy on diurnal and seasonal basis at a location become an important indicator of the overall performance of installed SPV system. The option of solar tracking involves the relative movement of the interception device in a way that maximizes effective gain of solar insolation. The strategy of tracking in based on instantly improving the angle of incidence to ensure a larger magnitude of incoming solar energy to the collector. The study has investigated tracking as,

E-W axis Tracking: Slope = zenith angle. (Solar altitude track)—**Mode 1**
Incidence angle minima (Maximize beam irradiance)—**Mode 2**
N-S axis Tracking: Slope equating hour angle (Constant speed track)—**Mode 3**
Maximum beam irradiance—**Mode 4**

The daily solar irradiance cumulatively added up for the month and averaged, as referred in Fig. 2. The four track modes are compared during 12 months of the year indicated minimum incidence angle tracking gives maximum value against the minimum observed for mode 1 tracking of solar collector. The choice of tracking gets recommended only when nett gain exceeds the auxiliary energy consumed (Fig. 3).

The sun-path diagram (Fig. 4) helps track seasonal changes with respect to solar azimuth angle and solar altitude angle. The computational tool can generate the sun-path diagram for a given location to orient collector so as to maximize the energy gain.

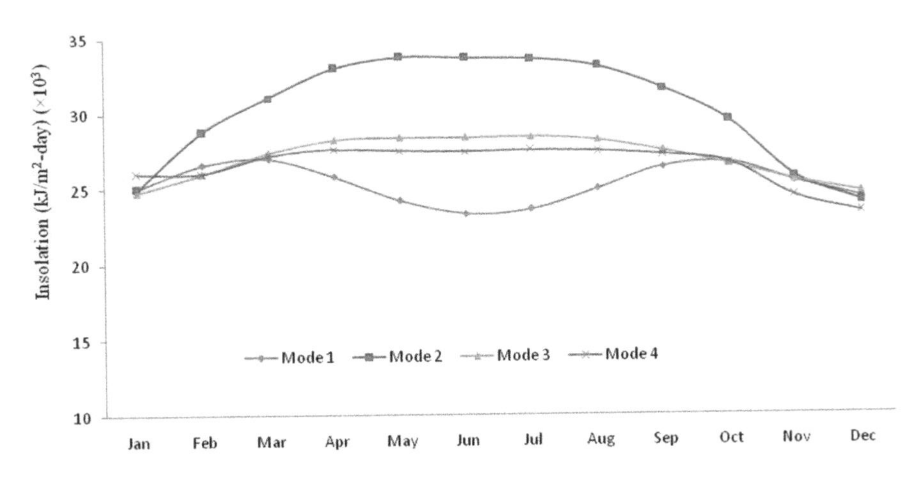

Fig. 3 Comparison of collector track modes

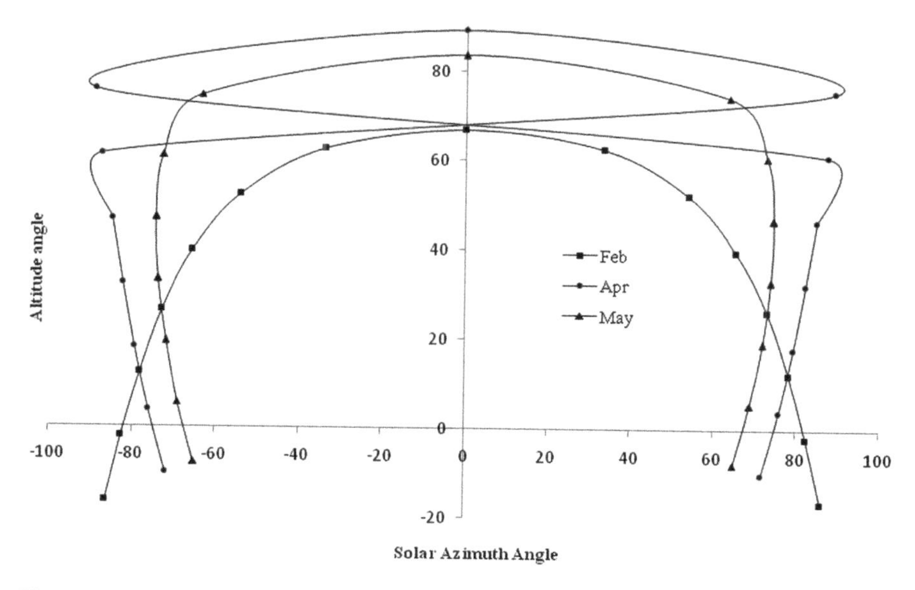

Fig. 4 Sun-path diagram for select months of year

4.3 Economics of SPV System

The economics of power generation is expressed in terms of the cost per kWh generated. The SPV system involves capital investment and operating cost that decides the cost of electricity. The location with a good sunshine obviously qualifies to have a better economics compared to location with lower solar insolation. The investment gets distributed for situations with higher generation and thereby reduces unit cot of electricity generated. Figure 5 shows that increased energy demand per day, reduced cost per kWh on account of the fact that solar module get designed for emergency loads. On the contrary, low-energy demand situations have a lower kW rating of SPV modules, but cost toward the balance of system that includes battery storage, relays and other control systems remains marginally less. This leads to incomplete usage of available solar potential; hence, the law of economics applicable to any other system components applies to SPV systems as well. Larger volume of generation reduced the unit cost of production as explained by learning curves of commodity pricing.

4.4 Simulation Studies on Solar Energy Accessibility

The magnitude of solar radiation at a location has dependence on latitude of location, day of the year, time, slope angle and surface azimuth angle. The solar radiation geometry literature provides reliable experimental and empirical correlations

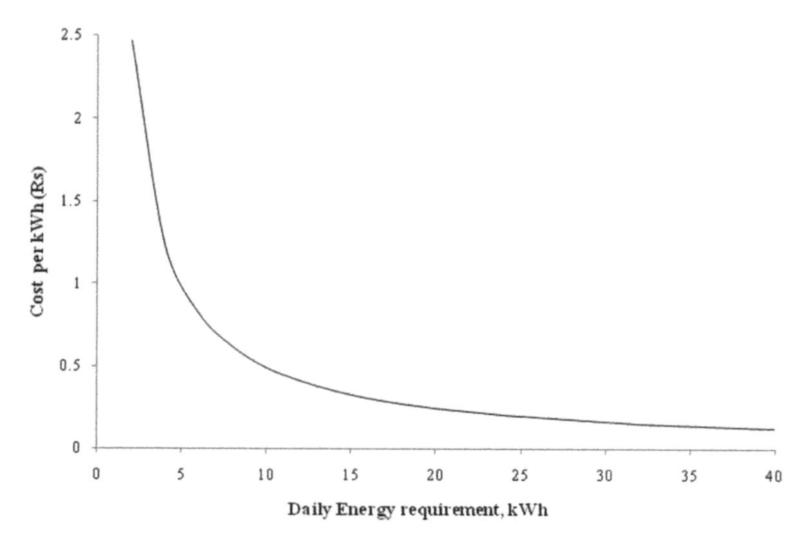

Fig. 5 Dependance of system load on cost per kWh

to predict the potential of solar insolation. These exercises are essential in design of solar energy conversion devices. Figure 6a–c highlights variation of solar energy available at location-A with respect to collector at different time of day, slope angle, surface azimuth and latitudinal variations.

As indicated in Fig. 6a, intensity of solar irradiation on horizontal and tilted surfaces indicated marginally higher values for former orientation with respect to a due south collector position. The intensity peaked at solar noon on account of unity value for air mass. Figure 6b indicated slope varied from zero to a maximum of 100° showed intensity to exhibit a decreasing trend for higher slope. Figure 6c indicated that surface azimuth angle of zero (due south position) was giving a marginally lesser insolation for a collector at location A. Similarly, a positive value of surface azimuth angle(east-wards alignment) gave a rise in solar irradiation against a drop for negative azimuth values. The similar variation in the intensity of solar insolation as a function of latitude angle can be obtained through relevant equations. The location closer to equator has higher potential compared to places closer to polar regions.

The consequent observation made through the experimental and computational studies strongly supported the approach of making prior investigations for installation of SPV systems. The tools used in the study helped in optimization of the system design and thereby made SPV power generation competitive to existing alternatives.

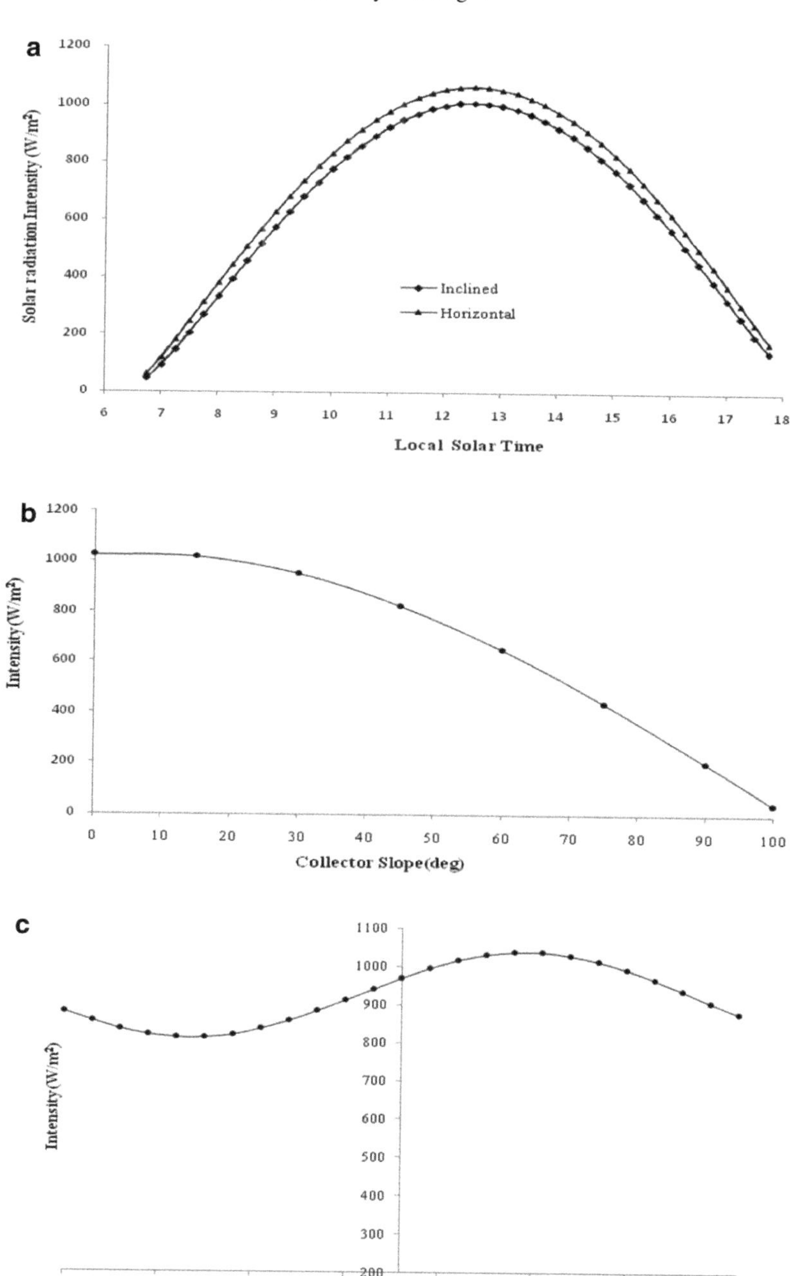

Fig. 6 **a** Solar insolation exhibiting diurnal variations. **b** Variation of solar insolation with slope. **c** Variation of solar insolation with collector surface azimuth angle

5 Conclusions

The studies performed on assessment of solar energy potential and subsequent marking of the location have resulted in the following broad observations for on site implementation.

Solar energy was established as viable option for both electric power generation and process energy needs on account of its huge potential and eco-friendly nature. The implementation of solar energy programme on a large scale can curtail substantial amount of carbon emissions.

The solar potential at the three identified locations A, B and C revealed that the equatorial region had a large solar potential compared to the temperate region. Solar energy showed wide variations based on geographical changes; however, India is blessed with abundant solar energy potential.

The strategies of solar tracking can help optimize the nett solar energy gain on account of time-based changes in collector orientation to map movement of sun in the horizon. The tracking strategies can be adopted to make intelligent solar collector that respond to diurnal-seasonal cycles along with climate induced aberration.

The future solar energy equipment are expected to be more responsive to variability factors associated with solar energy, and hence, development of hybrid renewable energy options could drastically enhance system reliability and carbon footprint minimization.

References

1. Ulgen K., Hepbasli, A.: Comparison of solar radiation correlations for Izmir Turkey. J. Energy Res. **26**(5), 413–430 (2002)
2. Ricci, R., Vitali, D., Montelpare, S.: An innovative wind–solar hybrid street light: development and early testing of a prototype. Int. J. Low-Carbon Technol. **10**(4), 420–429 (2015)
3. Mousavi, M., Hizam, S.A., Gomes, C.: Estimation of hourly daily and monthly global solar radiation on inclined surfaces: Models Re-visited Energies **10**(1), 134 (2017)
4. Zhang, T., Zhu, Q.Z., He, W., Pei, G., Ji, J.: Annual performance comparison between solar water heating system and solar photovoltaic/thermal system—A case study in Shanghai City. Journal of Low-Carbon Technologies **12**(4), 420–429 (2017)
5. Harrison, S., Jiang, L.: An investigation into the energy performance gap between the predicted and measured output of photovoltaic systems using dynamic simulation modeling software—A case study. J. Low-Carbon Technolo. **13**(1), 23–29 (2018)
6. Mills, S.: Combining solar power with coal-fired power plants or co-firing natural gas Clean Energy **2**(1), 1–9 (2018)
7. Laseinde, T., Ramere, D.: Low-cost automatic multi-axis solar tracking system for performance improvement in vertical support solar panels using Arduino board. Int. J. Low-Carbon Technol. **14**(1), 76–82 (2019)
8. Gao, Y., Huang, H., Yuehong, Su., Riffat, S.B.: A parametric study of characteristics of concentrating PV modules. Int. J. Low-Carbon Technol. **5**(2), 57–62 (2010)
9. Cuce, D., Bali, T., Sekucoglu, S.A.: Effects of passive cooling on performance of silicon photovoltaic cells. Int. J. Low-Carbon Technol. **6**(4), 299–308 (2011)
10. Rahman, A.: Modeling of a maximum power point tracker for a stand-alone photovoltaic system using MATLAB/simulink. J. Low-Carbon Technol. **9**(3), 195–201

11. Mukherjee, S., Ghosh, P.B.: Estimation of carbon credit and direct carbon footprint by solar photovoltaic cells in West Bengal India. J. Low-Carbon Technol. **9**(1), 52–55 (2014)
12. Ong, K.S.: Review of solar, heat pipe and thermoelectric hybrid systems for power generation and heating. J. Low-Carbon Technol. **11**(4), 460–465 (2016)
13. Gao, Y., Ji, J., Guo, Z., Peng, Su.: Comparison of the Solar PV Cooling System and Other Cooling Systems. J. Low-Carbon Technol. **13**(4), 353–363 (2018)
14. Chen, H., Riffat, S.B.: Development of photovoltaic thermal technology in recent years: a review. Int. J. Low-Carbon Technol. **6**(1), 1–13
15. Maleki, A.: Modeling and optimum design of an off-grid PV/WT/FC/diesel hybrid system considering different fuel prices. J. Low-Carbon Technol. **13**(2), 140–147 (2018)
16. Baljit, S.S.S., Chan, H.Y., Zaidi, S.H., Sopian, K.: Performance study of a dual-fluid photovoltaic thermal collector with reflection and refraction solar concentrators. J. Low-Carbon Technol. **15**(1), 25–39 (2020)
17. Solanki, C.S.: Solar photovoltaics: fundamentals technologies and applications, 2nd edn. PHI Learning, pp 200–280 (2015)

Gait Abnormality Detection without Clinical Intervention Using Wearable Sensors and Machine Learning

Subhrangshu Adhikary, Ruma Ghosh, and Arindam Ghosh

Abstract The gait pattern varies from person to person. However, the gait of a normal healthy human differs substantially from that of an individual with an abnormal gait. The gait abnormalities could arise from various underlying health conditions such as rheumatoid arthritis, injuries, etc. The gaits of individuals could be studied to differentiate healthy gaits from abnormal gaits which can help to identify underline health conditions and for preventive medication. In this work, for the identification of abnormal gait, we have used wearable sensors, comprising of tri-axial accelerometer and tri-axial gyroscope that track's the motion signatures produced while walking. We have collected such motion signatures of healthy persons and persons with walking abnormalities over a period of time. Later performed classification on collected data using different machine learning algorithms to segregate the cases with abnormality which can be used for facilitating remote detection of gait abnormalities using wearable sensors. With our approach, we have successfully classified different cases with an accuracy of over 90% . In the future, implementing such models in the real-world scenario could be very beneficial for the detection of gait diseases remotely with the help of smart devices like smartphones without clinical intervention.

Keywords Gait abnormality · IoT · Machine learning · Biomechanics · Biomedical signal processing

S. Adhikary · R. Ghosh · A. Ghosh (✉)
Department of Computer Science and Engineering,
Dr. B.C. Roy Engineering College, Durgapur, India
e-mail: arindam202@gmail.com

S. Adhikary
e-mail: subhrangshu.adhikary@spiraldevs.com

R. Ghosh
e-mail: ruma.ghosh@bcrec.ac.in

P. Muthukumar et al. (eds.), *Innovations in Sustainable Energy and Technology*,
Advances in Sustainability Science and Technology,
https://doi.org/10.1007/978-981-16-1119-3_31

1 Introduction

The gait analysis of human to determine the quality of walking is among the most prominent fields of study; this is because gait is extensively affected by plenty of underlying neurological or physical deformity. Different types of deformities have a unique pattern which could be further exploited to detect any such situations. Such patterns could be differentiated visually with keen observation. To standardize the process, mostly wearable sensors and image processing techniques are often used to study the gait of an individual. Image processing techniques often need hours of visual data and hours of the resource-intensive training process. Image processing-based classification has other disadvantages like the training needs to be performed in several angles to maintain consistency and provide accurate results. To remember the pattern of the gait for each step, image processing techniques would also require memory-based learning techniques like a combination of convolution neural network and recurrent neural network comprising long short-term memory techniques which would make the model much complex and would not be practically feasible with the present set of technology. Therefore, we have focused to work with wearable sensors instead.

Previously, wearable sensors consisting of accelerometer and gyroscope have often been used to study the gait of an individual; however, most of these studies are mostly focused toward either feature extraction of the time series motion data or machine learning-based classification of the gait of a specific disability and normal gait [1, 2]. Figures 1 and 2 are the plot of sensor data of normal and abnormal subject, respectively, and shows a clear deviated irregular pattern of peaks and trough, and we were motivated by this to differentiate normal and abnormal gait pattern.

To conduct our experiment, with consent from the human subjects, we have tracked 15 healthy subjects and 15 subjects with different abnormal health conditions. We have used scaling methods to normalize the data and we have implemented some of the widely accepted machine learning classifiers (KNN, Random forest and more) to isolate abnormal gaits from normal ones. The system is implemented by using Android smartphones to collect data and computer to study those data. A plot of their dataset given in Figs. 1 and 2 show that both are visually differentiable like we can see that for 'Ay', in healthy gait, peaks are appearing near interval of 70–80 are almost up to 12.5 but for abnormal gait, the peaks are irregular, ranging mostly between 10.5 and 12. Similar irregularities are visible for other features as well. Therefore based on this observation, we can exploit the pattern to detect abnormality.

2 Related Works

Gait analysis has been widely performed in several research works. Some of the methods have been developed with the help of various algorithms. The information used for the study includes video data or sensor data. Attempts to identify different

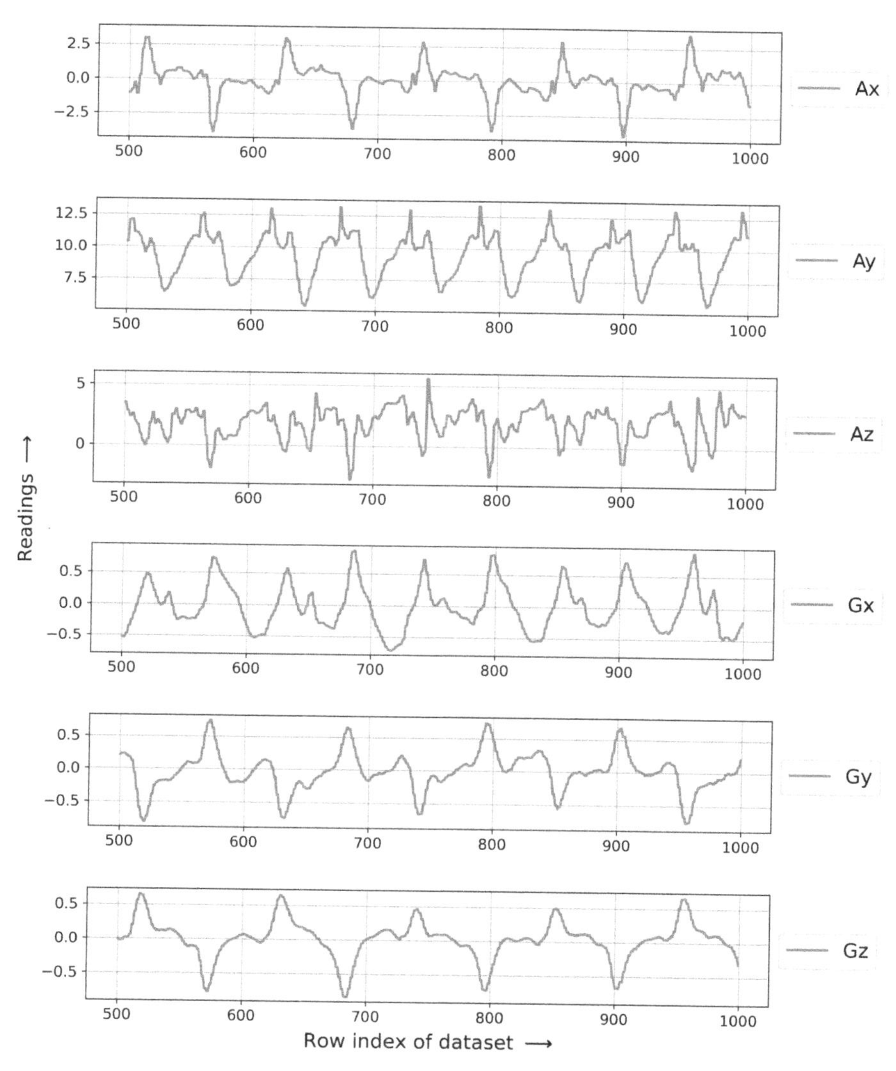

Fig. 1 Slice of gait sequence of a normal healthy individual

human activities such as walking, running has been demonstrated and the activities have been successfully classified with the application of certain machine learning algorithms like J48, Naive Bayes, SVM, MLP and attend high degree of accuracy. Shoe-integrated wireless sensors were used in an attempt to quantify gaits where people with healthy gaits and parkinsonian gaits were studied. In [3], uni-axial gyroscopic sensors were placed near the skin surface of the shank or things of subjects and angular velocity of their movement were recorded. A correlation between shank and thigh was verified. Smartphones embedded accelerometers were used to monitor

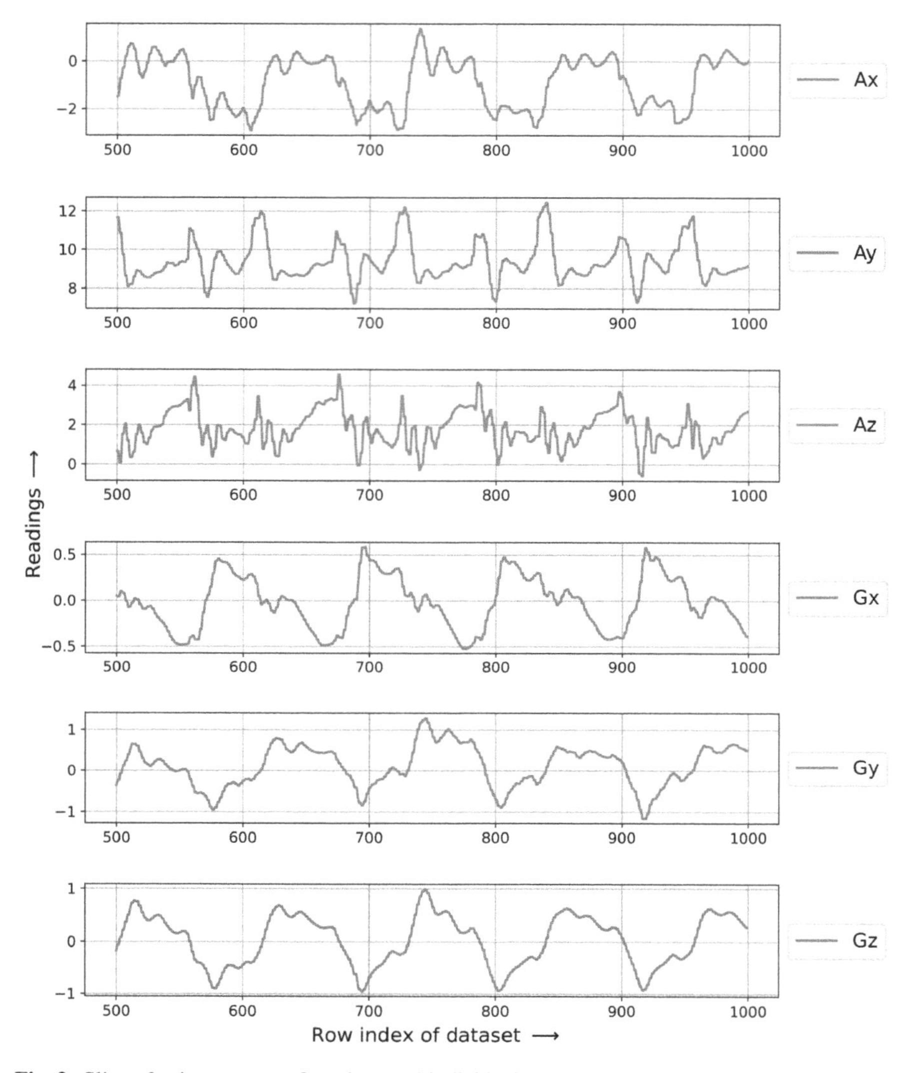

Fig. 2 Slice of gait sequence of an abnormal individual

the status of walking of 30 patients with chronic lung disease were studied while they walked for 6 min. They had evaluated a statistical model using the gait parameter to evaluate health status through lung function [4–6].

In [7], authors have attached sensors in the leg to measure their strides, swing and other gait features. Changing the foot, position and orientation is most likely to induce variation in the data which could reduce the consistency of the model. In this experiment, artificial techniques were used to simulate abnormal gait so its consistency to detect gait abnormality when the model is exposed to the production environment. In [8] was another recent experiment where sensors were placed in the foot, and they have used a threshold based detection system to study the gait. In [9], experiments have been performed where sensors were placed in both legs of the subject and observed the symmetry between the stride and swings of both legs. Authors in [10] have tried to detect cerebral palsy with an uncertainty based state-space model. Auto-encoder-based generative adversarial network technique was used to study gait in [11]. Auto-encoder is a technique popular in neural network algorithms where the training of the model is done with input and preferred output being the same, but the number of nodes in the hidden layer is set low so unwanted noise is filtered out and only significant features are passed on to the next layer in an attempt to generate output which appears like input but have lesser noise. The generative adversarial network attempts to create new sets of data based on previously known ones. These techniques were used to generate artificial gait data and study on them.

Till now all of these techniques have some flaw. In most cases, artificial approach was taken to generate defective gait which could produce inconsistency during deployment. On the other hand, the placement of sensors was mostly around the foot and so changing in orientation, position, height of the subject and other feature could impact the study of gait. This is why we have placed sensors on chest, with the same position and orientation so that we have consistent data for everyone.

3 Methodology

In this section, we will discuss our methods of implementing the gait abnormality detection model in details.

3.1 *Experimental Approach*

As mentioned earlier, smartphones contain various embedded sensors. We have included a tri-axial accelerometer and tri-axial gyroscope in our study. We have used Vivo V3 with Qualcomm Snapdragon 616 MSM8939 chipset, Octa-core (1.5 GHz, Quad-core, Cortex A53 + 1.2 GHz, Quad-core, Cortex A53) processor, Adreno 405 GPU (Fig. 3).

With the help of Android Studio 3.3.2, we have programmed a mobile application, which has two buttons, one to start and another to stop. In this, there is a text field where we provide the name by which the data file would be created. To remove redundancies and overwriting over a pre-existing file, we have added a timestamp to the file name. When the start button is clicked, a new file is created in mobile's local storage with the name provided in the text box concatenated with the current timestamp. We have used a collection rate of 100 Hz $\pm$ 1. Data recording stops when the stop button is pressed and the data file is saved in the local storage.

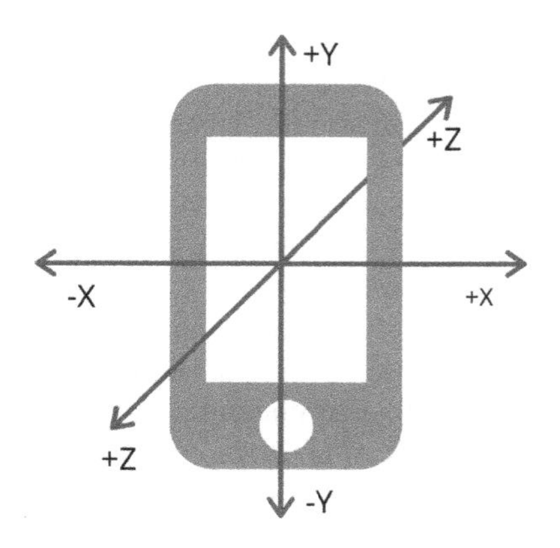

Fig. 3 Displayed the axes orientation of smartphone Tri-axial accelerometer and Tri-axial gyroscope

The data file is generated in a .csv extension. It consists of 7 columns respectively as x, y and z axes of the accelerometer, x, y and z axes of the gyroscope and time elapsed since the beginning of recording in milliseconds.

3.2 Data Collection

We have prepared a belt with which we fit the smartphone at the chest right above the sternum. After we have placed the smartphone, the volunteers were asked to walk on a straight road with their comfortable pace. Volunteers were asked to walk for up to 32 s (Fig. 4).

After collection of data, we have trimmed the datasets, first 2 s and last 2 s from each dataset were removed so that we can avoid initial acceleration and final retardation. So we have filtered datasets of 28 s for both healthy individuals and unhealthy individuals. The tests were performed on 15 healthy volunteers and 15

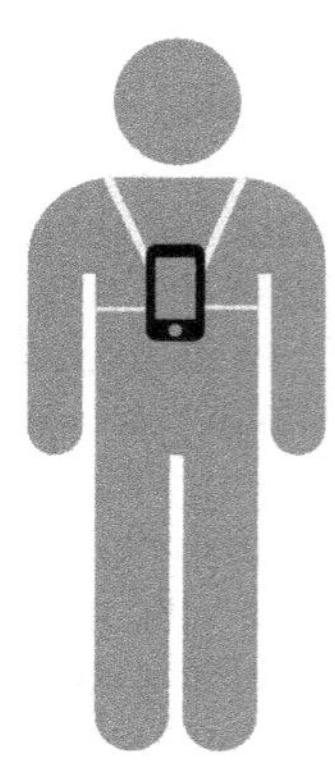

Fig. 4 Our setup to attach smartphone on chest

unhealthy volunteers. Unhealthy volunteers have one of the health problems like hemiplegia, osteoarthritis, rheumatoid arthritis, knee ligament fracture.

3.3 Classification Technique

We have pre-processed the dataset using Min-Max scaler method to transform the dataset to fit within the range of 0 and 1 as our algorithms perform best in this range. To scale dataset X, the following method has been implemented:

$$X_{\text{scaled}} = \frac{X - X_{\min}}{X_{\max} - X_{\min}} \tag{1}$$

Followed by normalization, we have tested four different supervised machine learning algorithms to classify the gaits. We will discuss the performances of all these classifiers in Sect. 4. The algorithms are described below.

K-Nearest Neighbour is a supervised machine learning algorithm which classifies classes based on majority wins approach. We find KNN as a suitable machine learning algorithm for this experiment, this is because the data points of the normal gait would accumulate toward mean distribution space; however, the data points of the abnormal gait appear to spread across a wider spectrum. Random forest classifier (RFC) is a popular ensemble-based learning algorithm which reaches its decision by voting from many different algorithms. We choose this classifier because it is immune to overfitting. Next on the list, we have decision tree classifier (DTC) is a supervised machine learning algorithm which belongs to an ensemble classifier family. We chose DTC for the classification purpose because missing data does not affect much on outcomes of DTC and also low pre-processing is required to train DTC model making is a suitable algorithm to test the validity of other classifiers. Finally, we have used extra tree classifier (ETC), in which the model samples the dataset without replacement and nodes are split on random splits regardless of the fact whether it is the best split or not. Thus, an extra tree classifier produces very low variance in its outcome [12–17].

4 Results and Discussion

While walking, when a person walks forward, motions are initiated in multiple direction and orientation. There are some to and from forward motion, some sidewise movements and some up and down motions. Now while we walk, when one foot is away from the ground and the other in on the ground; this is when our body reaches farthest from the ground. On the other hand, when both of our feet are on the ground, this is when we are closest to the ground.

When going away from the ground, we gain a linear acceleration against the gravity and so according to the orientation of the Android device we have placed, the magnitude of 'Ay' goes down. And conversely, when both our feet touch the ground, this is when we accelerate along with the acceleration due to gravity and thus 'Ay' increases. So from our data, each time 'Ay' has touched the peak is when both feet touch the ground and on the other hand, 'Ay' hits the trough when one leg is in the air and the other is on the ground.

The data points for both normal and abnormal data are first trimmed from edges and labelled as discussed in Sect. 3. Then we have combined the two different classes in 1:1 ratio combining to 84,000 individual data points which are then randomly split into two parts in 4:1 ratio for training and testing purpose. The dataset is then trained with the classifiers, and the following performances are observed.

4.1 Performance Analysis

When the normalized data passed through the classifiers, we have obtained the highest classification accuracy of 93.8% with KNN and almost similar accuracy of 93.79% was observed with RFC, followed by DTC, ETC with 89.18% and 85.65% respectively. As the raw dataset is balanced, that is the number of rows of normal data exactly equals the number of rows of abnormal data, the precision and recall would equal accuracy of the classifier. Now the next parameter to test the performance of our classifiers is resource optimization. ETC was the fastest to train by training the model in 122.6 ms followed by KNN, DTC and RFC being the slowest which took 16,202.7 ms. Based on the generation of output, DTC and ETC both performed similarly by training the model within 7.433 and 7.527 s, respectively. Now as the accuracy of KNN and RFC was higher compared to DTC and ETC, therefore we can say that KNN performs the best among the classifiers when training time is concerned, and RFC performs the best when the time to generate output is our concern. The performances have been recorded in Tables 1 and 2.

Table 1 Performance of raw sensor data classification

Performance metric	KNN	RFC	DTC	ETC
Accuracy (%)	97.03	94.95	90.70	87.19
Precision	0.97	0.949	0.907	0.871
Recall	0.97	0.949	0.907	0.871
Time to build model (ms)	176.3	16202.7	779.03	122.6
Time to predict outcomes (ms)	1092.9	527.99	7.433	7.527

Table 2 Health status detection results breakdown

Classifier	Actual and predicted normal (%)	Actual abnormal, Predicted normal (%)	Predicted abnormal, Actual normal (%)	Actual and predicted abnormal (%)
KNN	48.92	1.72	1.23	48.10
RFC	48.27	3.15	1.89	46.67
DTC	45.63	4.76	4.52	45.06
ETC	43.85	6.50	6.30	43.33

After the classification, we can check the confusion matrix (Table 2) produced by the classifiers while producing outcomes to test the reliability of the models. Out of total of 16800 tested data points, detection with KNN was the most reliable as it has accurately classified 16302 data with 290 and 208 false-positive and false-negative errors, followed by RFC with 15952, DTC with 15238 and ETC with 14648 accurate predictions. 48.92% of the data was truly detected as normal, 48.10% of the data was truly detected as abnormal, 1.7% abnormal subjects were detected as normal and 1.23% normal subjects were detected as abnormal. From this, we can verify our statement as mentioned in Sect. 3. That the normal gaits are clustered toward its mean distribution where KNN detected all neighbouring points as normal and discarded all remaining data points marking it as abnormal which are spread over a wider spectrum. Similarly, RFC produced second highest accurate predictions as RFC decides by gathering votes from multiple other classifiers. DTC and ETC, on the other hand, were considerably good but not as good as KNN and RNN; this is because these are made of datasets having high number of missing points; however, our datasets have no such missing points, and therefore, these two classifiers could not outperform KNN and RFC as we have expected it to happen.

5 Conclusion

Smartphones are now an integral part of our lives. Smartphones with its good computational power and presence of different sensors make it a suitable device to track our movements regularly. Identifying abnormal gaits while carrying our smartphones could alert the patient so that they can consult a doctor.

With this paper, we have discussed collecting accelerometer and gyroscope data, feature scaling and then we have used this to train the model and test the results and discussed its accuracy, precision, recall and optimization metrics.

The experimental results reveal that our model could successfully isolate normal gaits and abnormal gaits with a maximum accuracy of 97.03% with K-Nearest Neighbour Classifier. Random forest classifier also provided close results in obtaining 94.95% accuracy. RFC proves to be 91 times slower than KNN for training;

however, RFC performed twice as fast as KNN to predict results. Based on this, KNN and RFC could be interchangeably used based on resources availability with similar reliability.

The model could further be developed to detect specific abnormal pattern associated with particular disease reliably to accurately diagnose disease without clinical intervention using widely available gadgets such as a smartphone.

Acknowledgements We would like to thank Prabhat Accu Clinic, Prova X-Ray and G.S. Enterprise based in Uttar Dinajpur, West Bengal, for providing human subjects with consent to conduct the experiment.

References

1. Nishiguchi, S., Yamada, M., Nagai, K., Mori, S., Kajiwara, Y., Sonoda, T., Yoshimura, K., Yoshitomi, H., Ito, H., Okamoto, K., et al.: Telemed. e-Health **18**(4), 292 (2012)
2. Gafurov, D., Helkala, K., Søndrol, T.: JCP **1**(7), 51 (2006)
3. Tong, K., Granat, M.H.: Med. Eng. Phys. **21**(2), 87 (1999)
4. Yin, X., Shen, W., Samarabandu, J., Wang, X.: In: 2015 IEEE 19th International Conference on Computer Supported Cooperative Work in Design (CSCWD), pp. 582–587. IEEE (2015)
5. Bamberg, S.J.M., Benbasat, A.Y., Scarborough, D.M., Krebs, D.E., Paradiso, J.A.: IEEE Trans. Inf. Technol. Biomed. **12**(4), 413 (2008)
6. Juen, J., Cheng, Q., Prieto-Centurion, V., Krishnan, J.A., Schatz, B.: Telemed. e-Health **20**(11), 1035 (2014)
7. Chattopadhyay, S., Nandy, A.: In: TENCON 2018-2018 IEEE Region 10 Conference, pp. 0623–0628. IEEE (2018)
8. Dhokai, R.: A personalized gait abnormality detection system. Ph.D. thesis, Dhirubhai Ambani Institute of Information and Communication Technology (2018)
9. Duhaylungsod, C.R.E., Magbitang, C.E.B., Mercado, J.F.I.R., Osido, G.E.D., Pecho, S.A.C., dela Cruz, A.R.: In: 2017IEEE 9th International Conference on Humanoid, Nanotechnology, Information Technology, Communication and Control, Environment and Management (HNICEM), pp. 1–4. IEEE (2017)
10. Chakraborty, S., Thomas, N., Nandy, A.: In: International Conference on Computational Science, pp. 536–549. Springer (2020)
11. Nguyen, T.N., Meunier, J.: Pattern Anal. Appl. **22**(4), 1597 (2019)
12. Choi, S., Youn, I.H., LeMay, R., Burns, S., Youn, J.H.: In: 2014 International Conference on Computing, Networking and Communications (ICNC), pp. 1091–1095. IEEE (2014)
13. Barnich, O., Van Droogenbroeck, M.: Pattern Recogn. Lett. **30**(10), 893 (2009)
14. Gwak, M., Sarrafzadeh, M., Woo, E.: In: Proceedings, APSIPA Annual Summit and Conference, vol. 2018, pp. 12–15 (2018)
15. Dupuis, Y., Savatier, X., Vasseur, P.: Image Vision Comput **31**(8), 580 (2013)
16. Açıcı, K., Erdaş, Ç.B., Aşuroğlu, I., Toprak, M.K., Erdem, H., Oğul, H.: In: International Conference on Engineering Applications of Neural Networks, pp. 609–619. Springer (2017)
17. Kirkwood, C., Andrews, B., Mowforth, P.: J. Biomed. Eng. **11**(6), 511 (1989)

Non-invasive Group Activity Identification Using Neural Networks

Arindam Ghosh, Dibyendu Das, Ruma Ghosh, Sandip Mondal, and Mousumi Saha

Abstract Human activity recognition has been a topic of discussion for long. Proper identification would lead to better automation, surveillance, and more comfortable living. Although single person activity is comparatively easier to detect and significant research have been done recently, the challenge lies in identification of group activity. Existing research for identification of group activity is either computer vision-based or wearable device-based. However, computer vision might breach privacy and heavy in computation whereas wearable devices are based on people's will and might be discomforting as well. Therefore, in our proposed work we have used linear data, i.e., distance of the person from the sensor to predict group activity with high accuracy, at the same time it is non-invasive, and is automated. The proposed work has been conducted using customized multiple ultrasonic sensor grids on 10 individuals in a closed lab environment. The evaluation of experimental results shows an accuracy of 95.48% in group activity identification using neural networks.

Keywords Group activity · IoT · Machine learning · Sensors · Neural network

1 Introduction

As people tend to stay in closed rooms more often than outdoors, monitoring their activities can help us implement room automation in the future. Atomic activities might not prove that useful, in this case, because rooms tend to occupy groups of people. Thus, the need for group activity detection arises. The work that has been done related to human activity identification, can be mostly divided into two main phases. The first being environment placed sensors and wearable sensors. Wearable sensors [1] include mobile sensors, accelerometer, gyroscope, proximity sensors etc.

A. Ghosh (✉) · D. Das · R. Ghosh
Dr. B C Roy Engineering College, Durgapur, India
e-mail: arindam202@gmail.com

S. Mondal · M. Saha
National Institute of Technology Durgapur, Durgapur, India

P. Muthukumar et al. (eds.), *Innovations in Sustainable Energy and Technology*, Advances in Sustainability Science and Technology,
https://doi.org/10.1007/978-981-16-1119-3_32

These are generally worn by the individual or placed right above the individual. Though this method has the ease of deployment and is easy to use. They are limited by the very fact that it depends upon the will of the user. Also they mostly can detect atomic activities and have to be synchronized with proper network to a central machine, this also creates a problem of unknown number of sensing devices, since we do not know how many people would be present in the room.

Environment placed sensors solve the issues faced with wearable sensors, since they are independent of the users, and a fixed set of tools that can determine activities. This approach involves devices like cameras [2], PIR (pyro-electric) [3], acoustic sensor [4], thermal sensors [5], etc. Most of the related work in group activity recognition has been done using image recognition. But that not only is pervasive but requires high computational power.

Some work regarding ultrasonic sensors have been like object detection done, but as per our knowledge no group activity identification has been done using ultrasonic sensors, which uses neural networks and performs with an accuracy above 90%. Our proposed work uses an Decision Tree Classification for atomic activity recognition from labeled data-set. Thereafter, the output of the model is used in a neural network for determining the group activities based on distance reading of the person from the ultrasonic sensors.

Organization: The proposed work has been organized in six sections. Section 1 consists of Introduction, Sect. 2 explains the lab setup for receiving the data through ultrasonic sensors, Sect. 3 explains the noise reduction and machine learning models used, Sect. 4 details the Experimental results, Sect. 5 gives the conclusion and Sect. 6 mentions the references

2 System Architecture

In this section, we have illustrated the sensor arrangement for capturing activity data and processing technique used for recognizing group activity.

2.1 Working of Ultrasonic Sensors

We used the ultrasonic sensor, HC-SR04, in the suggested framework. Among the large range of currently present sensors, the HC-SR04 sensors are low-cost industrial sensors used to measure short distances. It consumes 5 V of power. It consists of two components—Transmitter and receiver. The distance of the object from the sensors is determined by using a sequence of ultrasonic pulses. The transmitter emits ultrasonic wave, if an obstacle comes in the path of the ultrasonic wave, it gets reflected back and is received by the sensor. The time elapsed between sending and receiving of this ultrasonic signal along with the speed of sound in air, helps us determine the distance between the object and the sensor using following equation $S = \frac{T}{2*29}$. Here, S is the distance, t is the time elapsed between sending and receiving of ultrasonic signal. Sound travels at approximately 340 ms^{-1}. This corresponds to about 29.412 μs/cm. Since the sound travels away from the sensor, then it bounces off of a surface and returns back, the time that is elapsed is for both way travel. Thus, we divide the time by 2, to determine the time required for one way travel. The sensors that have been used in the experiment have a distance range of 4 to 400 cm. HCSR04 operates in the 40 KHz frequency range. The sensors also have in-built noise filters which prevent disturbance due to close proximity.

2.2 *System Deployment*

Our suggested approach utilizes grids of homogeneous ultrasonic sensors. Four ultrasonic HC-SRO4 sensors have been placed 55 cm apart at the four corner of the sensor grid. Having considered the sensing coverage of each sensor, the given distance is optimized to minimize the overlapping of sensing regions. Such panels are suspended at a height of 2.5 m above the ground, facing downwards, covering a $72 \times 72\,\text{cm}^2$ of field of view. The real scenario of deployment and the sensor grid is shown in Fig. 1a, b. An region of $4 \times (72 \times 72)\,\text{cm}^2$ covers the four sensor grids. Because only one person can sit or stand at any given time under a given sensor, we can identify activity for a group of people at any given time, considering data from all the sensors used in the grid.

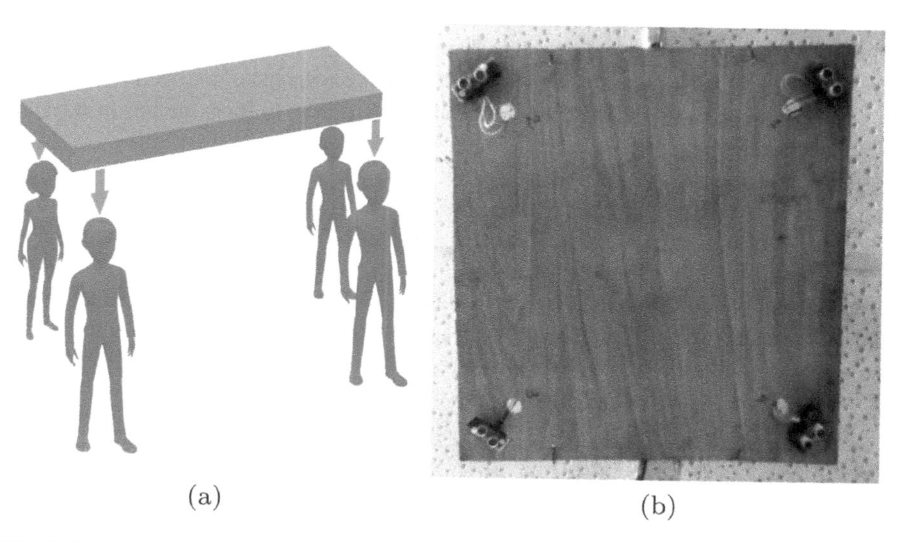

Fig. 1 Deployment overview

3 Data Processing Technique for Activity Recognition

The concept of this suggested approach is to facilitate automated recognition of group behavior. Every particular ultrasonic sensor measures distance and helps detect atomic behaviors such as sitting, standing and dropping. When obtained from multiple sensors, this data lets us simultaneously assess group behavior inside the room.

3.1 *Pre-processing of Sensor Data*

Ultrasonic sensors have varying measuring ranges that suggest diverse scenarios originating from several grids. The probability of random noise in records, however, is very high. Sound might be reflected by various undesirable objects such as walls, tables and chairs. To prevent this, certain thresholds have been used to eliminate such a situation. After a variety of studies, these threshold values have been calculated and are the most effective in reducing noise.

3.2 *Identification of Atomic Activities*

Since height varies among different users, so sensor reading also varies for activities like sitting, standing, fall, etc. Thus, a range of values has been taken, this range of values have been identified and used with labels for feeding a Decision Tree Classification model. The dataset has been classified under four labels which are empty, walking, standing and sitting. Serial dependencies have not been considered in this classification model but this classification will help to form the group activity identification from atomic activities.

After determining the various ranges from the given data, for all the four atomic activities, these ranges are further used for various group activity identification like walking, standing, sitting, gathering, separating, etc. These states are determined by adding up all the labels output for each sensor of each atomic states to get a single label instead of four. The data is then reduced using principal component analysis and then clusters are formed using K-means clustering, on clustering we find 3 of the states are redundant and can be merged with the rest of the states. The output data is then fed to the neural networks to identify the group activity.

3.3 *Recognition of Group Activities*

The group-level features in our implementation can be broadly classified into six categories. These are as follows:

1. **Walking in group**: When two or more individuals walk in the same direction in close vicinity.
2. **Standing in group**: When two or more individuals are in close proximity to each other and in standing posture.
3. **Sitting in group**: When two or more individuals are in close proximity to each other and in sitting posture.
4. **Gather in group**: When two or more individuals come from various directions and converge at a point.
5. **Separating from a group**: When two or more individuals were standing in close proximity and then diverge.
6. **Empty space**: When no individual is present under the grid.

As discussed above, the sum of labels from atomic states has been made which were identifying group activities into nine classes, however after reducing no. of columns to two using Principal Component Analysis(PCA), and thereafter using K-means over the group data to visualize the clusters, we identify that three of these classes are redundant and can be merged with the existing classes, respectively. Thus, the following changes are made from Fig. 2a, b.

Prediction of a single label, instead of four labels, results in better accuracy. The NxN matrix is flattened to a vector. Thereafter, it is passed through a six-layered neural network, with 512 neurons in the first five layers and 6 at the penultimate layer. The structure of used neural network has been shown in Fig. 3.

Each input is fed to each neuron of the first layer. The first layer of neurons is connected to the next layer via channels. These channels have weights associated with them. Each neuron in the hidden layer has some bias associated with them. Based on the sum of the product of weights and inputs with the bias a score is decided. Based on this score, an activation function is used to decide which neurons should be activated. This process is continued for all subsequent layers. The output layer predicts the output. The output is in the form of probability for the six classes present, indicating how likely that class can be an output for the given input. If this value is found to be wrong based on already present labels, then the weights are adjusted and the computation is performed again. After the process of training has been done, the index with the maximum probability (argmax) is given as output.

4 Experimental Results

The experiment was carried out in two stages, the first being the identification of atomic states and the second the identification of group states. The experiment has been performed in a closed lab environment. Where a Decision Tree algorithm has been used to identify the atomic states, a neural network has been used to identify the group states.

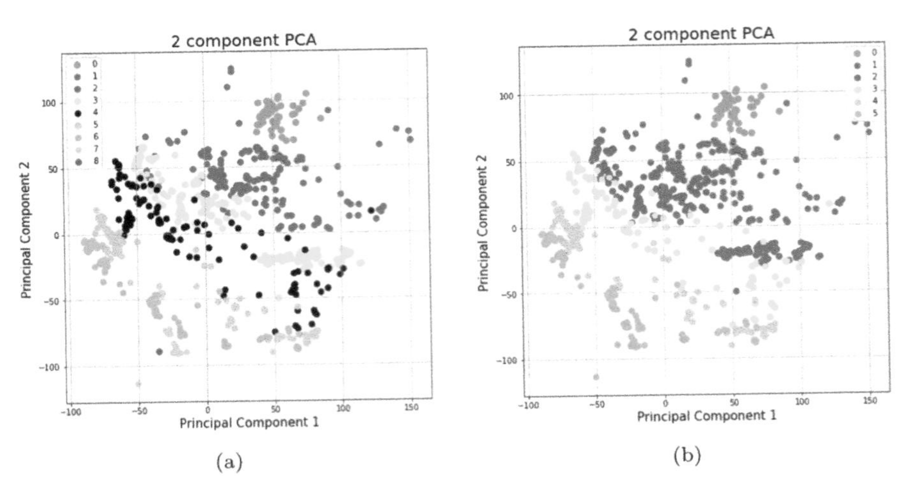

Fig. 2 Clusters for group states

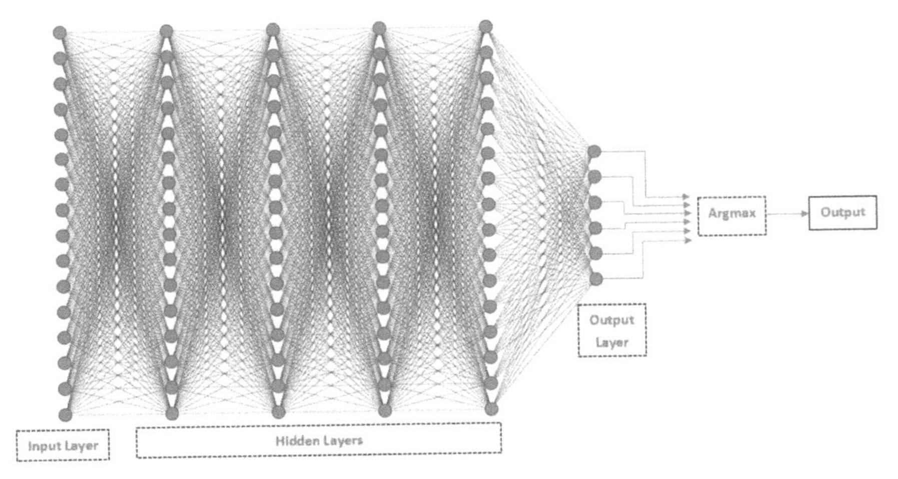

Fig. 3 Neural network architecture for group activity recognition

4.1 Results of Atomic Activity Identification

We use Decision Tree to identify the range of values for the various distance data that is received by the ultrasonic sensors to infer atomic activity. The highest accuracy has been found in case of sitting, i.e., 97% since sitting is a static position and height in case of sitting varies very little. The lowest accuracy has been detected during walking, i.e., 95% because not only walking occurs in standing position where there is a lot of variance in height but also it is a dynamic activity which is difficult to detect. This data is then fed into a sequential neural network to identify the group activities.

Table 1 Illustration of experimental result of atomic activity

Activity	Precision	Recall	$F1$ score	Accuracy
Empty	0.97	0.99	0.98	0.96
Walking	0.56	0.33	0.42	0.95
Standing	0.75	0.87	0.81	0.96
Sitting	0.67	0.40	0.50	0.97

Table 2 Illustration of experimental result of group activity

Activity	Precision	Recall	$F1$ Score	Accuracy
Empty space	1	1	1	1
Standing in group	1	0.92	0.96	0.98
Sitting in group	0.94	1	0.97	0.98
Gather in group	0.92	0.9	0.91	0.97
Separating from group	0.93	0.94	0.93	0.97
Walking in group	0.97	0.96	0.96	0.98

4.2 *Results of Group Activity Identification*

The input is determined at the start of the experiment, the final state is given in the form of a label, namely from 0 to 5, i.e., Empty Space, Standing in Group, Sitting in Group, Gather in group, Separating from Group, and Walking in Group. Individual activities were performed within a range of 110 cm and the group interaction zone is 140 cm. This has been determined with respect to the different experiments conducted in the test area. The group activity with the help of neural networks have been identified with an accuracy of 95.48% and has an F1-score of 95.68%. A slight lower accuracy has been observed in case of gathering and separating from group because these were very dynamic activities. In case of standing or sitting in a group, all remain still; in case of walking, all are in a state of motion; however, in case of gathering and separating from group, some remain still while others remain in motion which in some cases gets resulted in either being predicted standing or sitting or walking, thus resulting in lower accuracy compared to others (Tables 1 and 2).

5 Conclusion

Our proposed work has tried to identify group activity recognition using ultrasonic sensors and neural networks in a closed lab environment and has achieved an accuracy of 95.48% and f1-score of 95.68%. This can be further extended in future to identify complex intergroup and intragroup activities for better automation. Accuracy of prediction of dynamic activities like gather and separation could also be improved.

References

1. Trabelsi, D., Mohammed, S., Chamroukhi, F., Oukhellou, L., Amirat, Y.: An unsupervised approach for automatic activity recognition based on hidden markov model regression. IEEE Trans. Autom. Sci. Eng. **10**(3), 829–835 (2013)
2. Cheng, Z., Qin, L., Huang, Q., Jiang, S., Tian, Q.: Group activity recognition by Gaussian processes estimation. IEEE 3228–3231 (2010)
3. Raykov, Y.P., Ozer, E., Dasika, G., Boukouvalas, A., Little, M.A.: Predicting room occupancy with a single passive infrared (PIR) sensor through behavior extraction. In: Proceedings of the 2016 ACM International Joint Conference on Pervasive and Ubiquitous Computing, pp. 1016–1027 (2016)
4. Mokhtari, G., Zhang, Q., Nourbakhsh, G., Ball, S., Karunanithi, M.: Bluesound: a new resident identification sensor using ultrasound array and ble technology for smart home platform. IEEE Sensors J. **17**(5), 1503–1512 (2017)
5. Taniguchi, Y., Nakajima, H., Tsuchiya, N., Tanaka, J., Aita, F., Hata, Y.: Estimation of human posture by multi thermal array sensors. IEEE 3930–3935 (2014)
6. Ghosh, A., Sanyal, A., Chakraborty, A., Sharma, P.K., Saha, M., Nandi, S., Saha, S.: On automatizing recognition of multiple human activities using ultrasonic sensor grid. In: 2017 9th International Conference on COMSNETS, pp. 488 –491 (2017)
7. Brand, M., Oliver, N., Pentland, A.: Coupled hidden Markov models for complex action recognition. IEEE 994–999 (1997)
8. Young-Ji Kim, S.-W. L., Cho, N.-G.: Group activity recognition with group interaction zone. In: 2014 22nd International Conference on Pattern Recognition (ICPR), pp. 3517–3521 (2014)

Author Index

P. Muthukumar et al. (eds.), *Innovations in Sustainable Energy and Technology*, Advances in Sustainability Science and Technology,
https://doi.org/10.1007/978-981-16-1119-3

Lightning Source UK Ltd.
Milton Keynes UK
UKHW020104190522
403175UK00002B/8